Lecture Notes in Economics and Mathematical Systems 616

Founding Editors:

M. Beckmann
H.P. Künzi

Managing Editors:

Prof. Dr. G. Fandel
Fachbereich Wirtschaftswissenschaften
Fernuniversität Hagen
Feithstr. 140/AVZ II, 58084 Hagen, Germany

Prof. Dr. W. Trockel
Institut für Mathematische Wirtschaftsforschung (IMW)
Universität Bielefeld
Universitätsstr. 25, 33615 Bielefeld, Germany

Editorial Board:

A. Basile, A. Drexl, H. Dawid, K. Inderfurth, W. Kürsten

Alberto Cambini • Laura Martein

Generalized Convexity and Optimization

Theory and Applications

 Springer

Prof. Alberto Cambini
Prof. Laura Martein
University of Pisa
Department of Statistics and Applied Mathematics
Via Ridolfi, 10
56124 Pisa
Italy
acambini@ec.unipi.it
lmartein@ec.unipi.it

ISBN 978-3-540-70875-9   e-ISBN 978-3-540-70876-6

DOI 10.1007/978-3-540-70876-6

Lecture Notes in Economics and Mathematical Systems ISSN 0075-8442

Library of Congress Control Number: 2008933853

Cover design: WMXDesign GmbH, Heidelberg

Printed on acid-free paper

9 8 7 6 5 4 3 2 1

springer.com

To Giovanna and Paolo

Foreword

In the latter part of the twentieth century, the topic of generalizations of convex functions has attracted a sizable number of researchers, both in mathematics and in professional disciplines such as economics/management and engineering. In 1994 during the 15th International Symposium on Mathematical Programming in Ann Arbor, Michigan, I called together some colleagues to start an affiliation of researchers working in generalized convexity. The international Working Group of Generalized Convexity (WGGC) was born. Its website at www.genconv.org has been maintained by Riccardo Cambini, University of Pisa.

Riccardo's father, Alberto Cambini, and Alberto's long-term colleague Laura Martein in the Faculty of Economics, University of Pisa, are the co-authors of this volume. My own contact with generalized convexity in Italy dates back to my first visit to their department in 1980, at a time when the first international conference on generalized convexity was in preparation. Thirty years later it is now referred to as GC1, an NATO Summer School in Vancouver, Canada. Currently WGGC is preparing GC9 which is to take place in Kaohsiung, Taiwan. As founding chair and also current chair of WGGC, I am delighted to see the continued interest in generalized convexity of functions, augmented by the topic of generalized monotonicity of maps.

Eight international conferences have taken place in this research area, in North America (2), Europe (5) and Asia (1). We thought it was now time to return to Asia since our membership has shifted towards Asia.

As an applied mathematician I have taught mostly in management schools. However, I am currently in the process of joining an applied mathematics department. One of the first texts I will try out with my mathematics students is this volume of my long-term friends from Pisa. I recommend this volume to anyone who is trying to teach generalized convexity/generalized monotonicity in an applied mathematics department or in a professional school. The volume is suitable as a text for both. It contains proofs and exercises. It also provides sufficient references for those who want to dig deeper as graduate students and as researchers. With dedication and much love the authors have written a book that is useful for anyone with a limited background in basic mathematics. At the same time, it also leads to more advanced mathematics.

The classical concepts of generalized convexity are introduced in Chaps. 2 and 3 with separate sections on non-differentiable and differentiable functions. This has not been done in earlier presentations. Chapter 4 deals with the relationship of optimality conditions and generalized convexity. One of the reasons for a study of generalized convexity is that convexity usually is just a convenient sufficient condition. In fact most of the time it is not necessary. And it is a rather rigid assumption, often not satisfied in real-world applications. That is the reason why economists have replaced it by weaker assumptions in more contemporary studies. In fact, some of the progress in this research area is due to the work of economists. I am glad that the new book emphasizes economic applications.

In Chap. 5 the transition from generalized convexity to generalized monotonicity occurs. Historically, this happened only around 1990 when I was working with the late Stepan Karamardian after joining the University of California at Riverside. He was a former PhD student of George Dantzig at the University of California at Berkeley. We collaborated on the last two papers he published, both on generalized monotonicity. (a new research area) We had opened up together.

In 2005 Nicolas Hadjisavvas, Sandor Komlosi and I completed the first *Handbook of Generalized Convexity and Generalized Monotonicity* with contributions from many leading experts in the field, including Alberto Cambini and Laura Martein, a proven team of co-authors who in their unique colorful way have left an imprint in the field. The new book is further evidence of their style.

Chapters 6 and 7 are devoted to specialized results for quadratic functions and fractional functions. With this the authors follow the outline of the first monograph in this research area, *Generalized Concavity* by Mordecai Avriel, Walter E. Diewert, Siegfried Schaible and Israel Zang in 1988. Chapter 8 contains algorithmic material on solving generalized convex fractional programs. It defeats the objection sometimes raised that the area of generalized convexity lacks algorithmic contributions. It is true that there could be more results in this important direction on a topic which by nature is theoretical. Perhaps the presentation in Chap. 8 will motivate others to take up the challenge to derive more results with a computational emphasis.

Today *Generalized Concavity* (1988) is available to us as the first volume on the topic, together with the comprehensive *Handbook of Generalized Convexity and Generalized Monotonicity* (2005), an edited volume of 672 pages, written by 16 different researchers including Alberto Cambini and Laura Martein. In addition, the published proceedings of GC1–GC8 are available from reputable publishing houses. The proceedings of GC9 will appear partially in the prestigious Taiwanese Journal of Mathematics.

As somebody who has participated in all the conferences, GC1–GC8, and who is co-organizing GC9 together with Jen-Chih Yao, Kaohsiung and who has been involved in most publications mentioned before, I congratulate the authors for having produced such a fine volume in this growing area of research. Like me they stumbled into it when no monographs on the topic were available. I can see the usefulness of the book for teaching and research for generations to come. Its technical level makes it suitable for undergraduate and graduate students. The level is pitched wisely. The book is more accessible than the Handbook as it assumes less background knowledge about the topic. This is not surprising as the purpose of the Handbook is different. The new book can serve as an up-to-date link to the Handbook. It also saves the reader from going through the earlier proceedings with more dated results.

As someone who, like the authors, has not departed from the area of generalized convexity in his career, I can highly recommend this excellent new volume in our community of researchers. WGGC has been the background for most recent publications in our field of study. It is the excitement of working in teams which has been promoted by WGGC. A sense of community very common in Italy is the background of this new volume. It made me happy when I reviewed the manuscript first. I hope that many readers will come to the same conclusion. My thanks and congratulations go to the authors for a job well done.

I want to thank the authors for having taken the time to write *Generalized Convexity and Optimization with Economic Applications* and for their diligent effort to produce an up-to-date text and wish the book much success among our growing community of researchers.

Riverside, California,   *Siegfried Schaible*
June 2008   *Chair of WGGC*

Contents

1

Convex Functions

1.1 Introduction

Convex and concave functions have many important properties that are useful in Economics and Optimization. In this Chapter the basic properties of convex and concave functions are explained, including some fundamental results involving these functions. In particular, the role of convexity and concavity in Optimization is stressed. Since a function f is concave if and only if $-f$ is convex, any result related to a convex function can easily be translated for a concave function. For this reason only the proofs related to convex functions are presented. For the sake of completeness, the corresponding results for the concave case are summarized in Appendix B.

1.2 Convex Sets

From a geometrical point of view, a set $S \subseteq \Re^n$ is convex if, for any two points in S, the line segment connecting these two points lies entirely in S (see Fig. 1.1).

Fig. 1.1. Convex and not convex set

Formally, we have the following definition.

Definition 1.2.1. *A set $S \subseteq \Re^n$ is convex if*

$$x_1, x_2 \in S \Rightarrow \lambda x_1 + (1 - \lambda)x_2 \in S, \ \forall \lambda \in [0, 1]. \tag{1.1}$$

The point $x = \lambda x_1 + (1 - \lambda)x_2$, $\lambda \in [0, 1]$, is said to be a *convex combination* of x_1 and x_2. By $[x_1, x_2] = \{x \in S : x = \lambda x_1 + (1 - \lambda)x_2, \ \lambda \in [0, 1]\}$ we shall denote the closed line segment joining x_1 and x_2.

By convention, the empty set and the singleton set (a set consisting of a single point) are considered convex sets. The following are simple examples of convex sets:

- The whole set $\Re^n$;
- The line through x_0 and direction u: $r = \{x \in \Re^n : x = x_0 + tu, \ t \in \Re\}$;
- The hyperplane $H = \{x \in \Re^n : \alpha^T x = \beta\}$, $\alpha \in \Re^n, \alpha \neq 0, \beta \in \Re$;
- The closed half-spaces associated with H: $H^+ = \{x \in \Re^n : \alpha^T x \geq \beta\}$, $H^- = \{x \in \Re^n : \alpha^T x \leq \beta\}$.

Theorem 1.2.1. *The intersection of an arbitrary family of convex sets is convex.*

Proof. See Exercise 1.2.   □

Definition 1.2.2. *A convex combination of finitely many points $x_i \in \Re^n$, $i = 1, ..., k$, is a point x of the form*

$$x = \sum_{i=1}^{k} \lambda_i x_i, \ \sum_{i=1}^{k} \lambda_i = 1, \ \lambda_i \geq 0, \ i = 1, ..., k.$$

The following theorem characterizes a convex set in terms of convex combinations of its points.

Theorem 1.2.2. *A set $S \subseteq \Re^n$ is convex if and only if every convex combination of finitely many points of S belongs to S.*

Proof. Suppose that S is convex. The proof proceeds by induction on the number k of points. For $k = 2$ the thesis is true by definition. Assuming that every convex combination of k points of S belongs to S, we must prove that every convex combination of $k + 1$ points $x_1, ..., x_k, x_{k+1} \in S$ is a point of S.

Let $z = \sum_{i=1}^{k+1} \lambda_i x_i$, $\sum_{i=1}^{k+1} \lambda_i = 1$, $\lambda_i \geq 0$, $i = 1, ..., k+1$. If $\lambda_{k+1} = 0$ or $\lambda_{k+1} = 1$, then $z \in S$ by assumption. In any other case we can re-write z in the form

$$z = \mu \sum_{i=1}^{k} \frac{\lambda_i}{\mu} x_i + \lambda_{k+1} x_{k+1}, \ \mu = \sum_{i=1}^{k} \lambda_i = 1 - \lambda_{k+1} > 0.$$

The induction assumption implies that the convex combination of k points

$$\bar{x} = \sum_{i=1}^{k} \frac{\lambda_i}{\mu} x_i$$ belongs to S so that we have $z = \mu \bar{x} + (1 - \mu)x_{k+1}$, that is z is a convex combination of two points of S and so $z \in S$.

The sufficiency follows by noting that we can consider, in particular, every convex combination of two points in S so that S is convex by definition. $\quad\square$

1.2.1 Topological Properties of Convex Sets

A key theorem is the following.

Theorem 1.2.3. *Let $S \subseteq \Re^n$ be a convex set with $intS \neq \emptyset$. Let $x_1 \in clS$ and $x_2 \in intS$. Then, $\lambda x_1 + (1-\lambda)x_2 \in intS$ for all $\lambda \in [0,1)$.*

Proof. The assumption $x_2 \in intS$ implies the existence of a ball $B(x_2, \epsilon)$ of radius $\epsilon > 0$ and center x_2 such that $B(x_2, \epsilon) = \{x : \| x - x_2 \| < \epsilon\} \subset S$. We prove that each point $y = \lambda x_1 + (1-\lambda)x_2$, $\lambda \in (0,1)$ is an interior point showing that the ball $B(y, (1-\lambda)\epsilon) \subset S$, i.e., every point z such that $\| z - y \| < (1-\lambda)\epsilon$ belongs to S. Set $R = \frac{(1-\lambda)\epsilon - \|z-y\|}{\lambda}$. Since $x_1 \in clS$ there exists a point $z_1 \in S$ such that $\| z_1 - x_1 \| < R$. Let $z_2 = \frac{z-\lambda z_1}{1-\lambda}$. We have

$$\| z_2 - x_2 \| = \tfrac{1}{1-\lambda} \| z - \lambda z_1 - (1-\lambda)x_2 \| \leq \tfrac{1}{1-\lambda} \| z - \lambda z_1 - (y - \lambda x_1) \| \leq$$
$$\leq \tfrac{1}{1-\lambda} \| z - y \| + \lambda \| z_1 - x_1 \| < \tfrac{1}{1-\lambda}(\| z - y \| + \lambda \tfrac{(1-\lambda)\epsilon - \|z-y\|}{\lambda}) = \epsilon.$$

Consequently, $z_2 \in S$. By definition of z_2 we have $z = \lambda z_1 + (1-\lambda)z_2$, i.e., z is a convex combination of two points of S and thus $z \in S$. The proof is complete. $\quad\square$

Theorem 1.2.4. *Let $S \subseteq \Re^n$ be a convex set with $intS \neq \emptyset$. Then, the following conditions hold:*
(i) clS is convex;
(ii) $intS$ is convex;
(iii) $cl(intS) = clS$;
(iv) $int(clS) = intS$.

Proof. (i) Let $x_1, x_2 \in clS$ and let $z \in intS$. By Theorem 1.2.3, $\lambda x_1 + (1-\lambda)z \in intS$ for all $\lambda \in [0,1)$ so that $\mu x_2 + (1-\mu)(\lambda x_1 + (1-\lambda)z) \in intS$ for all $\mu \in [0,1)$. Taking the limit as λ approaches 1, we have $\mu x_2 + (1-\mu)x_1 \in clS$.
(ii) This follows directly from Theorem 1.2.3 by noting that the interior point x_1 is obtained for $\lambda = 1$.
(iii) Since $intS \subseteq S$, we have $cl(intS) \subseteq clS$. Consider now $z \in clS$ and let $x \in intS$. By Theorem 1.2.3, $z + \lambda(x - z) \in intS$, $\forall \lambda \in (0,1]$; consequently, $z + \frac{1}{n}(x - z) \in intS$ for all n so that taking the limit as n approaches $+\infty$, we have $z \in cl(intS)$ and thus $clS \subseteq cl(intS)$.
(iv) Since $S \subseteq clS$, we have $intS \subseteq int(clS)$. Let $z \in int(clS)$; then, there exists $\epsilon > 0$ such that the closed ball $\bar{B}(z, \epsilon) = \{x : \| x - x_2 \| \leq \epsilon\}$ is contained in clS. Let $x \in intS$ and put $y = z + \epsilon\frac{z-x}{\|z-x\|} \in B$. By simple calculations, setting $\lambda = \frac{\epsilon}{\epsilon + \|z-x\|}$, we have $z = \lambda x + (1-\lambda)y$, so that $z \in intS$ by Theorem 1.2.3. Consequently $int(clS) \subseteq intS$ and thus $int(clS) = intS$. $\quad\square$

Remark 1.2.1. Property (iii) of Theorem 1.2.4 implies that every boundary point of S is a limit point of a sequence of interior points of S.

The following theorem points out that every interior point of a convex set S may be expressed as a convex combination of two points of S, one of which is arbitrary.

Theorem 1.2.5. *Let $S \subseteq \Re^n$ be a convex set with $intS \neq \emptyset$. Then, $z \in intS$ if and only if for every $x \in S$ there exists $\mu > 1$ such that $x + \mu(z - x) \in S$.*

Proof. See Exercise 1.5. □

1.2.2 Relative Interior of Convex Sets

The properties stated in the previous theorems are established assuming that the set of interior points of a convex set is nonempty; sometimes such an assumption may appear to be a restrictive condition. For instance, a line on the plane or a triangle in the ordinary space or, in general, a convex set which lies entirely in a linear manifold, does not have interior points. In order to extend the previous results to every convex set it is necessary to introduce the concept of the relative interior of a convex set.

Let S be a convex set and let W be the smallest linear manifold containing S. Then, the *relative interior* of S, denoted by riS, is the set of all interior points of S with respect to the topology induced by $\Re^n$ on W; in others words, a point $x_0 \in riS$ if and only if there exists a ball B of radius ϵ and center x_0 such that $B \cap S \subset W$.

Obviously, $riS = intS$ if and only if $W = \Re^n$. In contrast to $intS$ the relative interior has the fundamental property that $riS \neq \emptyset$ for every nonempty convex set (see Exercise 1.9).

Properties stated in Theorems 1.2.3, 1.2.4 and 1.2.5 may be restated in terms of the relative interior of a convex set.

Theorem 1.2.6. *Let $S \subseteq \Re^n$ and let $x_1 \in clS$ and $x_2 \in riS$. Then, $\lambda x_1 + (1 - \lambda)x_2 \in riS$ for all $\lambda \in [0, 1)$.*

Theorem 1.2.7. *Let $S \subseteq \Re^n$ be a nonempty convex set. Then, the following conditions hold:*
(i) $riS \neq \emptyset$;
(ii) riS is convex;
(iii) $cl(riS) = clS$;
(iv) $ri(clS) = riS$;
(v) $z \in riS$ if and only if for every $x \in S$ there exists $\mu > 1$ such that $x + \mu(z - x) \in S$.

1.2.3 Extreme Points and Extreme Directions

A point x belonging to a convex set $S \subseteq \Re^n$ is said to be an extreme point of S if it is not possible to express x as a convex combination of two distinct points of S.

The following example points out that the set of extreme points may be empty, finite, or infinite.

Example 1.2.1.
- A line does not have extreme points while a closed half-line has only one extreme point;
- A rectangle has four extreme points in its vertices;
- Every boundary point of a ball is an extreme point of the ball.

Regarding the existence of an extreme point, we have the following theorem (see [234]).

Theorem 1.2.8. *The set of all extreme points of a compact convex set S is nonempty. Furthermore, every $x \in S$ may be expressed as a convex combination of finitely many extreme points of S.*

The last statement of Theorem 1.2.8 cannot be extended to an unbounded convex set. For instance, a point of a closed half-line starting from x_0 cannot be expressed as a convex combination of its only extreme point x_0. This motivates the introduction of the concepts of recession direction and extreme direction. Consider, firstly, the following theorem.

Theorem 1.2.9. *Let $S \subseteq \Re^n$ be a closed convex set. Then, S is unbounded if and only if there exists a half-line contained in S. Furthermore, if the half-line $x = x_0 + td, t \geq 0$ is contained in S then, for every $y \in S$, the half-line $x = y + kd, k \geq 0$ is contained in S.*

Proof. Obviously, the existence of a half-line contained in S implies the unboundedness of S. Viceversa, the unboundedness of S implies the existence of a sequence $\{x_n\} \subset S$ such that $\lim_{n \to +\infty} \| x_n \| = +\infty$. Let $x_0 \in S$; without loss of generality we can suppose that the sequence $\left\{ \frac{x_n - x_0}{\|x_n - x_0\|} \right\}$ converges to a point $d \in \Re^n \backslash \{0\}$. In order to reach the thesis it is sufficient to prove that the half-line $x_0 + td, t \geq 0$, is contained in S. The convexity of S implies $x_0 + \lambda(x_n - x_0) \in S$ for all $\lambda \in [0, 1]$. For any fixed $t > 0$ choose $\lambda_n = \frac{t}{\|x_n - x_0\|}$; the sequence $\left\{ x_0 + \frac{t}{\|x_n - x_0\|}(x_n - x_0) \right\}$ is contained in S so that its limit, given by $x_0 + td$, belongs to the closure of S for all $t \geq 0$. The thesis is achieved since $clS = S$.

The last statement of the theorem still needs to be proven. Since the half-line $x = x_0 + td, t \geq 0$, is contained in S, we have $x_n = x_0 + nd \in S$ for all n. The convexity of S implies that, for every $y \in S$, $y + \lambda(x_n - y) = y + \lambda(x_n - x_0) + \lambda(x_0 - y) = y + \lambda nd + \lambda(x_0 - y) \in S$ for all $\lambda \in [0, 1]$. For any fixed $t \geq 0$ choose $\lambda_n = \frac{t}{n}$; the sequence $y + \frac{t}{n}nd + \frac{t}{n}(x_0 - y)$ converges to $y + td$ so that $y + td \in clS = S$ for all $t \geq 0$. The proof is complete. $\square$

A direction $d \in \Re^n$ such that for every $y \in S$, the half-line $x = y + kd, k \geq 0$ is contained in S, is called a recession direction.

Theorem 1.2.9 establishes that the set of recession directions of a closed convex set S is nonempty if and only if S is unbounded.

A recession direction d is said to be an extreme direction if it is not possible to express d as a convex combination of two distinct recession directions.

Regarding the existence of an extreme point and extreme direction for an unbounded closed convex set, we have the following theorem (see [234]).

Theorem 1.2.10. *An unbounded closed convex set containing no lines has at least one extreme point and one extreme direction.*

The following fundamental representation theorem holds (see [234]).

Theorem 1.2.11. *Let $S \subseteq \Re^n$ be a closed convex set containing no lines. Then, $x \in S$ if and only if x can be expressed as the sum $x = y + d$, where y is a convex combination of extreme points of S and d is a positive linear combination of extreme directions.*

A polyhedron, defined as the intersection of finitely many closed half-spaces, is a special convex set having a finite number of extreme points and extreme directions. The extreme points of a polyhedron are also called vertices of the polyhedron. A bounded polyhedron is called a polytope.

1.2.4 Supporting Hyperplanes and Separation Theorems

Theorems of separation play a fundamental role in Optimization. We will limit ourselves to presenting some basic results which will be utilized later.

Let S be a convex subset of $\Re^n$ and let x_0 be a boundary point of S.

A supporting half-space to S at x_0 is a closed half-space containing S.

A supporting hyperplane to S at x_0 is the boundary of a supporting half-space to S at x_0.

In other words, the hyperplane $H_{x_0} = \{x \in \Re^n : \alpha^T x = \alpha^T x_0\}$ is a supporting hyperplane to S at x_0 if either $S \subseteq H_{x_0}^+ = \{x \in \Re^n : \alpha^T x \geq \alpha^T x_0\}$ or else $S \subseteq H_{x_0}^- = \{x \in \Re^n : \alpha^T x \leq \alpha^T x_0\}$.

Without loss of generality we can assume that $S \subseteq H_{x_0}^+$ by replacing α with $-\alpha$ if necessary.

Definition 1.2.3. *Let S, T be two subsets of $\Re^n$.*
A hyperplane $H = \{x \in \Re^n : \alpha^T x = \beta\}$ is said to separate S and T if $\alpha^T x \geq \beta$, $\forall x \in S$, and $\alpha^T x \leq \beta$, $\forall x \in T$.

In Fig. 1.2, a supporting hyperplane and a separating hyperplane are depicted. Some fundamental results related to the existence of a supporting hyperplane and to the existence of a separation hyperplane are found in the following theorems whose proofs can be found in any text-book (see references at the end of this Chapter).

Theorem 1.2.12. *(Separation of a convex set and a point)*
Let S be a closed convex subset of $\Re^n$ and let $y_0 \notin S$. Then, there exist $\alpha \in \Re^n \backslash \{0\}$, $x_0 \in S$ such that $\alpha^T x \geq \alpha^T x_0$ for all $x \in S$ and $\alpha^T y_0 < \alpha^T x_0$.

Theorem 1.2.13. *(Existence of a supporting hyperplane at a boundary point)*
Let S be a closed convex subset of $\Re^n$ and let x_0 be a boundary point of S.
Then, there exists $\alpha \in \Re^n \backslash \{0\}$ such that $\alpha^T x \geq \alpha^T x_0$ for all $x \in S$.

Supporting hyperplane   Separating hyperplane

Fig. 1.2. Separating hyperplanes

Theorem 1.2.14. *(Separation of two sets)*
Let S_1 and S_2 be nonempty convex sets in $\Re^n$. Then, there exists a hyperplane
which separates S_1 and S_2 if and only if $riS_1 \cap riS_2 = \emptyset$.

The following corollary shows that every closed convex set can be represented
as the intersection of closed half-spaces.

Corollary 1.2.1. *Let S be a closed convex subset of $\Re^n$. Then, S is the*
intersection of all its supporting half-spaces, i.e., $S = \bigcap\limits_{x_0 \in S} H_{x_0}^+$.

Proof. Obviously, S is contained in the intersection of all half-spaces $H_{x_0}^+$. Let
$y \in \bigcap\limits_{x_0 \in S} H_{x_0}^+$ and suppose that $y \notin S$. Then, from Theorem 1.2.12, there exists
a supporting hyperplane at a point x_0 belonging to the boundary ∂S of S such
that $y \notin H_{x_0}^+$ and this is a contradiction.   □

1.2.5 Convex Cones and Polarity

A cone (with vertex at zero) in $\Re^n$ is a nonempty set C satisfying the following
property:
$$x \in C, \ k \geq 0 \Rightarrow kx \in C.$$
A convex cone is a cone which is convex as a set.

Half-lines, lines, subspaces, and half-spaces through the origin are examples of
convex cones. The union of disjoint closed convex cones generates non-convex
cones.
A cone is convex if and only if it is closed under the operations of addition and
multiplication by a non-negative scalar as is shown in the following theorem.

Theorem 1.2.15. *A set $C \subseteq \Re^n$ is a convex cone if and only if the following properties hold:*
(i) $x \in C$, $k \geq 0 \Rightarrow kx \in C$;
(ii) $x, y \in C \Rightarrow x + y \in C$.

Proof. If C is a convex cone, (i) follows by definition of a cone. Furthermore, if $x, y \in C$, the convexity of C implies $\frac{1}{2}x + \frac{1}{2}y \in C$ and thus $2 \cdot (\frac{1}{2}x + \frac{1}{2}y) = x + y \in C$, so that (ii) holds.
Assume now the validity of (i) and (ii) and let $x, y \in C$. Then, from (i) C is a cone so that $\lambda x \in C$, $\lambda \geq 0$, $(1 - \lambda)y \in C$, $\lambda \leq 1$, and from (ii) $\lambda x + (1 - \lambda)y \in C$, $0 \leq \lambda \leq 1$, i.e., C is convex. □

Corollary 1.2.1 may be specified in the case where S is a closed convex cone obtaining the following corollary.

Corollary 1.2.2. *Let C be a closed convex cone in $\Re^n$. Then, C is the intersection of all its supporting half-spaces at the origin.*

Proof. Taking into account Corollary 1.2.1, it is sufficient to prove that a supporting hyperplane H_{x_0} to C at $x_0 \in \partial C$ passes through the origin. We have $\alpha^T x \geq \alpha^T x_0, \forall x \in C$. Since $kx_0 \in C$ for all $k > 0$ we have $k\alpha^T x_0 \geq \alpha^T x_0, \forall k > 0$, that is $(k - 1)\alpha^T x_0 \geq 0, \forall k > 0$ and this last inequality holds if and only if $\alpha^T x_0 = 0$. □

In some problems we are interested in the existence of a strict supporting hyperplane to C at the origin, i.e., in the existence of a supporting hyperplane H such that $H \cap C = \{0\}$. In order to fully illustrated this important aspect, we shall first introduce, the notion of polarity.

Definition 1.2.4. *Let C be a cone in $\Re^n$. Then, the positive polar of C, denoted by C^+, is given by $C^+ = \{\alpha \in \Re^n : \alpha^T c \geq 0, \ \forall \ c \in C\}$.*

The opposite of C^+ is referred to as the negative polar of C and is denoted by C^-. Equivalently, $C^- = \{\alpha \in \Re^n : \alpha^T c \leq 0, \ \forall \ c \in C\}$.

Remark 1.2.2. It follows immediately from the definition that polarity is order-inverting, i.e., if $C_1 \subset C_2$ then $C_1^+ \supset C_2^+$.

The following theorem states the structure of C^+ and, in addition, it characterizes the elements of a closed convex cone in terms of its positive polar and viceversa.

Theorem 1.2.16. *Let C be a closed convex cone in $\Re^n$. Then:*
(i) C^+ is a closed convex cone;
(ii) $c \in C$ if and only if $\alpha^T c \geq 0$ for all $\alpha \in C^+$;
(iii) $c \in \text{int}C$ if and only if $\alpha^T c > 0$ for all $\alpha \in C^+ \backslash \{0\}$;
(iv) $C = C^{++}$, where C^{++} is the positive polar of C^+;
(v) $\alpha \in \text{int}C^+$ if and only if $\alpha^T c > 0$ for all $c \in C \backslash \{0\}$.

Proof. (i) Let $\alpha_1, \alpha_2 \in C^+$. Then, we have $\alpha_1^T c \geq 0$, $\alpha_2^T c \geq 0$ for all $c \in C$, so that $(\alpha_1 + \alpha_2)^T c \geq 0$ and $k\alpha_1^T c \geq 0$, for all $k \geq 0$. It follows, from Theorem 1.2.15, that C^+ is a convex cone. Consider now a sequence $\{\alpha_n\} \subset C^+$ converging to an element α. By means of the continuity of the scalar product, and by taking the limit in $\alpha_n^T c \geq 0$, we obtain $\alpha^T c \geq 0$ for all $c \in C$. Consequently, C^+ is a closed cone.

(ii) Taking into account the definition of C^+, we must prove that the condition $\alpha^T c \geq 0$, $\forall \alpha \in C^+$ implies $c \in C$. If not, by Theorem 1.2.12 and by Corollary 1.2.2, there exist $\gamma \in \Re^n \backslash \{0\}$, $c_0 \in \partial C$, such that $\gamma^T x \geq \gamma^T c_0 = 0$, $\forall x \in C$ and $\gamma^T c < 0$. The former inequality implies $\gamma \in C^+$ contradicting the latter which implies $\gamma \notin C^+$.

(iii) Let $c \in intC$. From (ii) we have $\alpha^T c \geq 0$ for all $\alpha \in C^+$. Assume, by contradiction, the existence of $\alpha \in C^+, \alpha \neq 0$, such that $\alpha^T c = 0$. Since c is an interior point, there exists $\epsilon > 0$ such that $c + \epsilon\, d \in C$ for every direction d of unitary norm. It follows that $\alpha^T(c + \epsilon\, d) = \epsilon\alpha^T d \geq 0$ for all d and this is absurd since, by choosing $d^* = -\frac{\alpha}{\|\alpha\|}$, we have $\epsilon\alpha^T d^* < 0$. Assume now $\alpha^T c > 0$ for all $\alpha \in C^+ \backslash \{0\}$. We must prove that $c \in intC$. If not, taking into account (ii), we have $c \in \partial C$ so that, by Theorem 1.2.13, there exists $\gamma \in \Re^n \backslash \{0\}$ such that $\gamma^T x \geq \gamma^T c = 0$, $\forall x \in C$, which contradicts the assumption.

(iv) By applying (ii) to the polar cone C^+, we have $\alpha \in C^+$ if and only if $z^T \alpha \geq 0$ for all $z \in C^{++}$. By comparing this last inequality with (ii), the thesis is achieved.

v) This follows by applying (iii) to the polar cone C^+, taking into account that $C^{++} = C$. $\qquad\square$

Remark 1.2.3. The proof of (i) of Theorem 1.2.16 points out that C^+ is a closed convex cone even if C is not closed and/or convex.

From (v) of Theorem 1.2.16, the existence of a strict supporting hyperplane to C at the origin is equivalent to the condition $intC^+ \neq \emptyset$. As we will see, this last condition is strictly related to the non-existence of lines contained in C. A closed cone which does not contain lines, i.e., $c \in C$ implies $-c \notin C$, is called a pointed cone. Equivalently, C is a pointed cone if and only if $C \cap (-C) = \{0\}$. The set $C \cap (-C)$ is called the lineality space of C and it is denoted by $\ell(C)$. The following theorem holds, where $dim\ C^+$ denotes the dimension of C^+, i.e., the maximum number of linearly independent vectors contained in C^+ or, equivalently, the dimension of the smallest subspace containing C^+.

Theorem 1.2.17. *Let C be a closed convex cone in $\Re^n$. Then:*
(i) $\ell(C)$ is the largest subspace contained in C;
(ii) $dim\ \ell(C) + dim\ C^+ = n$;
(iii) $intC^+ \neq \emptyset$ if and only if $\ell(C) = \{0\}$.

Proof. (i) Let $c \in C \cap (-C)$; $c \in C$ implies $kc \in C$ for all $k \geq 0$, while $c \in -C$ implies $-kc \in C$ for all $k \geq 0$, so that $kc \in C \cap (-C)$ for all $k \in \Re$.

Furthermore, Theorem 1.2.15 implies that $C \cap (-C)$ is closed with respect to the addition. It follows that $\ell(C)$ is a subspace. Let $W \subset C$ be a subspace; since $w \in W$ implies $-w \in W$, we have $w, -w \in C$ or equivalently, $w \in C \cap (-C)$. Consequently $W \subseteq \ell(C)$ so that $\ell(C)$ is the largest subspace contained in C.

(ii) Let $c \in \ell(C)$ and $\alpha \in C^+$. Since $c, -c \in C$, we have $\alpha^T c \geq 0$, $\alpha^T(-c) \geq 0$, so that $\alpha^T c = 0$ for all $c \in \ell(C)$. It follows that $\alpha \in [\ell(C)]^\perp$, i.e., $C^+ \subseteq [\ell(C)]^\perp$. Let V be the smallest subspace containing C^+. If $V \subset [\ell(C)]^\perp$, from $C^+ \subseteq V \subset [\ell(C)]^\perp$ we have $C^{++} = C \supseteq V^+ = V^\perp \supset \ell(C)$ and this contradicts (i). Consequently, $V = [\ell(C)]^\perp$ so that $dim\ C^+ = dim\ V = dim\ [\ell(C)]^\perp$ Since $dim\ \ell(C) + dim\ [\ell(C)]^\perp = n$, (ii) follows.

(iii) It follows from (ii) by noting that $int C^+ \neq \emptyset$ if and only if $dim\ V = dim\ C^+ = n$ or, equivalently, if and only if $dim\ \ell(C) = 0$, i.e., $\ell(C) = \{0\}$.   □

The following corollary, which is a direct consequence of (v) of Theorem 1.2.16 and of (iii) of Theorem 1.2.17, states a necessary and sufficient condition for the existence of a strict supporting hyperplane to a cone at the origin.

Corollary 1.2.3. *Let C be a closed convex cone in $\Re^n$. Then, there exists $\alpha \in \Re^n$ such that $\alpha^T c > 0$ for all $c \in C, c \neq 0$, if and only if C is pointed.*

By noting that Theorem 1.2.17 implies $ri C^+ = int C^+$ with respect to the topology induced by $\Re^n$ on the subspace $[\ell(C)]^\perp$, we have the following theorem which generalizes (v) of Theorem 1.2.16.

Theorem 1.2.18. *Let C be a closed convex cone in $\Re^n$. Then $\alpha \in ri C^+$ if and only if*
(i) $\alpha^T c = 0$ for all $c \in \ell(C)$;
(ii) $\alpha^T c > 0$ for all $c \in C \backslash \ell(C)$.

1.3 Convex Functions

From a geometrical point of view, a function f is convex provided that the line segment connecting any two points of its graph lies on or above the graph. The function f is strictly convex provided that the line segment connecting any two points of its graph lies above the graph (see Fig. 1.3).

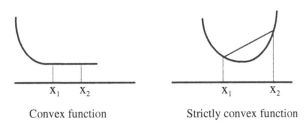

| Convex function | Strictly convex function |

Fig. 1.3. Examples of convex functions

From an analytical point of view, we have the following definitions.

Definition 1.3.1. *Let f be a function defined on a convex set $S \subseteq \Re^n$.*
(i) The function f is said to be convex on S if for every $x_1, x_2 \in S$

$$f(\lambda x_1 + (1 - \lambda)x_2) \leq \lambda f(x_1) + (1 - \lambda)f(x_2), \quad \forall \lambda \in [0, 1]. \qquad (1.2)$$

(ii) The function f is said to be strictly convex on S if for every $x_1, x_2 \in S$

$$f(\lambda x_1 + (1 - \lambda)x_2) < \lambda f(x_1) + (1 - \lambda)f(x_2), \quad \forall \lambda \in (0, 1). \qquad (1.3)$$

A function f defined on a convex set $S \subseteq \Re^n$ is concave if and only if $-f$ is convex on S. It follows that all the results related to convex functions that we are going to establish can be easily stated in terms of concave functions. For the sake of completeness, and also taking into account that concave functions are more common in Economics, in Appendix B we shall give a summary of the main properties of concave functions.
Simple examples of convex and concave functions are given below.

Example 1.3.1.
1. An affine function $f(x) = a^T x + b$, $x \in \Re^n$ is both convex and concave (not strictly);
2. The function $f(x) = x + |x|$, $x \in \Re$ is convex (not strictly);
3. The function $f(x) = ax^2 + bx + c$, $x \in \Re$ is strictly convex if $a > 0$ and it is strictly concave if $a < 0$.

Obviously, a strictly convex function is convex, too; the converse statement is not true as it follows from (i) or (ii) of Example 1.3.1.
The inequalities (1.2), (1.3), may be extended to any weighted average of its values at a finite number of points as is shown in the following theorem.

Theorem 1.3.1. *(Jensen's inequality)*
(i) A function f is convex on S if and only if for every $x_1, ..., x_n \in S$,

$$f\left(\sum_{i=1}^n \lambda_i x_i\right) \leq \sum_{i=1}^n \lambda_i f(x_i), \lambda_i \geq 0, \ i = 1, ..., n, \ \sum_{i=1}^n \lambda_i = 1. \qquad (1.4)$$

(ii) A function f is strictly convex on S if and only if for every $x_1, ..., x_n \in S$,

$$f\left(\sum_{i=1}^n \lambda_i x_i\right) < \sum_{i=1}^n \lambda_i f(x_i), \lambda_i \geq 0, \ i = 1, ..., n, \ \sum_{i=1}^n \lambda_i = 1. \qquad (1.5)$$

Proof. (i) Suppose that f is convex. The proof proceeds by induction on the number n of points. For $n = 2$ the thesis is true by definition. By assuming that (1.4) is verified for every convex combination of n points, we must prove that

$$f\left(\sum_{i=1}^{n+1} \lambda_i x_i\right) \leq \sum_{i=1}^{n+1} \lambda_i f(x_i), \lambda_i \geq 0, \ i = 1, ..., n+1, \ \sum_{i=1}^{n+1} \lambda_i = 1.$$

If $\lambda_i = 0$ for some i, the thesis follows by means of the induction assumption, otherwise we have (see the proof given in Theorem 1.2.2)

$$\sum_{i=1}^{n+1} \lambda_i x_i = \mu\bar{x} + (1-\mu)x_{n+1}, \ \mu = \sum_{i=1}^{n} \lambda_i, \ \bar{x} = \sum_{i=1}^{n} \frac{\lambda_i}{\mu} x_i.$$

By taking into account the convexity of f and the induction assumption, we have

$$f\left(\sum_{i=1}^{n+1} \lambda_i x_i\right) = f(\mu\bar{x} + (1-\mu)x_{n+1}) \le \mu f(\bar{x}) + (1-\mu)f(x_{n+1}) \le$$

$$\le \mu\left(\frac{1}{\mu}\sum_{i=1}^{n} \lambda_i f(x_i)\right) + \lambda_{n+1}f(x_{n+1}) = \sum_{i=1}^{n+1} \lambda_i f(x_i).$$

The reverse statement follows by noting that if (1.4) is verified for every n of points it is in particular verified for $n = 2$, so that by definition f is convex. (ii) This follows analogously. □

Associated with a convex function are the epigraph and the lower level sets defined, respectively, as follows:

$$epif = \{(x, z) : x \in S, \ z \ge f(x)\}; \ S_{\le\alpha} = \{x \in S : \ f(x) \le \alpha\}.$$

A convex function is characterized by the convexity of its epigraph as is shown in the following theorem.

Theorem 1.3.2. *Let f be a function defined on a convex set $S \subseteq \Re^n$. Then: (i) f is convex if and only if $epif$ is a convex set; (ii) f is strictly convex if and only if $epif$ is a convex set and it does not contain any line segment.*

Proof. (i) Let f be convex. If $(x_1, z_1), (x_2, z_2) \in epif$, then $z_1 \ge f(x_1)$, $z_2 \ge f(x_2)$, so that, for every $\lambda \in [0, 1]$, we have $\lambda(x_1, z_1) + (1-\lambda)(x_2, z_2) = (\lambda x_1 + (1-\lambda)x_2, \lambda z_1 + (1-\lambda)z_2) \in epif$ since $(\lambda z_1 + (1-\lambda)z_2) \ge \lambda f(x_1) + (1-\lambda)f(x_2) \ge f(\lambda x_1 + (1-\lambda)x_2)$.
Assume now the convexity of $epif$ and let $x_1, \ x_2 \in S$.
Since $(x_1, f(x_1)) \in epif$, $(x_2, f(x_2)) \in epif$, we have $\lambda(x_1, f(x_1)) + (1-\lambda)(x_2, f(x_2)) = (\lambda x_1 + (1-\lambda)x_2, \lambda f(x_1) + (1-\lambda)f(x_2)) \in epif$, $\forall \lambda \in [0, 1]$.
On the other hand $(\lambda x_1 + (1-\lambda)x_2, \lambda f(x_1) + (1-\lambda)f(x_2)) \in epif$ if $\lambda f(x_1) + (1-\lambda)f(x_2) \ge f(\lambda x_1 + (1-\lambda)x_2)$, i.e., if f is convex.
(ii) The proof is similar to the one given in (i). □

Regarding the lower level sets of a convex function we have the following theorem.

Theorem 1.3.3. *Let f be a convex function defined on a convex set $S \subseteq \Re^n$. Then, $S_{\le\alpha}$ is convex for every $\alpha \in \Re$.*

Proof. The thesis is true by convention if $S_{\leq\alpha} = \emptyset$ or $S_{\leq\alpha}$ is a singleton set. The points x_1, x_2 belong to $S_{\leq\alpha}$ if and only if $f(x_1) \leq \alpha$, $f(x_2) \leq \alpha$ so that, by means of the convexity of f, we have $f(\lambda x_1 + (1 - \lambda)x_2) \leq (\lambda f(x_1) + (1 - \lambda)f(x_2) \leq \lambda\alpha + (1 - \lambda)\alpha) = \alpha$. It follows that $\lambda x_1 + (1 - \lambda)x_2 \in S_{\leq\alpha}$.   □

Remark 1.3.1. The necessary condition for a function to be convex stated in Theorem 1.3.3 is not generally sufficient. For instance, any increasing nonlinear concave single-variable function has convex lower sets but it is not convex. This fact has led to the introduction of a new class of functions, as we will see in the next chapter.

1.3.1 Algebraic Structure of the Convex Functions

The class of convex functions defined on a convex set is closed with respect to the addition and with respect to the non-negative scalar multiplication. More precisely, we have the following theorem.

Theorem 1.3.4. *Let $f_1, f_2, ..., f_m$ be functions defined on a convex set $S \subseteq \Re^n$ and set $f(x) = \sum_{i=1}^{m} \alpha_i f_i(x)$, $\alpha_i \geq 0$. Then:*

(i) If f_i, $i = 1, .., m$, are convex on S, then f is convex on S.
(ii) If f_i, $i = 1, .., m$, are strictly convex on S, then f is strictly convex on S.

Proof. See Exercise 1.24.   □

1.3.2 Composite Function

Another important property is related to the composition product.

Theorem 1.3.5. *Let $f : S \to \Re$ be a convex function defined on a convex set $S \subseteq \Re^n$ and let $g : A \to \Re$ be a non-decreasing convex function, with $f(S) \subseteq A$. Then the composite function $h(x) = g(f(x))$ is convex on S. Furthermore, if f is strictly convex and g is an increasing convex function, then h is strictly convex.*

Proof. See Exercise 1.25.   □

Let us note that the requirement of the convexity of g is essential to guaranteeing the convexity of the composite function. For instance, the function $h(x) = x$ is convex, the function $g(x) = x^3$ is an increasing non-convex function and the composite function $f(x) = g(h(x)) = x^3$ is not convex.
Theorems 1.3.4, 1.3.5 and the analogous ones for concave functions are sometimes useful in constructing convex or concave functions.

Example 1.3.2.
1. The function $f(x) = e^{a^T x + b}$, $x \in \Re^n$ is convex since the affine function $a^T x + b$ is convex and the exponential function is an increasing convex function.
2. The function $f(x) = (a^T x + b)^2$ is convex on $S = \{x \in \Re^n : a^T x + b > 0\}$ since the affine function $a^T x + b$ is convex and the square function is an increasing convex function on the set of positive real numbers.

Example 1.3.3.
1. The power $f(x) = x^\alpha$, $x \geq 0$, is strictly concave for $0 < \alpha < 1$ and it is strictly convex for $\alpha < 0$ and for $\alpha > 1$;
2. $f(x) = \log x$, $x > 0$, is strictly concave.

Example 1.3.4. If f is a positive concave function, then $z(x) = \log f(x)$ is concave since the logarithm function is increasing and concave.

Example 1.3.5. If f is a positive concave function, then $\frac{1}{f}$ is convex. In fact, $z(x) = \log \frac{1}{f(x)} = -\log f(x)$ is convex as the opposite of the concave function $\log f(x)$. It follows that $e^{z(x)} = \frac{1}{f(x)}$ is convex.

1.3.3 Differentiable and Twice Differentiable Convex Functions

A convex function is continuous on the interior of its domain but not necessarily differentiable. For instance, the convex function $f(x) = |x|$ is continuous on $\Re$ but it is not differentiable at $x = 0$.
From a geometrical point of view, a differentiable function is convex if and only if its graph lies on or above the tangent in any point of the graph; it is strictly convex if its graph lies above the tangent in any point of the graph.
From an analytical point of view, the convexity of a function of one variable may be characterized by means of its first and second derivatives, according to the following properties:
• Let I be an open interval of the real line. A differentiable function f is convex on I if and only if for every $x_0 \in I$ we have

$$f(x) \geq f(x_0) + f'(x_0)(x - x_0), \ \forall x \in I. \tag{1.6}$$

• Let I be an open interval of the real line. A twice differentiable function f is convex on I if and only if

$$f''(x) \geq 0, \ \forall x \in I. \tag{1.7}$$

The extensions of (1.6) and (1.7) to functions of more variables are given below.

Theorem 1.3.6. *Let f be a differentiable function defined on a nonempty open convex set $S \subseteq \Re^n$. Then, f is convex on S if and only if for every $x_0 \in S$*

$$f(x) \geq f(x_0) + (x - x_0)^T \nabla f(x_0), \quad \forall x \in S. \tag{1.8}$$

Proof. Assume the convexity of f. Then, for every x_0, $x \in S$ with $x \neq x_0$, we have $f(x_0 + \lambda(x - x_0)) \leq f(x_0) + \lambda(f(x) - f(x_0))$, $\forall \lambda \in [0, 1]$ or, equivalently,

$$\frac{f(x_0 + \lambda(x - x_0)) - f(x_0)}{\lambda} \leq f(x) - f(x_0).$$

It follows that $\lim_{\lambda \to 0^+} \frac{f(x_0 + \lambda(x - x_0)) - f(x_0)}{\lambda} = (x - x_0)^T \nabla f(x_0) \leq f(x) - f(x_0)$ so that (1.8) is verified.

Conversely, let $x_1, x_2 \in S$. Setting $x = x_2$, $x_0 = \lambda x_1 + (1 - \lambda)x_2$ in (1.8), we obtain

$$f(x_2) \geq f(\lambda x_1 + (1 - \lambda)x_2) + \lambda(x_2 - x_1)^T \nabla f(\lambda x_1 + (1 - \lambda)x_2). \qquad (1.9)$$

Setting $x = x_1$, $x_0 = \lambda x_1 + (1 - \lambda)x_2$ in (1.8), we obtain

$$f(x_1) \geq f(\lambda x_1 + (1 - \lambda)x_2) + (1 - \lambda)(x_1 - x_2)^T \nabla f(\lambda x_1 + (1 - \lambda)x_2). \quad (1.10)$$

By multiplying (1.9) and (1.10) by $(1 - \lambda)$ and λ, respectively, and by adding them up we obtain $(1 - \lambda)f(x_2) + \lambda f(x_1) \geq f(\lambda x_1 + (1 - \lambda)x_2)$, $\forall \lambda \in [0, 1]$, i.e., the convexity of the function. $\qquad \square$

The following theorem points out that there is a strict and useful connection between the convexity of a function and the convexity of its restriction along lines.

Theorem 1.3.7. *Let f be a function defined on a convex set $S \subseteq \Re^n$. Then, f is convex (strictly convex) on S if and only if the restriction of f to each line segment contained in S is a convex (strictly convex) function.*

Proof. Let $\varphi(t) = f(x_0 + tu)$, $t \in I$, be the restriction of f on a line through $x_0 \in S$. Assuming the convexity of f, we must prove that $\varphi(t)$ is convex, too, i.e., $\varphi(\lambda t_1 + (1 - \lambda)t_2) \leq \lambda \varphi(t_1) + (1 - \lambda)\varphi(t_2)$, $\lambda \in [0, 1]$. Setting $x_1 = x_0 + t_1 u$, $x_2 = x_0 + t_2 u$, we have $\lambda x_1 + (1 - \lambda)x_2 = x_0 + (\lambda t_1 + (1 - \lambda)t_2)u$ so that $f(\lambda x_1 + (1 - \lambda)x_2) = f(x_0 + (\lambda t_1 + (1 - \lambda)t_2)u) = \varphi(\lambda t_1 + (1 - \lambda)t_2)$. The convexity of f implies $f(\lambda x_1 + (1 - \lambda)x_2) \leq \lambda f(x_1) + (1 - \lambda)f(x_2)$ or, equivalently, $\varphi(\lambda t_1 + (1 - \lambda)t_2) \leq \lambda \varphi(t_1) + (1 - \lambda)\varphi(t_2)$.

Assume now that every restriction of f on a line segment is convex; we must prove that $f(\lambda x_1 + (1 - \lambda)x_2) \leq \lambda f(x_1) + (1 - \lambda)f(x_2)$, $\lambda \in [0, 1]$. We have $f(\lambda x_1 + (1 - \lambda)x_2) = f(x_2 + \lambda(x_1 - x_2)) = \varphi(\lambda)$. The convexity of $\varphi(\lambda)$ implies $\varphi(\lambda \cdot 1 + (1 - \lambda) \cdot 0) \leq \lambda \varphi(1) + (1 - \lambda)\varphi(0)$, i.e., the thesis. $\qquad \square$

The previous theorem is a simpler way to obtain a characterization that is easier to check than (1.8) when the function is continuously twice differentiable.

Theorem 1.3.8. *Let f be a continuously twice differentiable function on a nonempty open convex set $S \subseteq \Re^n$. Then, f is convex on S if and only if its Hessian matrix $H(x)$ is positive semidefinite at each point x of S.*

Proof. Let $x_0 \in S$, $d \in \Re^n$ and consider the restriction $\varphi(t) = f(x_0 + td)$ with t such that $x_0 + td \in S$. The thesis follows from Theorem 1.3.7 and from (1.7), taking into account that $\varphi''(t) = d^T H(x_0 + td)d$.   □

1.4 Convexity and Homogeneity

In this section we will point out the connection between homogeneity and convexity.

Definition 1.4.1. *Let $C \subseteq \Re^n$ be a convex cone. A function $f : C \to \Re$ is said to be homogeneous of degree $\alpha \in \Re$ if for every $x \in C$*

$$f(tx) = t^\alpha f(x), \ \forall t > 0.$$

In particular, a function f is homogeneous of degree zero, if for every $x \in C$ $f(tx) = f(x)$, $\forall t > 0$; while a function f is said to be linearly homogeneous if it is homogeneous of degree one, i.e., for every $x \in C$, $f(tx) = tf(x)$, $\forall t > 0$. Homogeneous functions appear frequently in Economics. For instance:

• The demand function $D(p, R) = \frac{R}{p}$, where $p > 0$ is the price of a good and $R > 0$ is the income, is homogeneous of degree zero; this means that the demand does not change if both the income and the price double, triple, ..., as may happen when there is a change in the monetary unit;

• The production function $f(L, K) = L^{\frac{1}{3}} K^{\frac{2}{3}}$, where $L > 0$ is the labour and $K > 0$ is the capital, is homogeneous of degree one; this means that the production doubles, triples, .. if both the labour and the capital double, triple,... This kind of property is expressed in Economics by saying that f exhibits constant returns to scale.

Some other examples will be presented in Sect. 2.4.

The following theorem states some properties of homogeneous functions.

Theorem 1.4.1.
(i) Let $f_1, f_2, ..., f_m$ be homogeneous functions of degree α_i, $i = 1, ..., m$, respectively, defined on a convex cone $C \subseteq \Re^n$. Then $z(x) = f_1(x) \cdot f_2(x) \cdot ... \cdot f_m(x)$ is homogeneous of degree $\alpha_1 + \alpha_2 + ... + \alpha_m$.
(ii) Let $f_1, f_2, ..., f_m$ be homogeneous functions of the same degree α defined on a convex cone $C \subseteq \Re^n$. Then $z(x) = (f_1(x) + f_2(x) + ... + f_m(x))^\beta$ is homogeneous of degree $\alpha\beta$.

Proof. See Exercise 1.34.   □

The following theorem shows that linear homogeneity plus subadditivity produces convexity.

Theorem 1.4.2. *Let f be a linearly homogeneous function defined on a convex cone $C \subseteq \Re^n$. Then, f is convex if and only if for every $x, y \in C$*

$$f(x + y) \leq f(x) + f(y).$$

Proof. Linear homogeneity and convexity imply subadditivity since $f(x+y) = f\left(2(\frac{x+y}{2})\right) = 2f(\frac{x+y}{2}) \leq 2(\frac{1}{2}f(x) + \frac{1}{2}f(y))$.

On the other hand subadditivity and linear homogeneity imply convexity since $f(\lambda x + (1-\lambda)y) \leq f(\lambda x) + f((1-\lambda)y) = \lambda f(x) + (1-\lambda)f(y)$. $\qquad\square$

1.5 Minima of Convex Functions

In this section we point out the role of convexity in Optimization. Knowing whether or not a local minimum is also global is one of the most important questions in Optimization. The assumption of convexity gives a positive answer to this question as is stated in the following theorem.

Theorem 1.5.1. *Let $S \subseteq \Re^n$ be a convex set and let f be a convex function on S. Then:*
(i) A local minimum point is also global;
(ii) The set S^* of all minimum points is a convex set;
(iii) S^* has at most one element if f is strictly convex.

Proof. (i) Let $x_0 \in S$ be a local minimum point of f and assume, by contradiction, the existence of $x^* \in S$ such that $f(x^*) < f(x_0)$. The convexity of S implies $x = \lambda x^* + (1-\lambda)x_0 \in S$, $\forall \lambda \in [0,1]$ and the convexity of the function implies $f(x) \leq \lambda f(x^*) + (1-\lambda)f(x_0) < \lambda f(x_0) + (1-\lambda)f(x_0) = f(x_0)$, $\forall \lambda \in (0,1)$. In particular, when λ approaches to zero, we have $f(x) < f(x_0)$ for every point x belonging to a neighbourhood of x_0, contradicting the local minimality assumption.

(ii) The thesis follows by convention if $S^* = \emptyset$. Let $x_0 \in S^*$ and consider the lower level set $S_{\leq f(x_0)} = \{x \in S : f(x) \leq f(x_0)\}$. Obviously we have $S_{\leq f(x_0)} = S^*$ so that the thesis follows by Theorem 1.3.3.

(iii) Assume, to get a contradiction, the existence of two elements $x_1, x_2 \in S^*$. Since $f(x_1) = f(x_2)$, by the strict convexity of f we have $f(\lambda x_1 + (1-\lambda)x_2) < \lambda f(x_1) + (1-\lambda)f(x_2) = f(x_1) = f(x_2)$, $\forall \lambda \in (0,1)$ thereby contradicting the global minimality assumption. $\qquad\square$

Knowing whether or not a critical point is a local minimum is another important question. Once again, the assumption of convexity gives a positive answer to this question.

Theorem 1.5.2. *Let $S \subseteq \Re^n$ be a convex set and let f be a differentiable convex function on S. Then, a critical point $x_0 \in S$ is a global minimum point.*

Proof. See Exercise 1.43. $\qquad\square$

1.6 Exercises

1.1. Let B be a ball of center x_0 and radius R. Verify that:
1. If $z \in B$, then every point $x = x_0 + t(z - x_0)$, $t \in (0, 1)$, is an interior point of B;
2. If $z_1, z_2 \in B$, then every point $x = \lambda z_1 + (1 - \lambda)z_2, \lambda \in (0, 1)$, is an interior point of B;
3. B is a convex set;
4. Every boundary point of B is an extreme point.

1.2. Show Theorem 1.2.1.

1.3. Let S, T be convex sets. Prove that:
1. For every $\alpha, \beta \in \Re$, $\Gamma = \alpha S + \beta T$ is convex. In particular, the sum $S + T$ and the product for a scalar $kS, k \in \Re$ are convex sets.
2. The cartesian product $S \times T$ is convex.

1.4. Let S be a subset of $\Re^n$. The set of all finitely many convex combinations of points of S is called the convex hull of S and is denoted by $convS$. Show that:
(a) $convS$ is a convex set;
(b) $convS$ is the smallest convex subset containing S.

1.5. Prove Theorem 1.2.5.

1.6. Let $S \subseteq \Re^n$ be a convex set with $intS \neq \emptyset$. Show that there is at most one boundary point on every half-line starting from an interior point of S.

1.7. The points $x_1, x_2, ..., x_{k+1}$ of $\Re^n$ are said to be affinely independent if $x_2 - x_1, x_3 - x_1, ..., x_{k+1} - x_1$, are linearly independent. The convex hull of $S = \{x_1, x_2, ..., x_{k+1}\}$ is called a (k-dimensional) simplex. Show that:
1. The smallest linear manifold W containing $x_1, x_2, ..., x_{k+1}$ has dimension k;
2. The maximum dimension of a simplex is n.

1.8. Let S be a convex set of $\Re^n$ and let W be the smallest linear manifold containing S. Show that:
1. $dimW = k$ if and only if the largest simplex contained in S has dimension k;
2. $W = \Re^n$ if and only if W contains a n-dimensional simplex.

1.9. Let S be a nonempty convex set. Prove that $riS \neq \emptyset$.

1.10. Prove that any point of a compact convex set S may be expressed as a convex combination of the extreme points of S.

1.11. Prove that any point of a polytope may be expressed as a convex combination of its vertices.

1.12. Prove that the sum of two convex cones is a convex cone.

1.13. Let C be a convex cone in $\Re^n$ and let $affC$ be the smallest subspace containing C. Prove that:
1. $affC = C - C$;
2. $intC \neq \emptyset$ if and only if $affC = \Re^n$.

1.14. Let C be a convex cone with $intC \neq \emptyset$. Prove that $riC^+ \cap riC^- = \emptyset$.

1.15. Consider the cones C_1, C_2 in $\Re^n$. Prove that $C_1^+ \cap C_2^+ = (C_1 + C_2)^+$.

1.16. Let U, V be closed convex cones in $\Re^n$ such that $U \cap V = \{0\}$. Prove the existence of $\alpha \in \Re^n \setminus \{0\}$ such that $\alpha^T u \geq 0$, $\forall \, u \in U$, and $\alpha^T v \leq 0$, $\forall \, v \in V$.

1.17. Let U, V be closed convex cones in $\Re^n$ such that $U \cap V = \{0\}$. If U is pointed, prove the existence of $\alpha \in \Re^n \setminus \{0\}$ such that $\alpha^T u > 0$, $\forall \, u \in U \setminus \{0\}$, and $\alpha^T v \leq 0$, $\forall \, v \in V$.

1.18. Let U, V be closed convex pointed cones in $\Re^n$ such that $U \cap V = \{0\}$. Prove the existence of $\alpha \in \Re^n \setminus \{0\}$ such that $\alpha^T u > 0$, $\forall \, u \in U \setminus \{0\}$, and $\alpha^T v < 0$, $\forall \, v \in V \setminus \{0\}$.

1.19. Consider the linear programming problem

$$\inf c^T x, \quad x \in S = \{x \in \Re^n : Ax \leq b, \ x \geq 0\}.$$

Show that the infimum is $-\infty$ or that it is attained at a vertex of S.

1.20. Prove that a function is of the form $f(x) = a^T x + b$, $x \in \Re^n$ if and only if it is both convex and concave.

1.21. Let $S \subseteq \Re^n$ be a closed convex set. Show that the *distance function* $d(z) = \min\limits_{x \in S} \| z - x \|$ is convex.

1.22. (a) Let $f_i, i = 1, ..., m$, be convex functions defined on the convex set $S \subseteq \Re^n$. Show that the function $z(x) = \max\limits_{i \in \{1,..,m\}} \{f_i(x)\}$ is convex on S.
(b) Let $f_i, i = 1, ..., m$, be concave functions defined on the convex set $S \subseteq \Re^n$. Show that the function $z(x) = \min\limits_{i \in \{1,..,m\}} \{f_i(x)\}$ is concave on S.

1.23. Prove that the function $f(x_1, x_2) = x_1^\alpha x_2^\beta$, $x_1, x_2 > 0$, $\alpha, \beta \in \Re$ is concave if and only if $\alpha \in [0, 1]$, $\beta \in [0, 1]$, $\alpha + \beta \leq 1$ and it is convex if and only if $\alpha \leq 0$, $\beta \leq 0$.

1.24. Prove Theorem 1.3.4.

1.25. Prove Theorem 1.3.5.

1.26. Prove that (i) of Theorem 1.3.4 is equivalent to the following conditions:
1. The sum of two convex functions is a convex function;
2. The product between a convex function and a non-negative real number is a convex function.

1.27. Give an example showing that the product of two convex (concave) functions is not a convex (concave) function.

1.28. Let $f(x) = Ax + b$ be an affine function, where A is an $m \times n$ matrix, $b \in \Re^m$, and let $g : \Re^m \to \Re$ be a convex function. Show that $z(x) = g(Ax+b)$ is a convex function.
State and prove a similar result for concave functions.

1.29. Show that if $z(x) = \log f(x)$ is a convex function, then $f(x)$ is convex. Give an example showing that the reverse is not true.

1.30. Let f be a negative convex function. Show that $z(x) = \frac{1}{f(x)}$ is concave.

1.31. Give an example showing that the reciprocal of a positive convex function is not concave.

1.32. Let $f : \Re \to \Re$ be a differentiable convex function and let x_0 be such that $f'(x_0) > 0$. Show that $\lim_{x \to +\infty} f(x) = +\infty$.

1.33. Let $S \subseteq \Re^n$ be a nonempty open convex set and let f be a differentiable function on S. Prove that f is strictly convex if and only if for every $x_0 \in S$ we have $f(x) > f(x_0) + (x - x_0)^T \nabla f(x_0)$, $\forall x \in S$, $x \neq x_0$.

1.34. Prove Theorem 1.4.1.

1.35. Prove that a function of one variable is homogeneous of degree α if and only if it is of the form $f(x) = kx^\alpha$.

1.36. Let f be a linearly homogeneous function defined on a convex cone $C \subseteq \Re^n$. Prove that f is concave if and only if $f(x + y) \geq f(x) + f(y)$ for every $x, y \in C$.

1.37. A relevant result regarding homogeneous functions is Euler's Theorem: let f be a differentiable function defined on the open convex cone $C \subseteq \Re^n$. Then, f is homogeneous of degree α if and only if $x^T \nabla f(x) = \alpha f(x)$, $\forall x \in C$. By using Euler's theorem and the first characterization of convex functions, prove that a differentiable linearly homogeneous function is convex if and only if $x^T \nabla f(x_0) \leq f(x)$, $\forall x, x_0 \in int\Re^n_+$.

1.38. Use Euler's theorem to prove that the Hessian matrix of a twice differentiable linearly homogeneous function is singular at each point of the domain.

1.39. Find the set of all global minimum points of the function $f(x) = (a^T x + b) \log(a^T x + b)$, $a^T x + b > 0$, $a \in \Re^n \setminus \{0\}$.

1.40. Let $S \subseteq \Re^n$ be an open convex set and let f be a non-constant differentiable convex function on S. Show that f cannot have a global maximum. Can f have a local maximum? Can f have a local strict maximum?

1.41. Show that the quadratic form $f(x) = \frac{1}{2}x^T A x$ is convex if and only if $x_0 = 0$ is a local minimum.

1.42. Show that the convex quadratic function $f(x) = \frac{1}{2}x^T A x + a^T x + a_0$ has a global minimum point if and only if $rank A = rank[A, a]$.

1.43. Prove Theorem 1.5.2.

1.44. Let $S \subseteq \Re^n$ be a convex set and let $f : S \to \Re$ be a concave function. Prove that:
1. A local maximum point is also global;
2. The set S^* of all maximum points is a convex set;
3. S^* has at most one element if f is strictly concave;
4. If f is differentiable and $x_0 \in S$ is a critical point, then x_0 is a global maximum point.

1.7 References

Avriel M. [10], Bazaraa M. S., Sherali H. D., and Shetty C. M. [18], Carter M. [61], Ginsberg W. [124], Madden P. [191], Nikaido H. [216], Roberts A. W. [233], Rockafellar R. T. [234], Stoer J., and Witzgall C. [270], Takayama A. [274].

2

Non-Differentiable Generalized Convex Functions

2.1 Introduction

In several economic models convexity appears to be a restrictive condition. For instance, classical assumptions in Economics include the convexity of the production set in producer theory and the convexity of the upper level sets of the utility function in consumer theory.

On the other hand, in Optimization it is important to know when a local minimum (maximum) is also global. Such a useful property is not exclusive to convexity (concavity).

All this has led to the introduction of new classes of functions which have convex lower/upper level sets and which verify the local-global property, starting with the pioneer work of Arrow–Enthoven [7].

In this chapter we shall introduce the class of quasiconvex, strictly quasiconvex, and semistrictly quasiconvex functions and the inclusion relationships between them are studied. For functions in one variable, a complete characterization of the new classes is established. Examples of the most important generalized convex functions in Economics are given.

We shall focus our attention on generalized convexity since a function f is generalized concave if and only if $-f$ is generalized convex, so that every result for a generalized convex function can be easily re-stated in terms of generalized quasiconcave functions. For the sake of completeness in Appendix B we shall give a summary of the main properties of generalized concave functions.

2.2 Quasiconvexity and Strict Quasiconvexity

One way to generalize the definition of a convex function is to relax the convexity condition, and require, from a geometrical point of view, that the restriction of the function along a line joining any two points in the domain lies under at least one of the endpoints. A function that verifies such a condition is called quasiconvex. Figure 2.1 shows some examples of quasiconvex functions.

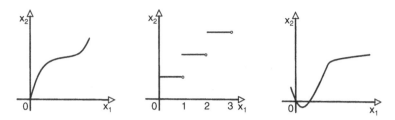

Fig. 2.1. Quasiconvex functions

From an analytical point of view, we have the following definition.

Definition 2.2.1. *Let f be defined on a convex set $S \subseteq \Re^n$. The function f is said to be quasiconvex on S if*

$$f(\lambda x_1 + (1 - \lambda)x_2) \leq \max\{f(x_1), f(x_2)\} \tag{2.1}$$

for every x_1, $x_2 \in S$ and for every $\lambda \in [0, 1]$ or, equivalently,

$$f(x_1) \geq f(x_2) \text{ implies that } f(x_1) \geq f(x_1 + \lambda(x_2 - x_1)) \tag{2.2}$$

for every $x_1, x_2 \in S$ and for every $\lambda \in [0, 1]$.

If the inequality in (2.1) is strict, the function is called strictly quasiconvex. Formally:

Definition 2.2.2. *A function f defined on a convex set $S \subseteq \Re^n$ is said to be strictly quasiconvex if*

$$f(\lambda x_1 + (1 - \lambda)x_2) < \max\{f(x_1), f(x_2)\} \tag{2.3}$$

for every x_1, $x_2 \in S$, $x_1 \neq x_2$, and for every $\lambda \in (0, 1)$ or, equivalently,

$$f(x_1) \geq f(x_2) \text{ implies that } f(x_1) > f(x_1 + \lambda(x_2 - x_1)) \tag{2.4}$$

for every $x_1, x_2 \in S$, $x_1 \neq x_2$, and for every $\lambda \in (0, 1)$.

It follows immediately, from the given definitions, that a strictly quasiconvex function is also quasiconvex; the converse is not true as is shown in the following example.

Example 2.2.1.

1. The function $f(x) = \begin{cases} \frac{|x|}{x} & x \neq 0 \\ 0 & x = 0 \end{cases}$ is quasiconvex but not strictly quasiconvex;

2. Every monotone function of one variable is quasiconvex and every increasing or decreasing function of one variable is strictly quasiconvex; for instance, the concave function $f(x) = \log x$, $x > 0$ is strictly quasiconvex.

As we have just pointed out, quasiconvexity may be viewed as a relaxation of convexity. Unfortunately, in the new larger class of functions some properties of convex functions are lost. For instance, as we can deduce from (1) in Example 2.2.1, a quasiconvex function may have interior discontinuity points, a local minimum which is not global and an interior global maximum.

The relationships between convexity, strict convexity, quasiconvexity and strict quasiconvexity are stated in the following theorem.

Theorem 2.2.1. *Let $S \subseteq \Re^n$ be a convex set.*
(i) If f is convex on S, then f is quasiconvex on S.
(ii) If f is strictly convex on S, then f is strictly quasiconvex on S.
(iii) If f is strictly quasiconvex on S, then f is quasiconvex on S.

Proof. (i) We have:
$f(\lambda x_1 + (1 - \lambda)x_2) \leq \lambda f(x_1) + (1 - \lambda)f(x_2) \leq \lambda \max\{f(x_1), f(x_2)\} + (1 - \lambda)$
$\max\{f(x_1), f(x_2)\} = \max\{f(x_1), f(x_2)\}$.
Similarly we can prove (ii), while (iii) follows directly by definition. □

Example 2.2.1 shows that the class of quasiconvex functions contains properly the class of convex functions and the class of strictly quasiconvex functions. Furthermore, there is not any inclusion relationship between the class of convex functions and the class of strictly quasiconvex ones. In fact, a convex function may have constant restrictions and a strictly increasing function of one variable is strictly quasiconvex but not necessarily convex.

The relationships between convexity, strict convexity, quasiconvexity and strict quasiconvexity are illustrated in Fig. 2.2 where the arrow ($\rightarrow$) reads "implies".

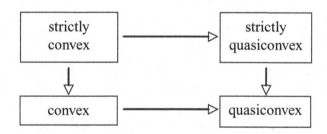

Fig. 2.2. Relationships between various types of convexity

An important case where quasiconvexity reduces to convexity is related to homogeneity. More exactly, the homogeneity assumption combined with quasiconvexity produces convexity as is shown in the following theorem.

Theorem 2.2.2. *Let f be a homogeneous function of degree $\alpha = 1$ defined on a convex set $S \subseteq \Re^n$.*
If $f(x) > 0$ for all $x \in S$, then f is quasiconvex if and only if it is convex.

Proof. From (i) of Theorem 2.2.1, convexity implies quasiconvexity without any other assumption. Now we will prove, firstly, that homogeneity combined with quasiconvexity implies subadditivity, i.e., for every $x_1, x_2 \in S$ we have $f(x_1 + x_2) \leq f(x_1) + f(x_2)$.

Let $y_1 = f(x_1) > 0$, $y_2 = f(x_2) > 0$. Since f is homogeneous of degree one, i.e., $f(tx) = tf(x), t > 0$, we have $f(\frac{x_1}{y_1}) = f(\frac{x_2}{y_2}) = 1$, so that the quasiconvexity of f implies $f((1-t)\frac{x_1}{y_1} + t\frac{x_2}{y_2}) \leq 1$ $\forall t \in (0,1)$. By choosing $t = \frac{y_2}{y_1+y_2}$, we have $1 - t = \frac{y_1}{y_1+y_2}$, and $f(\frac{x_1}{y_1+y_2} + \frac{x_2}{y_1+y_2}) = \frac{1}{y_1+y_2}f(x_1 + x_2) \leq 1$. Consequently, $f(x_1 + x_2) \leq (y_1 + y_2) = f(x_1) + f(x_2)$.

It remains to be proven that subadditivity and homogeneity imply convexity. We have $f(\lambda x_1 + (1-\lambda)x_2) \leq f(\lambda x_1) + f((1-\lambda)x_2) = \lambda f(x_1) + (1-\lambda)f(x_2)$. $\square$

While a convex function may be characterized by the convexity of its epigraph, a quasiconvex function may be characterized by the convexity of its lower level sets, as is shown in the following theorem.

Theorem 2.2.3. *A function f defined on a convex set $S \subseteq \Re^n$ is quasiconvex on S if and only if the lower level set $L_{\leq \alpha} = \{z \in S : f(z) \leq \alpha\}$ is convex for every $\alpha \in \Re$.*

Proof. Assume that f is quasiconvex and let $x, y \in L_{\leq \alpha}$. Since $f(x) \leq \alpha$, $f(y) \leq \alpha$, we have $f(\lambda x + (1-\lambda)y) \leq \max\{f(x), f(y)\} \leq \alpha$, so that $\lambda x + (1-\lambda)y \in L_{\leq \alpha}$.

In order to prove the converse statement, assume without loss of generality, that $\max\{f(x), f(y)\} = f(x)$ and consider the lower level set $L_{\leq \alpha}$ with $\alpha = f(x)$, that is $L_{\leq f(x)} = \{z \in S : f(z) \leq f(x)\}$; obviously $y \in L_{\leq f(x)}$. Since $L_{\leq f(x)}$ is convex, $x + \lambda(y - x) \in L_{\leq f(x)}$ for every $\lambda \in [0,1]$, i.e., $f(x + \lambda(y - x)) \leq f(x) = \max\{f(x), f(y)\}$. $\square$

By means of a simple application of the previous Theorem, we obtain the following result.

Theorem 2.2.4. *Let f be a quasiconvex function defined on a convex set $S \subseteq \Re^n$ and let S^* be the set of all global minimum points of f. Then, S^* is convex.*

Proof. If $S^* = \emptyset$ the thesis follows by convention. Taking into account that $S^* = \{x \in S : f(x) = m\} = \{x \in S : f(x) \leq m\}$, where m is the minimun value of f on S, the convexity of S^* follows from the convexity of the lower level sets of a quasiconvex function. $\square$

As happens for a convex function and for a strictly convex function, (2.1) and (2.3) may be extended to every convex combination of a finite number of points.

Theorem 2.2.5. *f is quasiconvex on a convex set $S \subseteq \Re^n$ if and only if, for $x^i \in S$, $i = 1, .., p$, we have*

$$f\left(\sum_{i=1}^{p}\lambda_i x^i\right) \leq \max_{i\in\{1,..,p\}} f(x^i), \ \sum_{i=1}^{p}\lambda_i = 1, \ \lambda_i \geq 0, \ i = 1,..,p. \qquad (2.5)$$

Furthermore, f is strictly quasiconvex on S if and only if the inequality in (2.5) is strict.

Proof. The validity of (2.5) for $p = 2$ is equivalent to the definition of a quasiconvex function.

Assume now that f is quasiconvex. We will prove that (2.5) holds by induction. Since (2.5) is true for $p = 2$, we must prove that the validity of (2.5) for every p elements implies that

$$f(\lambda_1 x^1 + ... + \lambda_p x^p + \lambda_{p+1}x^{p+1}) \leq \max_{i\in\{1,...,p+1\}} f(x^i)$$

with $\sum_{i=1}^{p+1}\lambda_i = 1, \ \lambda_i \geq 0, \ x^i \in S, \ i = 1,...,p+1.$

If $\lambda_{p+1} = 0$, the thesis follows by means of the induction assumption, otherwise, by setting $\lambda_0 = \lambda_1 + ... + \lambda_p$, we have $\lambda_0 + \lambda_{p+1} = 1$ so that $y = \frac{\lambda_1}{\lambda_0}x_1 + ... + \frac{\lambda_p}{\lambda_0}x^p$ is a convex combination of p points since $\sum_{i=1}^{p}\frac{\lambda_i}{\lambda_0} = 1$.

As a result, $y \in S$. On the other hand $\sum_{i=1}^{p+1}\lambda_i x^i = \lambda_0 y + \lambda_{p+1}x^{p+1}$; the thesis is reached by applying quasiconvexity to the points y, x^{p+1}.

The last statement follows similarly. $\qquad\square$

Another main difference between convex functions and quasiconvex functions is related to the algebraic structure. The class of convex functions is closed with respect to the addition while the sum of quasiconvex (strictly quasiconvex) functions is not in general quasiconvex (strictly quasiconvex). For instance, the functions $f(x) = x^3$, $g(x) = -3x$ are quasiconvex and strictly quasiconvex in $\Re$ since they are strictly monotone, but their sum $h(x) = f(x) + g(x) = x^3 - 3x$ is neither quasiconvex, nor strictly quasiconvex since $h(-2) = -2$, $h(0) = 0$, and $h(-1) = 2 > 0$.

Fortunately, in contrast to the convex case, increasing functions combined with quasiconvex functions produce quasiconvex functions, as is stated in the following theorems.

Theorem 2.2.6. *Let f be a quasiconvex function defined on a convex set $S \subseteq \Re^n$ and let $g : A \to \Re$ be a non-decreasing function, with $f(S) \subseteq A$. Then:*

(i) kf, $k > 0$ is quasiconvex on S;

(ii) $g \circ f$ is quasiconvex on S.

Proof. (i) For the positivity of k and the quasiconvexity of f we have
$(kf)(x_1 + \lambda(x_2 - x_1)) = kf(x_1 + \lambda(x_2 - x_1)) \leq k \ \max\{f(x_1), f(x_2)\} =$
$= \max\{kf(x_1), kf(x_2)\}.$

(ii) Let $x_1, x_2 \in S$ with $f(x_1) \geq f(x_2)$. Then, $f(x_1 + \lambda(x_2 - x_1)) \leq f(x_1)$, for all $\lambda \in [0, 1]$. Since g is a non-decreasing function, $g(f(x_1)) \geq g(f(x_2))$ implies $g(f(x_1 + \lambda(x_2 - x_1))) \leq g(f(x_1)), \forall \lambda \in [0, 1]$. $\square$

Theorem 2.2.7. *Let f be a strictly quasiconvex function defined on a convex set $S \subseteq \Re^n$ and let $g : A \to \Re$ be an increasing function, with $f(S) \subseteq A$. Then:*
(i) kf, $k > 0$ is strictly quasiconvex on S;
(ii) $g \circ f$ is strictly quasiconvex on S.

Proof. Similar to the proof given in Theorem 2.2.6. $\square$

Another useful composition theorem is the following.

Theorem 2.2.8. *Let $g(x) = Ax + b$ where A is an $m \times n$ matrix, $b \in \Re^m$, and let f be a quasiconvex function on a convex set $S \subseteq g(\Re^n)$. Then, $z(x) = f(Ax + b)$ is quasiconvex on S.*

Proof. We have $z(\lambda x_1 + (1 - \lambda)x_2) = f(\lambda(Ax_1 + b) + (1 - \lambda)(Ax_2 + b)) \leq max\{f(Ax_1 + b), f(Ax_2 + b)\} = max\{z(x_1), z(x_2)\}, \forall \lambda \in [0, 1]$. $\square$

Property (ii) of Theorem 2.2.6 leads to the following result which extends Theorem 2.2.2.

Theorem 2.2.9. *Let f be a homogeneous function of degree $\alpha \geq 1$ defined on a convex set $S \subseteq \Re^n$.*
If $f(x) > 0$ for all $x \in S$, then f is quasiconvex if and only if it is convex.

Proof. Taking into account Theorem 2.2.2, case $\alpha > 1$ remains to be considered. The function $g(x) = [f(x)]^{\frac{1}{\alpha}}$ is linearly homogeneous and quasiconvex since it is the composite function $g = z \circ f$, where $z(y) = y^{\frac{1}{\alpha}}$ is increasing and f is quasiconvex. It follows that $f(x) = [g(x)]^{\alpha}$ is convex as the composite function of a convex function and an increasing convex function. $\square$

Remark 2.2.1. Following the same line given in the proof of the previous theorem, it is easy to prove the following result.
Let f be a homogeneous function of degree α with $0 < \alpha \leq 1$, defined on a convex set $S \subseteq \Re^n$.
If $f(x) > 0$ for all $x \in S$, then f is quasiconcave if and only if it is concave.

Example 2.2.2. The function $f(x_1, .., x_n) = (\sum_{i=1}^{n} x_i^2)^{\beta}$ is convex for $\beta \geq \frac{1}{2}$.

In fact, $\sum_{i=1}^{n} x_i^2$ is convex so that, from Theorem 2.2.6, f is quasiconvex; on the other hand f is homogeneous of degree $\alpha = 2\beta$ and consequently, from Theorem 2.2.9, f is convex for $\beta \geq \frac{1}{2}$.

Remark 2.2.2. Theorems 2.2.3 and 2.2.6 are sometimes useful in identifying quasiconvex functions or in constructing new quasiconvex functions from existing ones. Examples will be given in Sect. 2.3.

The given definitions of generalized convexity point out that the behaviour of the function is strictly related to the behaviour of its restriction on every line segment. This connection is expressed in the following theorem.

Theorem 2.2.10. *Let f be a function defined on a convex set $S \subseteq \Re^n$. Then, f is quasiconvex (strictly quasiconvex) on S if and only if the restriction of f on each line segment contained in S is a quasiconvex (strictly quasiconvex) function.*

Proof. See Exercise 2.18. □

Remark 2.2.3. As we shall see, by means of Theorem 2.2.10, several results regarding generalized convexity of functions of several variables may be derived from the corresponding results for functions of one variable. For this reason we shall devote Sect. 2.5 to the study of generalized convex functions of one variable.

The non-constancy of a strictly quasiconvex function along a line allows us to characterize strictly quasiconvexity within quasiconvexity.

Theorem 2.2.11. *Let f be a function defined on a convex set $S \subseteq \Re^n$. Then, f is strictly quasiconvex on S if and only if (i) and (ii) hold:*
(i) f is quasiconvex on S;
(ii) Every restriction on a line segment is not constant.

Proof. (i) This follows from (iii) of Theorem 2.2.1 while (ii) follows directly from (2.3).
Assume now that (i) and (ii) hold. If f is not strictly quasiconvex, there exist x_1, $x_2 \in S$, $\bar{\lambda} \in (0,1)$, such that $f(x_1) \geq f(x_2)$ and $f(\bar{x}) = f(x_1 + \bar{\lambda}(x_2 - x_1)) = f(x_1)$; since f is not constant in $[x_1, \bar{x}]$, there exists $x_0 \in (x_1, \bar{x})$ such that $f(x_0) < f(x_1) = f(\bar{x})$. If $f(\bar{x}) = f(x_1) > f(x_2)$, the quasiconvexity of f on the line segment $[x_0, x_2]$ is contradicted. If $f(\bar{x}) = f(x_1) = f(x_2)$, there exists $x^* \in (\bar{x}, x_2)$ such that $f(x^*) < f(\bar{x}) = f(x_2)$ and this contradicts the quasiconvexity of f on the line segment $[x_0, x^*]$. □

As we have pointed out by means of Example 2.2.1, a quasiconvex function is not necessarily continuous. Under a continuity assumption, we have the useful property that quasiconvexity on an open convex set S is preserved on the closure of S. More precisely, we have the following result.

Theorem 2.2.12. *Let f be a continuous function on the closure of a convex set $S \subseteq \Re^n$ and quasiconvex on the interior of S. Then, f is quasiconvex on the closure of S.*

Proof. Let $x, y \in clS$; we must prove that $f(x + \lambda(y - x)) \leq \max\{f(x), f(y)\}$, $\forall \lambda \in [0, 1]$. If $x, y \in intS$ the inequality is true by assumption. Let $\{x_n\} \subset intS$, $\{y_n\} \subset intS$ be sequences converging to x and y, respectively, with the convention $x_n = x$ for all n ($y_n = y$ for all n) if $x \in intS$ ($y \in intS$). We have $f(x_n + \lambda(y_n - x_n)) \leq max\{f(x_n), f(y_n)\}$, $\forall \lambda \in [0, 1]$, so that the thesis follows by taking the limit for $n \to +\infty$ and taking into account the continuity of f. □

2.3 Semistrict Quasiconvexity

In this section we shall introduce a class of generalized convex functions which is intermediate between convex and quasiconvex ones. This class is obtained by strengthening quasiconvexity.

Definition 2.3.1. *A function f defined on a convex set $S \subseteq \Re^n$ is said to be semistrictly quasiconvex if*

$$f(\lambda x_1 + (1 - \lambda)x_2) < \max\{f(x_1), f(x_2)\} \tag{2.6}$$

for every $x_1, x_2 \in S$, with $f(x_1) \neq f(x_2)$ and for every $\lambda \in (0, 1)$ or, equivalently,

$$f(x_1) > f(x_2) \text{ implies that } f(x_1) > f(x_1 + \lambda(x_2 - x_1)) \tag{2.7}$$

for every $x_1, x_2 \in S, \ \lambda \in (0, 1)$.

As a direct consequence of the given definition, we have the following theorem.

Theorem 2.3.1. *Let f be a function defined on a convex set $S \subseteq \Re^n$.*
(i) If f is strictly quasiconvex on S, then f is semistrictly quasiconvex on S;
(ii) If f is convex on S, then f is semistrictly quasiconvex on S.

Since a constant function is semistrictly quasiconvex but not strictly quasiconvex, the class of strictly quasiconvex functions is properly contained in the class of semistrictly quasiconvex ones.

The following examples point out that there is not any inclusion relationship between the class of semistrictly quasiconvex functions and the one of quasiconvex functions.

Example 2.3.1. Consider the function $f(x) = \begin{cases} 1 & -1 \leq x \leq 1, \ x \neq 0 \\ 2 & x = 0 \end{cases}$

f is not quasiconvex since we have $f(0) = 2 > \max\{f(1), f(-1)\} = 1$; on the other hand, f is semistrictly quasiconvex, since, in order to apply the definition, we must necessarily consider the points $x_1 = 0$, $x_2 \neq 0$, so that $f(x_1 + \lambda(x_2 - x_1)) = f(\lambda x_2) = 1 < f(x_1) = 2$.

Example 2.3.2. Consider the function $f(x) = \begin{cases} x & 0 \le x \le 1, \\ 1 & 1 < x \le 2 \end{cases}$

f is a non-decreasing function and hence quasiconvex, but it is not semistrictly quasiconvex since we have $f(0) = 0 < f(2) = 1$ and $f(\frac{3}{2}) = 1 = f(2)$.

Let us note that in Example 2.3.1, the function f is not lower semicontinuous at $x_0 = 0$. We shall prove that a sufficient condition for a semistrictly quasiconvex function to be quasiconvex is the lower semicontinuity of the function.

Theorem 2.3.2. *If f is a lower semicontinuous and semistrictly quasiconvex function on a convex set S, then f is quasiconvex on S.*

Proof. Let x, $y \in S$ such that $f(x) \ge f(y)$. If $f(x) > f(y)$, the thesis follows from (2.7). Let $f(x) = f(y)$ and assume the existence of $\bar{\lambda} \in (0,1)$ such that $f(\bar{x}) = f(x + \bar{\lambda}(y - x)) > f(x)$. Since f is lower semicontinuous there exist $\epsilon > 0$ and a neighbourhood I of $\bar{x}$ such that $f(z) \ge f(\bar{x}) - \epsilon$, $\forall z \in I$. Since $f(\bar{x}) > f(x)$, by choosing $\epsilon < f(\bar{x}) - f(x)$, we have $f(z) > f(x)$, $\forall z \in I$ and, in particular, $\forall z \in I \cap [x, y]$. Let $z \in I \cap [x, y]$. If $z \in (x, \bar{x})$, then the semistrict quasiconvexity of f is contradicted on the segment $[z, y]$ if $f(z) < f(\bar{x})$ or in the segment $[x, \bar{x}]$ if $f(z) \ge f(\bar{x})$.
If $z \notin (x, \bar{x})$, the proof is analogous.   □

We can summarize the inclusion relationships between the various classes of convex and generalized convex functions by means of the diagram of Fig. 2.3 which assumes lower semicontinuity. All inclusions are proper.

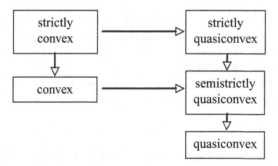

Fig. 2.3. Relationships between various types of convexity under lower semicontinuity

As happens in the quasiconvex case, the behaviour of a semistrictly quasiconvex function may be characterized by means of the behaviour of its restrictions on every line segment. Such a property allows us to characterize semistrict quasiconvexity within quasiconvexity and also to characterize strict quasiconvexity within semistrict quasiconvexity. More precisely, we have the following results.

Theorem 2.3.3. *Let f be a function defined on a convex set $S \subseteq \Re^n$. Then, f is semistrictly quasiconvex on S if and only if the restriction of f to each line segment contained in S is a semistrictly quasiconvex function.*

Theorem 2.3.4. *Let f be a lower semicontinuous function defined on a convex set $S \subseteq \Re^n$. Then, f is semistrictly quasiconvex if and only if (i) and (ii) hold:*
(i) f is quasiconvex;
(ii) Every local minimum is also global for each restriction on a line segment.

Proof. Assume that (i) and (ii) hold. If f is quasiconvex but not semistrictly quasiconvex, then there exist x_1, $x_2 \in S$, $\bar{\lambda} \in (0, 1)$ such that $f(x_1) > f(x_2)$ and $f(\bar{x}) = f(x_1 + \bar{\lambda}(x_2 - x_1)) = f(x_1)$. From (ii), f is not constant in $[x_1, \bar{x}]$, otherwise each point $\tilde{x} \in (x_1, \bar{x})$ is a local minimum which is not global for the restriction $\varphi(t) = f(x_1 + t(x_2 - x_1)), t \in [0, 1]$. Consequently, there exists $x_0 \in (x_1, \bar{x})$ such that $f(x_0) < f(x_1) = f(\bar{x})$ and this contradicts the quasiconvexity of f on the segment $[x_0, x_2]$.
With respect to the converse statement, (i) follows from Theorem 2.3.2 while (ii) is obvious. $\qquad\square$

The following theorem, whose proof can be found in [104, 209], shows that the class of continuous semistrictly quasiconvex functions is the wider class for which every local minimum is also global.

Theorem 2.3.5. *Let f be a continuous quasiconvex function defined on a convex set $S \subseteq \Re^n$. Then, f is semistrictly quasiconvex if and only if every local minimum point is also global for f on S.*

Theorem 2.3.6. *Let f be a lower semicontinuous quasiconvex function defined on a convex set $S \subseteq \Re^n$. Then, f is strictly quasiconvex if and only if the following conditions hold:*
(i) f is semistrictly quasiconvex;
(ii) Any restriction on a line segment attains its minimum at no more than one point.

Proof. See Exercise 2.19. $\qquad\square$

The composition Theorem 2.2.6 may easily be extended to the semistrictly quasiconvex case.

Theorem 2.3.7. *Let f be a semistrictly quasiconvex function defined on a convex set $S \subseteq \Re^n$ and let $g : A \to \Re$ be an increasing function, with $f(S) \subseteq A$. Then:*
(i) kf, $k > 0$ is semistrictly quasiconvex on S;
(ii) $g \circ f$ is semistrictly quasiconvex on S.

Composition theorems related to generalized convex functions are useful tools in identifying or in constructing generalized quasiconvex functions. Some examples are given below.

Example 2.3.3. Let f be a positive and convex function defined on a convex set $S \subseteq \Re^n$ and consider the increasing functions $h_1(y) = \log y$, $y > 0$, $h_2(y) = y^\alpha$, $y > 0$, $\alpha > 0$. From Theorem 2.3.7 and taking into account the inclusion relationships, we deduce the semistrict quasiconvexity of the functions $z(x) = \log f(x)$, and $z(x) = (f(x))^\alpha$, $\alpha > 0$.

Furthermore if f is positive and strictly convex, then $z(x) = \log f(x)$, and $z(x) = (f(x))^\alpha$, $\alpha > 0$ are strictly quasiconvex.

Example 2.3.4. Let f be a quasiconcave (strictly quasiconcave, semistrictly quasiconcave, respectively) function constant in sign defined on a convex set $S \subseteq \Re^n$. Then, the reciprocal function $z(x) = \frac{1}{f(x)}$ is quasiconvex (strictly quasiconvex, semistrictly quasiconvex, respectively).

In fact, since $h(t) = -\frac{1}{t}$ is an increasing function, the composite function $h(f(x)) = -\frac{1}{f(x)}$ is quasiconcave (strictly quasiconcave, semistrictly quasiconcave, respectively), so that $z(x) = \frac{1}{f(x)}$ is quasiconvex (strictly quasiconvex, semistrictly quasiconvex, respectively).

Note that the reciprocal of a constant in sign concave function is not convex but it is quasiconvex.

As is known, the class of convex functions is not closed with respect to the ratio. In the following theorem generalized convexity of a ratio is investigated.

Theorem 2.3.8. *Let f and g be functions defined on a convex set $S \subseteq \Re^n$, and let*

$$z(x) = \frac{f(x)}{g(x)}$$

Then, the following properties hold:
(i) If f is non-negative and convex, and g is positive and concave, then z is semistrictly quasiconvex;
(ii) If f is non-positive and convex, and g is positive and convex, then z is semistrictly quasiconvex;
(iii) If f is convex, and g is positive and affine, then z is semistrictly quasiconvex.

Proof. (i) We must prove that $z(x) = \frac{f(x)}{g(x)} < \frac{f(x_0)}{g(x_0)} = z(x_0)$ implies that $z((1 - \lambda)x_0 + \lambda x) < z(x_0)$, $\lambda \in (0, 1)$.

Taking into account the convexity of f and the concavity of g, together with their sign, we have $f((1 - \lambda)x_0 + \lambda x) \leq (1 - \lambda)f(x_0) + \lambda f(x) < (1-\lambda)f(x_0)+\lambda\frac{f(x_0)}{g(x_0)}g(x) = \frac{f(x_0)}{g(x_0)}((1-\lambda)g(x_0)+\lambda g(x)) \leq \frac{f(x_0)}{g(x_0)}g((1-\lambda)x_0+\lambda x)$.

It follows that $\frac{f((1-\lambda)x_0+\lambda x)}{g((1-\lambda)x_0+\lambda x)} < \frac{f(x_0)}{g(x_0)}$, i.e., $z((1 - \lambda)x_0 + \lambda x) < z(x_0)$.

(ii) This can be proven similarly.

(iii) This follows from (i) and (ii) by noting that an affine function is both convex and concave. $\qquad\square$

2.4 Generalized Convexity of Some Homogeneous Functions

In this section we shall characterize the quasiconcavity of some classes of homogeneous functions which appear frequently in Economics.

2.4.1 The Cobb–Douglas Function

One of the most important production or utility functions is the Cobb–Douglas function defined as:

$$f(x) = Ax_1^{\alpha_1} x_2^{\alpha_2}....x_n^{\alpha_n}, \ A > 0, \ x_i > 0, \ \alpha_i > 0, \ i = 1,...,n. \tag{2.8}$$

The main properties of the Cobb–Douglas function, which is homogeneous of degree $\alpha = \sum_{i=1}^{n} \alpha_i$ (see Theorem 1.4.1), are stated in the following theorem.

Theorem 2.4.1. *The Cobb–Douglas function (2.8) is quasiconcave and is concave if and only if $\alpha = \sum_{i=1}^{n} \alpha_i \leq 1$.*

Proof. Since $\log f(x) = \log A + \sum_{i=1}^{n} \alpha_i \log x_i$ is a concave function as a positive linear combination of concave functions, the function $f(x) = e^{\log f(x)}$ is quasiconcave. The last statement follows by means of the property that a positive homogeneous function of degree $\alpha \leq 1$ is concave if and only if it is quasiconcave. □

Remark 2.4.1. If $U : \Re_+^n \to \Re$ is a utility function which represents the preferences of the consumer and $\varphi : \Re \to \Re$ is an increasing function, then the monotone transformation $\varphi \circ U$ is a utility function which represents the same preferences. It follows that the Cobb–Douglas utility function

$$U(x) = Ax_1^{\gamma_1} x_2^{\gamma_2}....x_n^{\gamma_n}, \ A > 0, \ x_i > 0, \ \gamma_i > 0, \ i = 1,...,n, \ \sum_{i=1}^{n} \gamma_i = 1 \tag{2.9}$$

represents the same preferences as the Cobb–Douglas function (2.8). In fact, by setting $\beta = \frac{1}{\alpha_1 + ... + \alpha_n}$, we have that $\varphi(z) = z^\beta$ is an increasing function and furthermore $\varphi(f(x)) = (Ax_1^{\alpha_1} x_2^{\alpha_2}....x_n^{\alpha_n})^\beta = A^\beta x_1^{\alpha_1\beta} x_2^{\alpha_2\beta}....x_n^{\alpha_n\beta} = A^\beta x_1^{\gamma_1} x_2^{\gamma_2}....x_n^{\gamma_n}$ with $\sum_{i=1}^{n} \gamma_i = \beta \sum_{i=1}^{n} \alpha_i = 1$. This justifies the use of utility functions of the kind (2.9) found in consumer theory.

2.4.2 The Constant Elasticity of Substitution (C.E.S.) Function

Another important function in Economics is the C.E.S. function defined by

$$f(x) = (a_1 x_1^\beta + a_2 x_2^\beta + ... + a_n x_n^\beta)^{\frac{1}{\beta}}, \ a_i > 0, \ x_i > 0, \ i = 1, ..., n, \ \beta \neq 0 \quad (2.10)$$

The main properties of the C.E.S. function, which is linearly homogeneous, are stated in the following theorem.

Theorem 2.4.2. *The C.E.S. function (2.10) is quasiconcave if and only if $\beta \leq 1$ and it is convex if and only if $\beta \geq 1$.*

Proof. By setting $g(x) = a_1 x_1^\beta + a_2 x_2^\beta + ... + a_n x_n^\beta$, we have $f(x) = (g(x))^{\frac{1}{\beta}}$. When $\beta < 0$, g is convex as a linear combination of positive convex functions; it follows that $\frac{1}{g}$ is a quasiconcave function (see Example 2.3.4) so that f is quasiconcave as an increasing transformation of $\frac{1}{g}$. When $0 < \beta \leq 1$, g is concave as a positive linear combination of concave functions, so that f is quasiconcave as an increasing transformation of g. Finally, when $\beta \geq 1$, g is convex as a linear combination of convex functions, so that f is quasiconvex as an increasing transformation of g. The thesis follows from Theorem 2.2.2 since the C.E.S. function is linearly homogeneous. $\qquad\square$

2.4.3 The Leontief Production Function

Another important homogeneous function of degree α is the Leontief production function defined by

$$f(x) = \left(\min_i \left\{ \frac{x_i}{a_i} \right\} \right)^\alpha, \ x_i > 0, \ a_i > 0, \ i = 1, ..., n, \ \alpha > 0. \quad (2.11)$$

The following theorem holds.

Theorem 2.4.3. *The Leontief production function (2.11) is quasiconcave and it is concave if and only if $\alpha \leq 1$.*

Proof. The function $g(x) = \min_i \{ \frac{x_i}{a_i} \}$ is concave as the minimum of a finite number of concave functions, so that f is an increasing transformation of g and thus it is quasiconcave. The last statement follows from Remark 2.2.1. $\qquad\square$

2.4.4 A Generalized Cobb–Douglas Function

Consider the function $z(x) = \prod_{i=1}^{k} (f_i(x))^{\alpha_i}$, $\alpha_i > 0$, where $f_i(x)$, $i = 1, ..., k$, are positive concave functions on a convex set $S \subseteq \Re^n$. Since $\log z(x) = \sum_{i=1}^{k} \alpha_i \log f_i(x)$ is a concave function (as a positive linear combination of concave functions), the function $z(x) = e^{\log z(x)}$ is quasiconcave.

2.5 Generalized Quasiconvex Functions in One Variable

Theorems 2.2.10 and 2.3.3 suggest carrying on the study of generalized convexity for one real variable function with the aim of extending the obtained results for functions of several variables.

A complete characterization of quasiconvexity is given in the following theorem where the parenthesis $\rangle$ or $\langle$ indicates that the corresponding end point can or cannot belong to the interval.

Theorem 2.5.1. *Let φ be a function defined on the interval $[a, b] \subseteq \Re$. Then φ is quasiconvex if and only if one of the following conditions is verified:*
(i) φ is non-increasing or non-decreasing in $[a, b]$;
(ii) φ is non-increasing in $[a, b)$ but not in $[a, b]$ or φ is non-decreasing in $(a, b]$ but not in $[a, b]$;
(iii) there exists $t_0 \in (a, b)$ such that φ is non-increasing in $[a, t_0\rangle$ and non-decreasing in $\langle t_0, b]$, where at least one of the two intervals is closed.

Proof. It is easy to verify that the validity of (i) or (ii) or (iii) implies the quasiconvexity of φ. Assume now the quasiconvexity of φ. Set $\ell = \inf\{\varphi(t), t \in [a, b]\}$ (note that ℓ may be $-\infty$) and let $\{t_n\} \subset [a, b]$ be such that $\varphi(t_n) \to \ell$ when $t_n \to t_0$. The following exhaustive cases occur: $t_0 = a$, $t_0 = b$, $t_0 \in (a, b)$.
case $t_0 = a$.
We will prove that φ is non-decreasing in $(a, b]$. If not, there exist t_1, $t_2 \in (a, b]$ with $t_1 < t_2$, such that $\varphi(t_1) > \varphi(t_2) \geq \ell$. Since ℓ is the infimum value of φ, there exists $\bar{n}$ such that $a < t_{\bar{n}} < t_1 < t_2$ with $\varphi(t_{\bar{n}}) < \varphi(t_1)$ and this contradicts the quasiconvexity of φ applied to the interval $[t_{\bar{n}}, t_2]$.
If $\varphi(a) = \varphi(t_0) = \ell$ then φ is non-decreasing in $[a, b]$, otherwise φ is non-decreasing in $(a, b]$ but not in $[a, b]$ since $\varphi(a) > \ell$.
case $t_0 = b$.
By means of similar arguments, it is easy to prove that φ is non-increasing in $[a, b)$; furthermore, φ is non-increasing in $[a, b]$ if $\varphi(b) = \varphi(t_0) = \ell$.
case $t_0 \in (a, b)$.
We will prove that φ is non-increasing in $[a, t_0)$ and non-decreasing in $(t_0, b]$. If not, there exist $t_1, t_2, \hat{t}_1, \hat{t}_2 \in [a, b]$ such that $t_1 < t_2 < t_0 < \hat{t}_1 < \hat{t}_2$ with $\varphi(t_1) < \varphi(t_2)$ and $\varphi(\hat{t}_1) > \varphi(\hat{t}_2)$. Since ℓ is the infimum value of φ, there exists $\bar{n}$ such that $t_2 < t_{\bar{n}} < \hat{t}_1$ with $\varphi(t_{\bar{n}}) < \varphi(t_2)$, $\varphi(t_{\bar{n}}) < \varphi(\hat{t}_1)$ and this contradicts the quasiconvexity of φ in the intervals $[t_1, t_{\bar{n}}]$, $[t_{\bar{n}}, \hat{t}_2]$, respectively.
It remains to be proven that at least one of the two intervals $[a, t_0\rangle$, $\langle t_0, b]$ is closed.
Set $\ell_1 = \inf\{\varphi(t), t \in [a, t_0)\}$, $\ell_2 = inf\{\varphi(t), t \in (t_0, b]\}$. Let us note that at least one of the two infimum is finite, otherwise there exist $t_1 < t_0$, $t_2 > t_0$, such that $\varphi(t_1) < \varphi(t_0)$ and $\varphi(t_2) < \varphi(t_0)$, and this contradicts the quasiconvexity of φ in the interval $[t_1, t_2]$.
Now we will prove that $\varphi(t_0) \leq \max\{\ell_1, \ell_2\}$; if not, taking into account that the function is non-increasing in $[a, t_0)$ and $\ell_1 < \varphi(t_0)$, there exists $t_1 < t_0$ such that $\varphi(t_1) < \varphi(t_0)$. In a similar way, since the function is non-decreasing

in $(t_0, b]$ and $\ell_2 < \varphi(t_0)$, there exists $t_2 > t_0$ such that $\varphi(t_2) < \varphi(t_0)$. It follows that the function φ is not quasiconvex on the interval $[t_1, t_2]$ and this is absurd.

Finally, let us note that $\ell = \min\{\ell_1, \ell_2, \varphi(t_0)\}$. If $\ell = \ell_1$ then $\varphi(t_0) \le \ell_2$ and φ is non-decreasing in $[t_0, b]$. If $\ell = \ell_2$ then $\varphi(t_0) \le \ell_1$ and φ is non-increasing in $[a, t_0]$. If $\ell = \varphi(t_0)$, then φ is non-increasing in $[a, t_0]$ and non-decreasing in $[t_0, b]$.

The proof is complete.   □

Theorem 2.5.1 may be specified in the case where φ is a lower semicontinuous function.

Theorem 2.5.2. *Let φ be a lower semicontinuous function defined on the interval $[a, b] \subseteq \Re$. Then, φ is quasiconvex if and only if there exists $t_0 \in [a, b]$ such that φ is non-increasing in $[a, t_0]$ and non-decreasing in $[t_0, b]$, where one of the two subintervals may be reduced to a point.*

Proof. Assume that φ is quasiconvex. The lower semicontinuity of φ on $[a, b]$ implies the existence of its minimum value m. Set $A = \{t \in [a, b] : \varphi(t) = m\}$; A is a closed interval since φ is quasiconvex and lower semicontinuous. Let $t_0 \in A$. The function φ is non-increasing in $[a, t_0]$ when $t_0 \ne a$, since the existence of $t_1, t_2 \in [a, t_0)$, $t_1 < t_2$ with $\varphi(t_1) < \varphi(t_2)$ contradicts the quasiconvexity of φ in $[t_1, t_0]$.

In the same way it can be proven that φ is non-decreasing in $[t_0, b]$, $t_0 \ne b$, so that the thesis follows when $a < t_0 < b$.

When $a = t_0 = b$, φ is constant in $[a, b]$; when $a = t_0$ and $t_0 < b$, φ is constant in $[a, t_0]$ and non-decreasing in $[t_0, b]$.

When $a < t_0$ and $t_0 = b$, φ is non-increasing in $[a, t_0]$ and constant in $[t_0, b]$. In each case the thesis follows.

The converse statement follows immediately.   □

The given characterizations of quasiconvex functions can be specialized to the subclass of strictly quasiconvex functions as is stated in Corollary 2.5.1 and in Corollary 2.5.2.

Corollary 2.5.1. *Let $\varphi(t)$ be a function defined on the interval $[a, b] \subseteq \Re$. Then, $\varphi(t)$ is strictly quasiconvex if and only if one of the following conditions holds:*

(i) φ is increasing or decreasing in $[a, b]$;

(ii) φ is decreasing in $[a, b)$ but not in $[a, b]$ or increasing in $(a, b]$ but not in $[a, b]$;

(iii) there exists $t_0 \in (a, b)$ such that φ is decreasing in $[a, t_0)$ and increasing in $\langle t_0, b]$ where at least one of the two intervals is closed.

Corollary 2.5.2. *Let φ be a lower semicontinuous function defined on the interval $[a, b] \subseteq \Re$. Then, φ is strictly quasiconvex if and only if there exists*

$t_0 \in [a, b]$ *such that φ is decreasing in $[a, t_0]$ and increasing in $[t_0, b]$, where one of the two subintervals may be reduced to a point.*

Since the class of semistrictly quasiconvex functions is contained in the class of quasiconvex ones under the lower semicontinuity assumption, Theorem 2.5.2 may be specified as follows.

Corollary 2.5.3. *Let $\varphi(t)$ be a lower semicontinuous function defined on the interval $[a, b] \subseteq \Re$. Then, $\varphi(t)$ is semistrictly quasiconvex if and only if there exist $\alpha, \beta \in [a, b]$ such that φ is decreasing in $[a, \alpha]$, constant in $[\alpha, \beta]$ and increasing in $[\beta, b]$, where one or two subintervals may be reduced to a point.*

Proof. By referring to the proof given in Theorem 2.5.2, consider the closed interval $A = \{t \in [a, b] : \varphi(t) = m\}$ and let $\alpha = minA$, $\beta = maxA$. We must prove that $\varphi(t)$ is decreasing on the interval $[a, \alpha]$ and increasing on $[\beta, b]$. In this regard it is sufficient to note that the existence of $t_1, t_2 \in [a, \alpha]$ such that $t_1 < t_2 < \alpha$ and $\varphi(t_1) \geq \varphi(t_2)$, contradicts the semistrictly quasiconvexity of φ on the interval $[t_1, \alpha]$. Similarly, it can be proven that φ is increasing in $[\beta, b]$. $\square$

2.6 Exercises

2.1. Which of the following functions are quasiconvex, semistrictly quasiconvex or strictly quasiconvex?
(a) $f(x) = x \mid x \mid$;   (b) $f(x) = x \mid x \mid -x^2$;   (c) $f(x) = x \mid x \mid +x^2$;
(d) $f(x) = \begin{cases} \frac{1}{x-1} & 0 \leq x < 1 \\ 0 & x = 1 \\ \frac{1}{x+1} & x > 1 \end{cases}$

2.2. Sketch a graph of a continuous function and the graph of a discontinuous function of one variable which satisfies the following conditions:
decreasing in $[0, 1]$, constant in $[1, 2]$, decreasing in $[2, 3]$, constant in $[3, 4]$, increasing in $[4, 5]$, constant in $[5, 6]$ and increasing in $[6, 7]$.
Is this function quasiconvex, strictly quasiconvex or semistrictly quasiconvex?

2.3. Let $f : [a, b] \to \Re$ and let $x_0 \in (a, b)$. Assume that $f(x_0) = \alpha$,
$\lim_{x \to x_0^-} f(x) = -\infty$, $\lim_{x \to x_0^+} f(x) = -\infty$. Show that f is not quasiconvex.

2.4. Are the following functions quasiconvex, semistrictly quasiconvex or strictly quasiconvex?
(a) $f(x) = \log \sum_{i=1}^{n} x_i^2$;   (b) $f(x) = \log \sum_{i=1}^{n} x_i$, $x_i > 0$, $i = 1, ..., n$;
(c) $f(x) = (\sum_{i=1}^{n} e^{x_i})^\beta$, $\beta > 0$.

2.5. Let f be a function defined on the convex set $S \subseteq \Re^n$. Prove that:
(a) if f is positive and convex, then $-\frac{1}{f}$ is quasiconvex;
(b) if f is negative and quasiconvex (strictly quasiconvex) then $\frac{1}{f}$ is quasiconcave (strictly quasiconcave).

2.6. Let f, g be functions defined on a convex set $S \subseteq \Re^n$. Using the characterization of quasiconvex functions in terms of its lower level sets, prove that the function $z(x) = \frac{f(x)}{g(x)}$ is quasiconvex if f is non-negative and convex and g is positive and concave.

2.7. Let f and g be functions defined on a convex set $S \subseteq \Re^n$, and let $z(x) = \left(\frac{f(x)}{g(x)}\right)^\alpha$, $\alpha > 0$. If f is non-negative and convex, and g is positive and concave, then z is semistrictly quasiconvex.

2.8. Let f and g be functions defined on a convex set $S \subseteq \Re^n$, and let $z(x) = \frac{f(x)}{g(x)}$. Prove that z is strictly quasiconvex if one of the following conditions holds:
1. f is non-negative and strictly convex and g is positive and concave;
2. f is non-negative and convex and g is positive and strictly concave;
3. f is non-positive and strictly convex and $g(x)$ is positive and convex.

2.9. Apply Theorem 2.2.3 to show that the linear fractional function $f(x) = \frac{a^T x + a_0}{b^T x + b_0}$, $b^T x + b_0 > 0$, is both quasiconvex and quasiconcave.

2.10. Show that the following functions are quasiconvex:
(a) $f(x,y) = \log \frac{x^4}{y+1}$, $y > -1$;
(b) $f(x,y) = \log \frac{x^2 - xy + y^2}{4 - y^2}$, $-2 < y < 2$;
(c) $f(x,y) = \log(x^4 + y^2) - \log(y - 1)$, $y > 1$.

2.11. Show that the following functions are convex:
$$f(x_1, ..., x_n) = \left(\sum_{i=1}^n x_i^4\right)^{\frac{1}{3}}; \ g(x,y) = \frac{x^2}{y}; \ h(x) = \frac{(a^T x)^2}{b^T x}.$$

2.12. Show that the function $f(x,y) = xy$, x, $y \geq 0$ is quasiconcave and that the function $f(x,y) = \sqrt{xy}$, x, $y \geq 0$ is concave.

2.13. Show that the ratio $z(x) = \frac{f(x)}{g(x)}$ is semistrictly quasiconcave when f is non-negative and concave and g is positive and convex.

2.14. Show that the function $z(x) = f(x) \cdot g(x)$ is semistrictly quasiconcave when f is non-negative and concave and g is positive and concave.

2.15. Show that the function $f(x) = (a^T x + a_0)(b^T x + b_0)$ is semistrictly quasiconcave if $a^T x + a_0 \geq 0$ and $b^T x + b_0 > 0$ while it is semistrictly quasiconvex if $a^T x + a_0 \leq 0$ and $b^T x + b_0 > 0$.

2.16. Let f be a quasiconvex function defined on the convex set $S \subseteq \Re^n$. Prove that f cannot have an interior strict local maximum point.

2.17. Give an example which shows that a quasiconvex function may have an interior global maximum point which is not a local minimum.

2.18. Show Theorem 2.2.10.

2.19. Show Theorem 2.3.6.

2.20. Let f be a non-constant lower semicontinuous semistrictly quasiconvex function on the convex set $S \subseteq \Re^n$. Show that f, in contrast to the quasiconvex case, cannot have an interior global maximum.

2.21. Give an example which shows that the assumption of lower semicontinuity in Exercise 2.20 cannot be relaxed.

2.22. Show that, in contrast to the quasiconvex case, a non-constant lower semicontinuous semistrictly quasiconvex function cannot have an interior local maximum point which is not a local minimum.

2.23. Give a direct proof of the statement: if f is convex then f is semistrictly quasiconvex.

2.7 References

Arrow K. J., and Enthoven A. C. [7], Avriel M. [10], Avriel M., Diewert W. E., Schaible S., and Ziemba W. T. eds. [12], Avriel M., Diewert W. E., Schaible S., and Zang I. [13], Carter M. [61], De Finetti B. [92], Diewert W. E., Avriel M., and Zang I. [94], Elkin R. M. [104], Gerencsér L. [116], Ginsberg W. [124], Greenberg H. J., and Pierskalla W. P. [129], Madden P. [191], Mangasarian O. L. [193], Martos B. [211], Ponstein J. [223], Thompson W. A., and Parke D. W. [275].

3

Differentiable Generalized Convex Functions

3.1 Introduction

in this chapter we shall consider, under the differentiability assumption, the classes of generalized convex functions introduced in the previous chapter. Furthermore, a new class is defined: that of pseudoconvex functions, which is perhaps the most important of all.

Several first order and second order characterizations of quasiconvexity and pseudoconvexity are also established, some of which turn out to be useful tools in determining special classes of generalized convex functions.

By combining generalized convexity and generalized concavity, quasilinearity and pseudolinearity are introduced and studied.

In the last section, the notion of convexity and generalized convexity at a point is introduced. This represents a significant relaxation of the concept of generalized convexity, since the convexity of the domain of the function is not required.

3.2 Differentiable Quasiconvex and Pseudoconvex Functions

3.2.1 Differentiable Quasiconvex Functions

A differentiable function is convex if and only if its graph lies on or above the tangent at any point of the graph. Similarly, a differentiable function is quasiconvex if and only if any of its level sets lies on or below the tangent at any point of the level set. In order to express this geometrical property analytically, we shall begin to characterize the quasiconvexity of a differentiable function of one variable.

Theorem 3.2.1. *Let φ be a differentiable function[1] on an interval $I \subseteq \Re$. Then, φ is quasiconvex on I if and only if the following implication holds:*

$$t_1, t_2 \in I, \ \varphi(t_1) \geq \varphi(t_2) \Rightarrow \varphi'(t_1)(t_2 - t_1) \leq 0. \tag{3.1}$$

Proof. Let $t_1,\ t_2 \in I$ such that $t_1 < t_2$ ($t_1 > t_2$) and $\varphi(t_1) \geq \varphi(t_2)$. The quasiconvexity of φ implies $\varphi(t) \leq \varphi(t_1)$, $\forall t \in [t_1, t_2]$ ($\forall t \in [t_2, t_1]$) so that φ is locally non-increasing (locally non-decreasing) at t_1 and, consequently, (3.1) holds.

Assume now that (3.1) holds. If φ is not quasiconvex, there exist $t_1,\ t_2 \in I$ with $t_1 < t_2$, such that $M = \max\{\varphi(t),\ t \in [t_1, t_2]\} > \max\{\varphi(t_1), \varphi(t_2)\}$. Let $\bar{t} = \inf\{t \in [t_1, t_2] : \varphi(t) = M\}$; the continuity of φ implies the existence of $\epsilon > 0$ such that $\varphi(t) < M$, $\forall t \in (\bar{t} - \epsilon, \bar{t})$ and $\varphi(t) > \max\{\varphi(t_1), \varphi(t_2)\}$, $\forall t \in (\bar{t} - \epsilon, \bar{t})$. The Mean Value Theorem applied to the interval $[\bar{t} - \epsilon, \bar{t}]$ implies the existence of $t^* \in (\bar{t} - \epsilon, \bar{t})$ such that $\varphi'(t^*) > 0$. Consequently, we have $\varphi(t^*) > \varphi(t_2)$ with $\varphi'(t^*) > 0$ and this contradicts (3.1). □

Remark 3.2.1. Theorem 3.2.1 cannot be specified for a strictly or a semistrictly quasiconvex function in the sense that (3.1) cannot be improved. For instance, the function $\varphi(t) = -t^2$, $t \in [0, 1]$ is both strictly and semistrictly quasiconvex and we have $\varphi(0) = 0 > \varphi(1) = -1$ with $\varphi'(0) = 0$.

Theorem 3.2.1 allows us to obtain the following characterization of a differentiable quasiconvex function in more than one variable.

Theorem 3.2.2. *Let $S \subseteq \Re^n$ be a convex set and let f be a differentiable function on S. Then, f is quasiconvex on S if and only if the following implication holds:*

$$x_1,\ x_2 \in S,\ f(x_1) \geq f(x_2) \Rightarrow \nabla f(x_1)^T (x_2 - x_1) \leq 0. \tag{3.2}$$

Proof. Assume f is quasiconvex and let $x_1,\ x_2 \in S$ with $f(x_1) \geq f(x_2)$. Consider the restriction $\varphi(t) = f(x_1 + t(x_2 - x_1)), t \in [0, 1]$. We have $\varphi(0) \geq \varphi(1)$ so that Theorem 3.2.1 implies that $\varphi'(0) = (x_2 - x_1)^T \nabla f(x_1) \leq 0$ and (3.2) holds.

Assume now that (3.2) holds. If f is not quasiconvex, there exists a restriction $\varphi(t) = f(x_1 + t(x_2 - x_1))$, $x_1,\ x_2 \in S$, $t \in [0, 1]$, which is not quasiconvex (see Theorem 2.2.10) and consequently, from Theorem 3.2.1, there exist $t_1, t_2 \in [0, 1]$ such that $\varphi(t_1) \geq \varphi(t_2)$ and $\varphi'(t_1)(t_2 - t_1) > 0$. Set $\bar{x}_1 = x_1 + t_1(x_2 - x_1)$, $\bar{x}_2 = x_1 + t_2(x_2 - x_1)$; we have $f(\bar{x}_1) \geq f(\bar{x}_2)$ and $\varphi'(t_1) = (x_2 - x_1)^T \nabla f(\bar{x}_1) = \frac{1}{t_2 - t_1}(\bar{x}_2 - \bar{x}_1)^T \nabla f(\bar{x}_1)$. Consequently, $\varphi'(t_1)(t_2 - t_1) = (x_2 - x_1)^T \nabla f(\bar{x}_1) > 0$ and this contradicts (3.2). □

The following theorem points out that corresponding to a strict inequality on the left-hand side of (3.2) we also have a strict inequality on the right-hand side when S is open and x_1 is not a critical point.

---

[1] The differentiability of a function g on a set $X \subseteq \Re^n$ means that g is differentiable on an open set containing X.

Theorem 3.2.3. *Let f be a differentiable quasiconvex function on an open convex set $S \subseteq \Re^n$. Then, the following implication holds:*

$$x_1,\ x_2 \in S,\ f(x_1) > f(x_2),\ \nabla f(x_1) \neq 0 \Rightarrow \nabla f(x_1)^T (x_2 - x_1) < 0. \quad (3.3)$$

Proof. Assume the existence of $x_1, x_2 \in S$ such that $f(x_1) > f(x_2)$ and $(x_2 - x_1)^T \nabla f(x_1) \geq 0$. Since f is quasiconvex, we necessarily have $(x_2 - x_1)^T \nabla f(x_1) = 0$. For the continuity of f, there exists $\epsilon > 0$ such that $y = x_2 + \epsilon \nabla f(x_1) \in S$ with $f(y) < f(x_1)$ (note that $y \neq x_2$ since $\nabla f(x_1) \neq 0$). Consequently, for the quasiconvexity of f, we have $(y - x_1)^T \nabla f(x_1) \leq 0$. On the other hand, $y - x_1 = (y - x_2) + (x_2 - x_1)$, so that $(y - x_1)^T \nabla f(x_1) = (y - x_2)^T \nabla f(x_1) + (x_2 - x_1)^T \nabla f(x_1) = \epsilon \parallel \nabla f(x_1) \parallel^2 > 0$, and this is a contradiction. □

Let us note that assumption $\nabla f(x_1) \neq 0$ is essential for the validity of Theorem 3.2.3. In fact, the function $\varphi(t) = -t^2$, $t \in [0, 1]$, is quasiconvex, $\varphi(0) = 0 > \varphi(1) = -1$ but $\varphi'(0) = 0$.

Remark 3.2.2. As we have pointed out in Remark 3.2.1, it is not possible to obtain a characterization of a strictly or a semistrictly quasiconvex function in terms of the sign of the directional derivatives. Nevertheless, it is possible to characterize a strictly quasiconvex function when f does not have critical points as is stated in Theorem 3.2.4.

Theorem 3.2.4. *Let f be a differentiable function on an open convex set $S \subseteq \Re^n$ and assume that $\nabla f(x) \neq 0$, $\forall x \in S$. Then, f is strictly quasiconvex if and only if the following condition holds:*

$$x_1,\ x_2 \in S,\ f(x_1) \geq f(x_2) \Rightarrow \nabla f(x_1)^T (x_2 - x_1) < 0. \quad (3.4)$$

Proof. Assume that f is strictly quasiconvex. If $f(x_1) > f(x_2)$, the thesis follows from Theorem 3.2.3. If $f(x_1) = f(x_2)$ the strict quasiconvexity implies the existence of $t^* \in (0, 1)$ such that $f(x_1) > f(x^*)$ with $x^* = x_1 + t^*(x_2 - x_1)$. From Theorem 3.2.3 we have $(x^* - x_1)^T \nabla f(x_1) < 0$ or, equivalently, $t^*(x_2 - x_1)^T \nabla f(x_1) < 0$, so that (3.4) holds.
Assume now the validity of (3.4). From Theorem 3.2.2 the function is quasiconvex . If f is not strictly quasiconvex, there exist $x_1, x_2 \in S$, $t^* \in (0, 1)$, such that $f(x_1) \geq f(x_2)$, $f(x_1) = f(x^*)$ where $x^* = x_1 + t^*(x_2 - x_1)$. Consider the restriction $\varphi(t) = f(x_1 + t(x_2 - x_1))$, $t \in (0, 1)$. Equation (3.4) applied to points x^*, x_2 implies $\varphi'(t^*) < 0$. Consequently, there exists $t_1 \in (0, t^*)$ such that $\varphi(t_1) > \varphi(t^*)$ and this contradicts the quasiconvexity of $\varphi(t)$ on $[0, t^*]$. □

3.2.2 Pseudoconvex Functions

It is known that a critical point for a convex function is also a global minimum. This useful property does not hold for quasiconvex, strictly quasiconvex and semistrictly quasiconvex functions (for instance, the critical point for the strictly quasiconvex function $\varphi(t) = t^3$ is not a global minimum). This is the reason for introducing of the following class of pseudoconvex functions.

Definition 3.2.1. *A differentiable function f, defined on an open convex set $S \subseteq \Re^n$, is called pseudoconvex if*

$$x_1, \; x_2 \in S, \; f(x_1) > f(x_2) \Rightarrow \nabla f(x_1)^T (x_2 - x_1) < 0. \qquad (3.5)$$

If the inequality on the right hand-side of (3.5) still holds when $f(x_1) = f(x_2)$, the function is called *strictly pseudoconvex*. Formally,

Definition 3.2.2. *A differentiable function f, defined on an open convex set $S \subseteq \Re^n$, is called strictly pseudoconvex if*

$$x_1, \; x_2 \in S, \; x_1 \neq x_2, \; f(x_1) \geq f(x_2) \Rightarrow \nabla f(x_1)^T (x_2 - x_1) < 0. \qquad (3.6)$$

It follows, immediately, from the given definitions, that a strictly pseudoconvex function is pseudoconvex, too. The converse statement is not true; in fact, a constant function is pseudoconvex but it is not strictly pseudoconvex.

A function f is said to be pseudoconcave (strictly pseudoconcave) if and only if $-f$ is pseudoconvex (strictly pseudoconvex). Consequently, the results that we are going to establish can be easily adapted to the pseudoconcave case. For the sake of completeness, in Appendix B we shall summarize the main properties of pseudoconcave functions.

Remark 3.2.3. Note that, when $f(x_1) = f(x_2)$, $x_1, x_2 \in S$, $x_1 \neq x_2$, the strict pseudoconvexity implies that f is decreasing at x_1 in the direction $u = x_2 - x_1$ since the directional derivative $\nabla f(x_1)^T (x_2 - x_1)$ is negative. In particular, if x_1 is a local minimum, it is necessarily a strict local minimum.

The following theorem shows that a critical point for a pseudoconvex function is a global minimum as happens in the convex case.

Theorem 3.2.5. *Let f be a differentiable function on an open convex set $S \subseteq \Re^n$ and let $x_0 \in S$ be a critical point. If f is pseudoconvex, then x_0 is a global minimum for f. Furthermore, x_0 is unique if f is strictly pseudoconvex.*

Proof. Assume that there exists $y \in S$ such that $f(y) < f(x_0)$. Then $\nabla f(x_0) = 0$ implies that $\nabla f(x_0)^T (y - x_0) = 0$, which contradicts (3.5). It follows that x_0 is a global minimum for f and, furthermore, taking into account Remark 3.2.3, it is unique if f is strictly pseudoconvex. $\qquad \square$

Let us note that pseudoconvexity requires that (3.5) is satisfied at all points of the domain in contrast to quasiconvexity which implies that (3.5) is satisfied when $\nabla f(x_1) \neq 0$ (see Theorem 3.2.3). More exactly, we have the following theorem.

Theorem 3.2.6. *Let f be a differentiable function on an open convex set $S \subseteq \Re^n$.*
(i) If f is pseudoconvex on S, then f is quasiconvex on S;
(ii) If $\nabla f(x) \neq 0$, $\forall x \in S$, then f is pseudoconvex on S if and only if it is quasiconvex on S.

Proof. (i) Assume that f is not quasiconvex. Then, there exist $x_1, x_2 \in S$ with $f(x_1) \geq f(x_2)$ such that $\nabla f(x_1)^T (x_2 - x_1) > 0$. Consider the restriction $\varphi(t) = f(x_1 + t(x_2 - x_1))$, $t \in [0, 1]$. Since $\varphi'(0) = \nabla f(x_1)^T (x_2 - x_1) > 0$, φ attains its maximum value at an interior point $t_0 \in (0, 1)$, so that $\varphi(t_0) = f(x_0) > f(x_1) = \varphi(0) \geq f(x_2) = \varphi(1)$ and $\varphi'(t_0) = \nabla f(x_0)^T (x_2 - x_1) = 0$, where $x_0 = x_1 + t_0(x_2 - x_1)$. On the other hand, the pseudoconvexity of f, applied to points x_0, x_2, implies that $\nabla f(x_0)^T (x_2 - x_1)(1 - t_0) < 0$, and this is a contradiction.
(ii) It remains to be proven that a quasiconvex function is pseudoconvex when there are no critical points. This follows from Theorem 3.2.3. □

As for the classes of generalized convex functions introduced in the previous chapter, a function is pseudoconvex (strictly pseudoconvex) if and only if it is pseudoconvex (strictly pseudoconvex) over each restriction on a line segment. This property leads us to carry on the study of pseudoconvexity of a single-variable function with the aim of obtaining characterizations for functions of more than one variable.

Theorem 3.2.7. *Let φ be a differentiable function on an open interval $I \subseteq \Re$. Then, φ is pseudoconvex (strictly pseudoconvex) on I if and only if for every $t_0 \in I$ such that $\varphi'(t_0) = 0$, t_0 is a local minimum (strict local minimum) for φ.*

Proof. Assume the pseudoconvexity of φ. Let $t_0 \in I$ such that $\varphi'(t_0) = 0$ and assume that t_0 is not a local minimum for φ. Then, there exists $t^* \in I$ such that $\varphi(t^*) < \varphi(t_0)$. The pseudoconvexity of φ implies that $\varphi'(t_0)(t^* - t_0) < 0$ and this is absurd since t_0 is a critical point.
Assume now that every critical point is a local minimum. If φ is not pseudoconvex, there exist t_1, $t_2 \in I$ such that $\varphi(t_1) > \varphi(t_2)$ implies that $\varphi'(t_1)(t_2 - t_1) \geq 0$.
We can assume without loss of generality that $t_1 < t_2$ so that $\varphi'(t_1) \geq 0$. If $\varphi'(t_1) = 0$, t_1 is a local minimum point and thus there exist $\epsilon > 0$, $\bar{t} \in (t_1, t_1 + \epsilon)$ such that $\varphi(\bar{t}) \geq \varphi(t_1) > \varphi(t_2)$. If $\varphi'(t_1) > 0$ we can find $\bar{t} \in (t_1, t_2)$ with $\varphi(\bar{t}) > \varphi(t_1) > \varphi(t_2)$. In each case the maximum value of $\varphi(t)$ on $[t_1, t_2]$ is reached at an interior point. The largest maximizer point t_0 is such that $\varphi(t_0) > \varphi(t)$ for all $t \in (t_0, t_2]$ with $\varphi'(t_0) = 0$, and this contradicts the assumption that every critical point is a local minimum. The proof is complete, taking into account Remark 3.2.3. □

The following theorem is a direct consequence of Theorem 3.2.7.

Theorem 3.2.8. *Let f be a differentiable function on an open convex set $S \subseteq \Re^n$. Then, f is pseudoconvex (strictly pseudoconvex) on S if and only if for every $x_0 \in S$ and $u \in \Re^n$ such that $u^T \nabla f(x_0) = 0$, the function $\varphi(t) = f(x_0 + tu)$ attains a local minimum (strict local minimum) at $t = 0$.*

Remark 3.2.4. We have already remarked that a quasiconvex function may have local and also global interior maximum points, but cannot have a strict local maximum point and also a semistrict local maximum point defined as follows:

Let φ be defined on the open interval $I \subset \Re$.

A point $t_0 \in I$ is said to be a **semistrict local maximum point** for φ if there exist $t_1, t_2 \in I$ with $t_1 < t_0 < t_2$, such that $\varphi(\lambda t_1 + (1 - \lambda)t_2) \le \varphi(t_0)$ for every $\lambda \in [0, 1]$ and $\max\{\varphi(t_1), \varphi(t_2)\} < \varphi(t_0)$.

This special type of local maximum point, introduced by Dewert, Avriel, and Zang in [94], is stronger than the concept of a local maximum, but weaker than a strict local maximum.

By means of the non-existence of this kind of point, it is possible to characterize quasiconvexity in the non-differentiable and in the differentiable case (see Exercises 3.1, 3.4, 3.6, 3.25).

Referring to Theorem 3.2.6, we shall point out that within the class of quasiconvex fuctions, pseudoconvexity may be specified by means of its behaviour at a critical point. The following theorem holds.

Theorem 3.2.9. *Let f be a continuous differentiable function on an open convex set $S \subseteq \Re^n$. Then, f is pseudoconvex (strictly pseudoconvex) on S if and only if the following conditions hold:*
(i) f is quasiconvex on S;
(ii) If $x_0 \in S$, $\nabla f(x_0) = 0$, then x_0 is a local minimum (strict local minimum) for f.

Proof. If f is pseudoconvex, then (i) follows from Theorem 3.2.6, while (ii) follows from Theorem 3.2.5.

Assume now that (i) and (ii) hold. By applying Theorem 3.2.8 we must prove that if $x_0 \in S$ and $u \in \Re^n$ are such that $u^T \nabla f(x_0) = 0$, the function $\varphi(t) = f(x_0 + tu)$ attains a local minimum at $t = 0$. If $\nabla f(x_0) = 0$, the thesis follows from (ii); if $\nabla f(x_0) \ne 0$, the continuity of the gradient map implies the existence of an open neighbourhood $U(x_0)$ of x_0 such that $\nabla f(x_0) \ne 0$, $\forall x \in U(x_0)$. Theorem 3.2.6 implies that f is pseudoconvex on $U(x_0)$ and Theorem 3.2.7 implies that the function $\varphi(t) = f(x_0 + tu)$ attains a local minimum at $t = 0$. The proof is complete, taking into account Remark 3.2.3. □

Remark 3.2.5. In [82, 89] a more elaborate proof of Theorem 3.2.9 is given without the assumption of continuity of the gradient map.

Remark 3.2.6. Let us note that Theorems 3.2.6 and 3.2.9 cannot be extended to the closure of S. More exactly, a pseudoconvex function on an open set is not necessarily pseudoconvex on the closure of S. Consider for instance the quasiconvex function $f(x, y) = -xy, (x, y) \in int\Re_+^2$. Such a function is also pseudoconvex since its gradient does not vanish in $int\Re_+^2$. On the other hand, f is quasiconvex on $\Re_+^2$ (see Theorem 2.2.12) but it is not pseudoconvex on $\Re_+^2$ since the gradient vanishes at the origin which is not a global minimum.

Theorems 3.2.6 and 3.2.9 are sometimes useful in verifying the pseudoconvexity of a function.

Example 3.2.1. Consider the Cobb–Douglas function $f(x) = x_1^{\alpha_1} \cdot .... \cdot x_n^{\alpha_n}$, $x = (x_1, ..., x_n)$, $x_i > 0$, $\alpha_i < 0$, $i = 1, ..., n$. Since f is quasiconvex and $\nabla f(x) \neq 0$, f is also pseudoconvex.

Example 3.2.2. An affine function $f(x) = a^T x + b$ is pseudoconvex since it is constant if $a = 0$ and it is quasiconvex if $a = \nabla f(x) \neq 0$.

Theorem 3.2.10. *Consider the ratio $z(x) = \frac{f(x)}{g(x)}$ where f and g are differentiable functions defined on an open convex set $S \subseteq \Re^n$.*
(i) If f is convex and g is positive and affine, then z is pseudoconvex;
(ii) If f is non-negative and convex, and g is positive and concave, then z is pseudoconvex;
(iii) If f is positive and strictly convex, and g is positive and concave, then z is strictly pseudoconvex;
(iv) If f is non-negative and convex, and g is positive and strictly concave, then z is strictly pseudoconvex.

Proof. By Theorem 2.3.8 z is quasiconvex, so that, taking into account Theorem 3.2.9, it is sufficient to prove that a critical point x_0 (if one exists) is a local (strict local) minimum. We have $\nabla z(x) = \frac{\nabla f(x) \cdot g(x) - f(x) \cdot \nabla g(x)}{(g(x))^2}$ so that $\nabla z(x_0) = 0$ if and only if $\nabla f(x_0) = z(x_0)\nabla g(x_0)$. Consequently, $\nabla f(x_0)^T (x - x_0) = z(x_0)\nabla g(x_0)^T (x - x_0)$ and we have:
$z(x_0)\nabla g(x_0)^T (x - x_0) = z(x_0)(g(x) - g(x_0))$ if g is affine;
$z(x_0)\nabla g(x_0)^T (x - x_0) \geq z(x_0)(g(x) - g(x_0))$ if $z(x_0) \geq 0$, where the inequality is strict if g is strictly concave. In each case we have:
$f(x) \geq f(x_0) + \nabla f(x_0)^T (x - x_0) = f(x_0) + z(x_0)\nabla g(x_0)^T (x - x_0) \geq f(x_0) + z(x_0)(g(x) - g(x_0)) = z(x_0)g(x)$, where the first (second) inequality is strict if f (g) is strictly convex (strictly concave) and thus $\frac{f(x)}{g(x)} \geq z(x_0) = \frac{f(x_0)}{g(x_0)}$, $\forall x \in S$, where the inequality is strict if f (g) is strictly convex (strictly concave). $\square$

Another useful theorem in constructing new pseudoconvex functions from existing ones is the following.

Theorem 3.2.11. *Let* $f : S \subseteq \Re^n \to \Re$ *be a pseudoconvex (strictly pseudoconvex) function on an open convex set* S *and let* $\phi : \Re \to \Re$ *be a differentiable function such that* $\phi'(z) > 0, \forall z \in \Re$. *Then, the composite function* $\phi \circ f$ *is pseudoconvex (strictly pseudoconvex).*

Proof. ϕ is an increasing function so that $\phi \circ f$ is quasiconvex. Since $\nabla \phi(f(x)) = \phi'(f(x)) \nabla f(x)$, we have $\nabla \phi(f(x)) = 0$ if and only if $\nabla f(x) = 0$. It follows that x_0 is a local (strict local) minimum point for f and, consequently, it is a local (strict local) minimum for $\phi \circ f$. The thesis follows from Theorem 3.2.9. $\square$

Example 3.2.3. Let f be a pseudoconvex (strictly pseudoconvex) function on an open convex set $S \subseteq \Re^n$. Taking into account that the derivative of the exponential function $\phi(z) = e^z$ is positive, $g(x) = e^{f(x)}$ is pseudoconvex (strictly pseudoconvex) on S.
Furthermore, when f is positive on S, the functions $g(x) = \log f(x)$, $g(x) = \sqrt{f(x)}$, $g(x) = (f(x))^\alpha, \alpha > 0$, are pseudoconvex on S.

Remark 3.2.7. With the aim of extending the concept of pseudoconvexity to the non-differentiable case, in [217] and in [275] the following class of functions has been introduced. This reduces to a pseudoconvex function under the differentiability assumption:
For every $x_1, x_2 \in S$, there exists a positive number $b(x_1, x_2)$, depending on x_1, x_2, such that

$$f(x_1) > f(x_2) \Rightarrow f(x_1 + \lambda(x_2 - x_1)) \leq f(x_1) - (1 - \lambda)\lambda b(x_1, x_2), \ \forall \lambda \in (0, 1).$$

3.2.3 Relationships

In this subsection we shall complete the inclusion relationships between the various classes of convex and generalized convex functions.
The following theorem shows that the class of pseudoconvex functions is intermediate between the quasiconvex and semistrictly quasiconvex functions.

Theorem 3.2.12. *Let* f *be a differentiable function defined on an open convex set* $S \subseteq \Re^n$.
(i) If f *is pseudoconvex on* S, *then* f *is semistrictly quasiconvex on* S;
(ii) If f *is strictly pseudoconvex on* S, *then* f *is strictly quasiconvex on* S.

Proof. (i) Let $x_1, x_2 \in S$ with $f(x_1) > f(x_2)$ and set $\varphi(t) = f(x_1 + t(x_2 - x_1))$, $t \in [0, 1]$. If f is not semistrictly quasiconvex, then there exists $\bar{t} \in (0, 1)$ such that $f(x_1 + \bar{t}(x_2 - x_1)) \geq f(x_1) > f(x_2)$. Consequently, φ attains its maximum value at a point $t_0 \in (0, 1)$ with $\varphi(t_0) = f(x_1 + t_0(x_2 - x_1)) > f(x_2)$ and $\varphi'(t_0) = 0$. By applying pseudoconvexity to points $x_0 = x_1 + t_0(x_2 - x_1)$ and x_2, we have $\nabla f(x_0)^T(x_2 - x_0) = (1 - t_0)\nabla f(x_0)^T(x_2 - x_1) = (1 - t_0)\varphi'(t_0) < 0$, i.e., $\varphi'(t_0) < 0$, and this is a contradiction.

(ii) From (i), f is semistrictly quasiconvex on S so that, taking into account Theorem 2.3.6, it remains to be proven that every restriction on a line segment attains its minimum at no more than one point. If not, there exist at least two minimum points, x_1, $x_2 \in S$ with $f(x_1) = f(x_2)$. Strict pseudoconvexity implies that $\nabla f(x_1)^T(x_2 - x_1) < 0$, and this is a contradiction. □

Let us note that there is a strict inclusion relationship between the class of pseudoconvex (strictly pseudoconvex) functions and the semistrictly quasiconvex (strictly quasiconvex) ones. Consider, for instance, function $f(x) = x^3$ which is increasing and hence strictly quasiconvex; on the other hand, f is not pseudoconvex since $f'(0) = 0$ but $x_0 = 0$ is not a global minimum. Nevertheless, the classes of quasiconvex, semistricly quasiconvex and pseudoconvex functions collapse when there are no critical points as is stated in Theorem 3.2.6.

Remark 3.2.8. Let us note that there is not any inclusion relationships between the class of pseudoconvex functions and the strictly quasiconvex ones. In fact, a constant function is pseudoconvex but not strictly quasiconvex, while $f(x) = x^3$ is strictly quasiconvex but not pseudoconvex.

We can summarize the inclusion relationships between the various classes of convex and generalized convex functions by means of the diagram in Fig. 3.1, assuming differentiability.

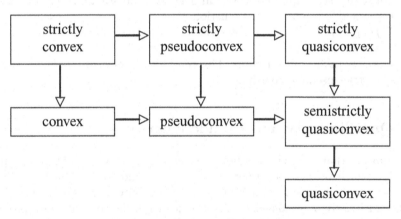

Fig. 3.1. Relationships between various types of convexity under differentiability

Examples in two variables showing that the inclusions are proper are given below.

Example 3.2.4.
1. $f(x, y) = x + y$, $(x, y) \in \Re^2$ is convex, strictly quasiconvex but not strictly convex.
2. $f(x, y) = (x + y)^3 + x + y$, $(x, y) \in \Re^2$ is pseudoconvex but not convex.

3. $f(x,y) = (x+y)^3$, $(x,y) \in \Re^2$ is semistrictly quasiconvex but neither pseudoconvex, nor strictly quasiconvex.

4. $f(x,y) = \begin{cases} -(x+y)^2 & x+y < 0 \\ 0 & x+y \geq 0 \end{cases}$ is quasiconvex but not semistrictly quasiconvex.

Finally, we shall prove that the class of strictly pseudoconvex functions is the intersection between the class of the pseudoconvex functions and the class of the strictly quasiconvex ones.

Theorem 3.2.13. *Let f be a continuous differentiable function defined on an open convex set $S \subseteq \Re^n$. Then, f is strictly pseudoconvex on S if and only if f is both pseudoconvex and stricly quasiconvex on S.*

Proof. Assume that f is both pseudoconvex and stricly quasiconvex on S. Taking into account Theorem 3.2.9, it remains to be proven that a local minimum point $x_0 \in S$ is strict. If not, there exist a neighbourhood I of x_0 and $x_1 \in I \cap S$ such that $f(x_1) = f(x_0)$. The strict quasiconvexity of f implies $f(x_0 + t(x_1 - x_0)) < f(x_0)$, $\forall t \in (0,1)$, and this contradicts the local optimality of x_0.

The converse statement is obvious. □

Remark 3.2.9. When $\nabla f(x) \neq 0$, $\forall x \in S$, since pseudoconvexity collapses to quasiconvexity, strict quasiconvexity implies pseudoconvexity. The converse is not true since, for instance, the function $f(x_1, x_2) = x_1 + x_2$ is pseudoconvex with $\nabla f(x_1, x_2) \neq 0$, but it is not strictly quasiconvex since it is constant on any level set.

Furthermore, taking into account Theorem 3.2.13, strict quasiconvexity collapses to strict pseudoconvexity.

3.3 Quasilinearity and Pseudolinearity

By requiring that a function is both convex and concave we obtain an affine function which verifies all properties of a convex and concave function. Similarly, we can require that a function is both quasiconvex and quasiconcave (or both semistrictly quasiconvex and semistrictly quasiconcave or both pseudoconvex and pseudoconcave) in order to obtain classes of functions for which all properties of a generalized convex function and generalized concave function hold.

3.3.1 Quasilinearity and Semistrict Quasilinearity

A function f is said to be quasilinear (semistrictly quasilinear) if it is both quasiconvex and quasiconcave (semistrictly quasiconvex and semistrictly quasiconcave).

Taking into account the various characterizations of a quasiconvex (semistrictly quasiconvex) and quasiconcave (semistrictly quasiconcave) function, we have the following theorems.

Theorem 3.3.1. *Let f be a function defined on a convex set $S \subseteq \Re^n$. Then, f is quasilinear on S if and only if one of the following conditions holds:*
(i) $x_1, x_2 \in S$, $\min\{f(x_1), f(x_2)\} \leq f(x_1 + \lambda(x_2 - x_1)) \leq \max\{f(x_1), f(x_2)\}$, for all $\lambda \in [0, 1]$;
(ii) The lower level sets and the upper level sets of f are convex;
(iii) Any restriction of f on a line segment is a non-increasing or a non-decreasing function.

Theorem 3.3.2. *Let f be a function defined on a convex set $S \subseteq \Re^n$. Then, f is semistrictly quasilinear on S if and only if one of the following conditions holds:*
(i) $x_1, x_2 \in S$, $f(x_1) \neq f(x_2)$, $\min\{f(x_1), f(x_2)\} < f(x_1 + \lambda(x_2 - x_1)) < \max\{f(x_1), f(x_2)\}$, $\lambda \in [0, 1]$;
(ii) Any restriction of f on a line segment is an increasing function or a decreasing function or a constant function.

Similarly to the quasiconvex case, a semistrictly quasilinear function is not necessarily a quasilinear function (see Example 2.3.2; the function is quasilinear but not semistrictly quasilinear). In order to mantain the inclusion relationships between the class of semistrictly quasilinear functions and the class of quasilinear functions, we must require continuity (see Theorem 2.3.2 and its analogous theorem for the quasiconcave case).

Theorem 3.3.3. *A continuous semistrictly quasilinear function is a quasilinear function.*

The converse statement of the above theorem is not true; for instance, the function of one variable $f(x) = x|x| - x^2$ is continuous and quasilinear but it is not semistrictly quasilinear.

Let us note that a continuous non-constant semistrictly quasilinear function is both strictly quasiconvex and strictly quasiconcave, so that there is no reason to introduce the class of the "strictly quasilinear functions".

Property (ii) of Theorem 3.3.1 implies that every level set Γ_k of a quasilinear function is convex since it is the intersection of the two convex sets $\Gamma_{\geq k}$, $\Gamma_{\leq k}$. Let us note that Γ_k is not necessarily the boundary of $\Gamma_{\geq k}$ or $\Gamma_{\leq k}$. For instance, consider the continuous quasilinear function

$$f(x) = \begin{cases} -x^2 & x \leq 0 \\ 0 & 0 < x \leq 2 \\ (x-2)^2 & x > 2 \end{cases}$$

We have $\Gamma_{k=0} = [0, 2]$, $\Gamma_{\geq 0} = [0, +\infty)$, $\Gamma_{\leq 0} = (-\infty, 2]$, so that the sets $\Gamma_{\geq 0}$, $\Gamma_{\leq 0}$ have different boundary points; note also that $int \Gamma_{k=0} \neq \emptyset$.

This kind of situation does not happen for a continuous semistrictly quasilinear function as is shown in the following theorem.

Theorem 3.3.4. *Let f be a non-constant continuous semistrictly quasilinear function on an open convex set $S \subseteq \Re^n$. Then, the following conditions hold:*
(i) $int\Gamma_k = \emptyset$ for every nonempty level set Γ_k;
(ii) A point of the level set Γ_k is a boundary point of upper level set $\Gamma_{\geq k}$ and a boundary point of lower level set $\Gamma_{\leq k}$.

Proof. (i) By contradiction, assume that $int\Gamma_k \neq \emptyset$ and let $x_0 \in int\Gamma_k$. For every $x \in S$, consider the line segment $[x_0, x]$; since $x_0 \in int\Gamma_k$, there exists $\tilde{x} \in (x_0, x)$ such that $[x_0, \tilde{x}] \subset \Gamma_k$. Consequently, the restriction of the function f along the line segment $[x_0, x]$ is constant in $[x_0, \tilde{x}]$ so that it is also constant (see (ii) of Theorem 3.3.2) in $[x_0, x]$. Then, we have $f(x_0) = f(x)$ for every $x \in S$, i.e., f is a constant function and this contradicts the assumption.
(ii) Let $x_0 \in \Gamma_k$. Since $int\Gamma_{k=0} = \emptyset$, for every neighbourhood I of x_0 there exists $\bar{x} \in I$ such that $f(\bar{x}) \neq f(x_0)$. Consider the restriction $\varphi(t) = f(x_0 + t(\bar{x} - x_0))$, $t \in (-\epsilon, \epsilon)$. From (ii) of Theorem 3.3.2, φ is increasing or decreasing so that (ii) holds. $\qquad\square$

The following theorem shows that the level sets of a continuous semistrictly quasilinear function are the intersections of its domain with hyperplanes.

Theorem 3.3.5. *Let f be a non-constant continuous function defined on an open convex set $S \subseteq \Re^n$. Then, f is semistrictly quasilinear if and only if every nonempty level set Γ_k can be expressed in the form $\Gamma_k = S \cap H_k$, where H_k is a hyperplane.*

Proof. Assume that f is a semistrictly quasilinear function. From Theorem 3.3.4 we have $int\Gamma_k = int\Gamma_{\geq k} \cap int\Gamma_{\leq k} = \emptyset$ so that the convexity of $\Gamma_{\leq k}$ and $\Gamma_{\geq k}$ implies the existence of a hyperplane H_k which separates $\Gamma_{\leq k}$ and $\Gamma_{\geq k}$, that is the existence of $\alpha_k \in \Re^n \backslash \{0\}$ such that $H_k = \{x \in \Re^n : \alpha_k^T(x - x_0) = 0\}$, $x_0 \in \Gamma_k$, $\alpha_k^T(x - x_0) \geq 0$, $\forall x \in \Gamma_{\geq k}$ and $\alpha_k^T(x - x_0) \leq 0$, $\forall x \in \Gamma_{\leq k}$. It follows that Γ_k is contained in the hyperplane H_k and thus $\Gamma_k \subseteq S \cap H_k$. On the other hand, $\alpha_k^T(x - x_0) > 0$, $\forall x \in \Gamma_{>k} = int\Gamma_{\geq k}$ and $\alpha_k^T(x - x_0) < 0$, $\forall x \in \Gamma_{<k} = int\Gamma_{\leq k}$, so that $x \in S \cap H_k$ implies $x \in \Gamma_k$.
Consider now the converse statement and assume, by contradiction, that f is not semistrictly quasilinear. Then, from (i) of Theorem 3.3.2 there exists a restriction φ of f on a line segment with end-points $x_0, x_1 \in S$, such that either $\min_{t\in[0,1]} \varphi(t) < \min\{f(x_0), f(x_1)\}$, or $\max_{t\in[0,1]} \varphi(t) > \max\{f(x_0), f(x_1)\}$, or both.
Consider the first case (the second case is similar). Let t^* be such that $\min_{t\in[0,1]} \varphi(t) = \varphi(t^*) < \min\{f(x_0), f(x_1)\}$, and let $x^* = x_0 + t^*(x_1 - x_0)$. Consider level set $\Gamma_{f(x^*)}$ and let H be a hyperplane such that $\Gamma_{f(x^*)} = S \cap H$. Since $x_0, x_1 \in \Gamma_{>f(x^*)}$, we have $x_0, x_1 \notin H$. On the other hand, $x^* \in H$ implies that x_0, x_1 are in opposite halfspaces and this is a contradiction. $\qquad\square$

Corollary 3.3.1. *Let f be a non-constant continuous function defined on $\Re^n$. Then, f is semistrictly quasilinear if and only if each of its nonempty level sets is a hyperplane.*

The characterization of quasilinearity in terms of its level sets is more complicated (for details see [211]). Nevertheless, in the differentiable case the convexity of the level sets suggests a characterization of quasilinearity based on the behaviour of the function at points belonging to the same level set. In this regard, the following theorem holds.

Theorem 3.3.6. *Let f be a differentiable function on an open convex set $S \subseteq \Re^n$. Then, f is quasilinear on S if and only if the following implication holds:*

$$x, y \in S, \ f(x) = f(y) \Rightarrow \nabla f(x)^T (y - x) = 0 \tag{3.7}$$

Proof. Let f be quasilinear. If $f(x) = f(y)$, then the convexity of the level set $\Gamma = \{z \in S : f(z) = f(x)\}$ implies that $[x, y] \subset \Gamma$ so that the restriction $\varphi(t) = f(x + t(y - x)), t \in [0, 1]$ is constant. Consequently, we have $\varphi'(0) = \nabla f(x)^T (y - x) = 0$.

Assume now that (3.7) holds. If f is not quasiconvex, taking into account (3.7), there exist $x, y \in S$ such that $f(x) > f(y)$ and $\nabla f(x)(y - x) > 0$; it follows that the restriction $\varphi(t) = f(x + t(y - x))$, $t \in [0, 1]$ has a maximum point t^* with $\varphi(t^*) > \varphi(0) > \varphi(1)$. The continuity of the function implies the existence of a point $\bar{t} \in (0, 1)$ such that $\varphi(\bar{t}) = \varphi(0)$. Setting $\bar{x} = x + \bar{t}(y - x)$ we have $f(\bar{x}) = f(x)$ so that $(\bar{x} - x)^T \nabla f(x) = 0$ and this is absurd since $(\bar{x} - x)^T \nabla f(x) = \bar{t}(y - x)^T \nabla f(x) > 0$. The quasiconvexity of f follows. Similarly, it can be proven that f is quasiconcave and thus the thesis follows. $\Box$

Remark 3.3.1. Consider the quasilinear function $f(x) = x^3$; setting $x = 0$, $y = 1$, we have $f'(x)(y - x) = 0$ with $f(x) \neq f(y)$. This means that the implication in (3.7) cannot be reversed. As we shall see in the next subsection, such a property characterizes the functions which are both pseudoconvex and pseudoconcave.

3.3.2 Pseudolinearity

A differentiable function f defined on an open convex set $S \subseteq \Re^n$ is said to be pseudolinear if it is both pseudoconvex and pseudoconcave.

Taking into account the results given for a pseudoconvex function and the analogous results for a pseudoconcave function, we have the following theorem.

Theorem 3.3.7. *Let f be a differentiable function on an open convex set $S \subseteq \Re^n$. Then, the following properties hold:*
(i) If f is pseudolinear on S, then $\nabla f(x) \neq 0$ for all $x \in S$ or f is a constant function;
(ii) If f is pseudolinear on S, then f is also semistrictly quasilinear on S;

(iii) f is pseudolinear on S if and only if the derivative of any of its non-constant restrictions on a line is constant in sign.

Proof. (i) Since for a pseudolinear function, a critical point is both a global minimum and a global maximum, the gradient of a non-constant pseudolinear function does not vanish on its domain.

(ii) This follows from Theorem 3.2.12 and from its analogous theorem in the pseudoconcave case.

(iii) This follows from Theorem 3.2.7 respectively and from its analogous theorem in the pseudoconcave case. □

Let us note that the converse statement of (ii) of Theorem 3.3.7 is not true. For instance, the semistricly quasilinear function $f(x) = x^3$ is not pseudolinear.

A pseudolinear function can be characterized by means of its behaviour at points belonging to the same level set strengthening condition (3.7) as is shown in the following theorem.

Theorem 3.3.8. *Let f be a differentiable function defined on an open convex set $S \subseteq \Re^n$. Then, f is pseudolinear on S if and only if the following double implication holds:*

$$x, y \in S, \ f(x) = f(y) \Longleftrightarrow \nabla f(x)^T (y - x) = 0 \qquad (3.8)$$

Proof. Since a pseudolinear function is also quasilinear, taking into account (3.7), it remains to be proven that $(y - x)^T \nabla f(x) = 0$ implies $f(x) = f(y)$. Setting $\varphi(t) = f(x + t(y - x)), t \in [0, 1]$, we have $\varphi'(0) = (y - x)^T \nabla f(x) = 0$, so that, for (i) of Theorem 3.3.7, φ is constant in $[0, 1]$ and this implies $f(x) = f(y)$.

Assume now that (3.8) holds. From Theorem 3.3.6 the function f is quasilinear. If $\nabla f(x) \neq 0$, $\forall x \in S$, quasilinearity implies pseudolinearity; if there exists a critical point $x_0 \in S$, from (3.8) f is constant and hence pseudolinear. □

Properties (i) and (ii) of Theorem 3.3.7, together with Theorem 3.2.6 and the analogous theorem for the pseudoconcave case, imply that the study of pseudolinearity is equivalent to the study of the subclass of semistrictly quasilinear functions having no stationary points.

Theorem 3.3.9. *Let f be a non-constant differentiable function on an open convex set $S \subseteq \Re^n$. Then, f is pseudolinear on S if and only if the following properties hold:*
(i) Each of the level sets of f is the intersection of S with a hyperplane;
(ii) $\nabla f(x) \neq 0$ for all $x \in S$.

Let us note that when a level set of a function f is contained in a hyperplane, the gradient of f is orthogonal at each point of the level set. From Theorem 3.3.9 it follows that the normalized gradient map, $x \to \frac{\nabla f(x)}{\|\nabla f(x)\|}$, is constant on each level set of a pseudolinear function. A direct proof of this last statement is given in the following theorem.

Theorem 3.3.10. *Let f be a function defined on an open convex set $S \subseteq \Re^n$ and assume $\nabla f(x) \neq 0$ for all $x \in S$. Then, f is pseudolinear on S if and only if its normalized gradient map is constant on each level set.*

Proof. Let f be pseudolinear. We must prove that the normalized gradient map is constant on each level set, i.e.,

$$f(x) = f(y) \Rightarrow \frac{\nabla f(x)}{\parallel \nabla f(x) \parallel} = \frac{\nabla f(y)}{\parallel \nabla f(y) \parallel} \tag{3.9}$$

Set $\Gamma_1 = \{d \in \Re^n : d^T \nabla f(x) = 0\}$, $\Gamma_2 = \{d \in \Re^n : d^T \nabla f(y) = 0\}$. We have $\Gamma_1 = \Gamma_2$. Indeed, if $d \in \Gamma_1$, from (3.8) it results that $f(x + td) = f(x) = f(y)$ for every t such that $x + td \in S$. From (3.8), it also follows that $(x + td - y)^T \nabla f(y) = 0$ and $(x - y)^T \nabla f(y) = 0$; consequently, $d^T \nabla f(y) = 0$ and thus $d \in \Gamma_2$. In an analogous way we can prove that $\Gamma_2 \subseteq \Gamma_1$. Since $\Gamma_1 = \Gamma_2$, it results that $\frac{\nabla f(x)}{\|\nabla f(x)\|} = \pm \frac{\nabla f(y)}{\|\nabla f(y)\|}$. Set $u = \frac{\nabla f(y)}{\|\nabla f(y)\|}$ and assume that $\frac{\nabla f(x)}{\|\nabla f(x)\|} = -u$; for a suitable $t \in (0, \epsilon)$ points $z_1 = x + tu$, $z_2 = y + tu$ are such that $f(z_1) < f(x)$, $f(z_2) > f(y)$. The continuity of f implies the existence of $\lambda \in (0, 1)$ such that $f(z) = f(x) = f(y)$ with $z = \lambda z_1 + (1 - \lambda) z_2$. From (3.8) we have $(z - y)^T u = 0$; on the other hand, $(z - y)^T u = (\lambda(x - y) + tu)^T u = t \parallel u \parallel^2 > 0$ so that $f(y) \neq f(z)$ and this is a contradiction. Consequently, we have $\frac{\nabla f(x)}{\|\nabla f(x)\|} = \frac{\nabla f(y)}{\|\nabla f(y)\|}$.

Assume now that (3.9) holds. Let $x, y \in S$ and set $\varphi(t) = f(x + t(y - x)), t \in [0, 1]$. If $\varphi'(t)$ is constant in sign, then $\varphi(t)$ is quasilinear on the line segment $[0, 1]$. Otherwise, from elementary Analysis, there exist $t_1, t_2 \in (0, 1)$ such that $\varphi(t_1) = \varphi(t_2)$ with $\varphi'(t_1)\varphi'(t_2) < 0$. Set $z_1 = x + t_1(y - x), z_2 = x + t_2(y - x)$. Since $f(z_1) = \varphi(t_1) = \varphi(t_2) = f(z_2)$, we have $\varphi'(t_2) = (1 - t_2)(y - x)^T \nabla f(z_2) = (1 - t_2)(y - x)^T \nabla f(z_1) \frac{\|\nabla f(z_2)\|}{\|\nabla f(z_1)\|} = \frac{1 - t_2}{1 - t_1} \varphi'(t_1) \frac{\|\nabla f(z_2)\|}{\|\nabla f(z_1)\|}$. Since $\varphi'(t_1)\varphi'(t_2) < 0$, we get a contradiction.

It follows that the restriction of the function over every line segment contained in S is quasilinear, so that f is quasilinear and also pseudolinear, since $\nabla f(x) \neq 0, \forall x \in S$. $\qquad \square$

Theorem 3.3.10 can be strengthened when f is defined on the whole space $\Re^n$, in the sense stated in the following theorem.

Theorem 3.3.11. *A non-constant function f is pseudolinear on the whole space $\Re^n$ if and only if its normalized gradient map is constant on $\Re^n$.*

Proof. Let f be pseudolinear on $\Re^n$ and assume that its normalized gradient map is not constant on $\Re^n$. Then, there exist $x_1, x_2 \in \Re^n$ such that $\frac{\nabla f(x_1)}{\|\nabla f(x_1)\|} \neq \frac{\nabla f(x_2)}{\|\nabla f(x_2)\|}$. From Theorem 3.3.10, we have $f(x_1) \neq f(x_2)$. Set $\Gamma_1 = \{d \in \Re^n : d^T \nabla f(x_1) = 0\}$ and $\Gamma_2 = \{d \in \Re^n : d^T \nabla f(x_2) = 0\}$. By noting that $x \in x_1 + \Gamma_1$ implies $(x - x_1) \in \Gamma_1$, we have $(x - x_1)^T \nabla f(x_1) = 0$

so that, from (3.8), $f(x) = f(x_1)$. Analogously, we have $f(x) = f(x_2)$ for all $x \in x_2 + \Gamma_2$. On the other hand, $\frac{\nabla f(x_1)}{\|\nabla f(x_1)\|} \neq \frac{\nabla f(x_2)}{\|\nabla f(x_2)\|}$ implies the existence of $\bar{x} \in (x_1 + \Gamma_1) \cap (x_2 + \Gamma_2)$, so that $f(x_1) = f(\bar{x}) = f(x_2)$, and this is a contradiction.

The converse statement follows from Theorem 3.3.10. $\qquad\qquad\square$

From a geometrical point of view, the previous theorem states that the level sets of a non-constant pseudolinear function, defined on the whole space $\Re^n$, are parallel hyperplanes; viceversa if the level sets of a differentiable function, with no critical points, are hyperplanes, then the function is pseudolinear.

In the following examples we shall use Theorem 3.3.9 for constructing pseudolinear functions and for verifying the pseudolinearity of given functions.

Example 3.3.1. The linear fractional function $(b \neq 0)$

$$f(x) = \frac{a^T x + a_0}{b^T x + b_0}, \quad b^T x + b_0 > 0$$

is pseudolinear.

It is easy to verify that the feasible level sets are open semi-hyperplanes. Furthermore, we have $\nabla f(x) = \frac{(b^T x + b_0)a - (a^T x + a_0)b}{(b^T x + b_0)^2}$ so that $\nabla f(x) \neq 0$ if a, b are not proportional. If $a = kb$, f reduces to $f(x) = k + \frac{a_0 - kb_0}{b^T x + b_0}$. In both cases (i) and (ii) of Theorem 3.3.9 hold and thus the linear fractional function is pseudolinear.

Example 3.3.2. Theorem 3.3.9 suggests constructing pseudolinear functions starting from a family of lines or hyperplanes.

Consider, for instance, the family of lines $y = \frac{kx+1}{\sqrt{k+1}}$. We have $(\sqrt{k+1}\, y)^2 = (kx+1)^2$, $k^2 x^2 + (2x - y^2)k + 1 - y^2 = 0$, $k = \frac{-2x + y^2 \pm |y|\sqrt{y^2 - 4x + 4x^2}}{2x^2}$, so that the given family of lines may be interpreted as the level sets of function

$$f(x, y) = \frac{-2x + y^2 + |\, y \,|\, \sqrt{y^2 - 4x + 4x^2}}{2x^2}.$$

From Theorem 3.3.9, the function is pseudolinear on each convex set S such that $\nabla f(x, y) \neq 0$, for all $(x, y) \in S$. For instance, f is pseudolinear on $S = \{(x, y) : x > 1, y > 0\}$.

Another way to construct a pseudolinear function from known functions is to apply the following theorem whose proof follows from Theorem 3.2.11 and its analogous theorem for pseudoconcave functions.

Theorem 3.3.12. *Let $f : S \subseteq \Re^n \to \Re$ be a pseudolinear function on an open convex set S and let $\phi : \Re \to \Re$ be a differentiable function such that $\phi'(z) > 0, \forall z \in \Re$, or $\phi'(z) < 0, \forall z \in \Re$. Then, the composite function $\phi \circ f$ is pseudolinear on S.*

Example 3.3.3. If $g(x)$ is a pseudolinear function on a convex set $S \subseteq \Re^n$, then the function $f(x) = e^{g(x)}$ is pseudolinear on S.

3.4 Twice Differentiable Generalized Convex Functions

In this section we shall present some characterizations of a twice differentiable generalized convex function.

3.4.1 Quasiconvex Functions

The Hessian matrix of a twice differentiable convex function is positive semidefinite or, equivalently, it has non-negative eigenvalues. As a result, the Hessian of a quasiconvex function which is not convex necessarily has some negative eigenvalue. The following theorem states that the Hessian cannot have two or more negative eigenvalues.

Theorem 3.4.1. *Let f be a twice continuously differentiable quasiconvex function defined on an open convex set $S \subseteq \Re^n$. Then, for every $x \in S$, the Hessian matrix $\nabla^2 f(x)$ has at most one negative eigenvalue.*

Proof. Let $x_0 \in S$ and suppose that $\nabla^2 f(x_0)$ has two (or more) negative eigenvalues. Denote with $v^1, v^2, ..., v^n$ a set of n orthogonal eigenvectors of $\nabla^2 f(x_0)$ and assume, without loss of generality, that the eigenvalues associated with v^1 and v^2 are negative. Let E be the subspace spanned by v^1 and v^2 and set $E^* = E \backslash \{0\}$. We have $u^T \nabla^2 f(x_0) u < 0$ for every $u \in E^*$. If $\nabla f(x_0) = 0$, then $u^T \nabla f(x_0) = 0$, $u^T \nabla^2 f(x_0) u < 0$ imply that x_0 is a strict local maximum for the restriction of f on the line through x_0 and direction u, and this contradicts the quasiconvexity of f. If $\nabla f(x_0) \neq 0$, the intersection between E and the orthogonal subspace to $\nabla f(x_0)$ has dimension equal to 1 or 2, so that there exists $u \in E^*$ such that $u^T \nabla f(x_0) = 0$, $u^T \nabla^2 f(x_0) u < 0$ and, once again, we get a contradiction. □

Theorem 3.4.1 establishes a necessary condition for a twice differentiable function to be quasiconvex. The following example shows that such a condition is not sufficient.

Example 3.4.1. Consider the function $f(x_1, x_2) = x_1^2 - x_2^2$, $x_1, x_2 > 0$. It is easy to verify that the Hessian matrix has one negative eigenvalue and one positive eigenvalue. The restriction of f on the half-line $x_2 = 2x_1 - 3$, $x_1 \geq \frac{3}{2}$, is given by $\varphi(x_1) = x_1^2 - (2x_1 - 3)^2 = -3x_1^2 + 12x_1 - 9$; such a function has a critical point at $x_1 = 2$ which is a strict local maximum, so that f is not quasiconvex.

Another necessary condition for a twice continuously differentiable function to be quasiconvex is stated in the following theorem.

Theorem 3.4.2. *Let f be a twice continuously differentiable quasiconvex function defined on an open convex set $S \subseteq \Re^n$. Then, the following property holds:*

$$x_0 \in S, \ u \in \Re^n, \ u^T \nabla f(x_0) = 0 \Rightarrow u^T \nabla^2 f(x_0) u \geq 0. \tag{3.10}$$

Proof. It sufficient to note that the condition $x_0 \in S$, $u \in \Re^n$, $u^T \nabla f(x_0) = 0$, $u^T \nabla^2 f(x_0) u < 0$ implies that x_0 is a strict local maximum point for the restriction of f on the line through x_0 and direction u, and this contradicts the quasiconvexity of f. $\qquad\square$

Condition (3.10) does not guarantee the quasiconvexity of f. In fact, in Example 3.4.1, condition (3.10) is verified for $x_0 = (2,1)^T$ and $u = (1,2)^T$ but f is not quasiconvex. Nevertheless, requiring the non-existence of critical points, condition (3.10) becomes sufficient too, as is shown in the following theorem (see also [219, 79, 82]).

Theorem 3.4.3. *Let f be a twice continuously differentiable function defined on an open convex set $S \subseteq \Re^n$ such that $\nabla f(x) \neq 0$ for all $x \in S$. Then, f is quasiconvex on S if and only if (3.10) holds.*

Proof. Taking into account Theorem 3.4.2, it remains to be proven that the condition (3.10) implies the quasiconvexity of f. Let $x_0, x_1 \in S$ such that $f(x_1) \leq f(x_0)$ and let $x(t) = tx_1 + (1-t)x_0$, $t \in [0,1]$. By contradiction, assume that f is not quasiconvex; the continuity of f implies the existence of $t_0 \in (0,1)$ such that $t_0 = \max\{t \in [0,1] : \varphi(t) = M\}$ where $M = \max_{t \in [0,1]} \{\varphi(t)\}$. Setting $\varphi(t) = f(x(t))$ we have $f(x(t)) < f(x(t_0))$, $\forall t > t_0$ and $\varphi'(t_0) = (x_1 - x_0)^T \nabla f(x(t_0)) = 0$.

Consider the function $\psi(\beta, \alpha) = f(\beta \nabla f(x(t_0)) + \alpha(x_1 - x_0) + x(t_0))$.

It results $\psi(0,0) = f(x(t_0))$ and $\frac{\partial \psi}{\partial \beta}(0,0) = \| \nabla f(x(t_0)) \|^2$. Since $\nabla f(x(t_0)) \neq 0$, we have $\frac{\partial \psi}{\partial \beta}(0,0) > 0$, so that, by means of the Implicit Function Theorem, there exists a differentiable function $\beta(\alpha)$ with α belonging to a suitable neighbourhood I of 0 such that $\beta(0) = 0$ and

$$f(z(\alpha)) = f(x(t_0)), \ \forall \alpha \in I \tag{3.11}$$

where $z(\alpha) = \beta(\alpha) \nabla f(x(t_0)) + \alpha(x_1 - x_0) + x(t_0)$.

From (3.11), through differentiation, we have

$$\nabla f(z(\alpha))^T [\beta'(\alpha) \nabla f(x(t_0)) + (x_1 - x_0)] = 0 \tag{3.12}$$

and

$$(\beta'(\alpha) \nabla f(x(t_0)) + (x_1 - x_0))^T \nabla^2 f(z(\alpha))(\beta'(\alpha) \nabla f(x(t_0)) + (x_1 - x_0)) +$$
$$+ \nabla f(z(\alpha))^T \beta''(\alpha) \nabla f(x(t_0)) = 0 \tag{3.13}$$

Set $\alpha = 0$ in (3.12); since $\nabla f(z(0))^T \nabla f(x(t_0)) = \| \nabla f(x(t_0)) \|^2$, we obtain

$$\beta'(0) \parallel \nabla f(x(t_0)) \parallel^2 + (x_1 - x_0)^T \nabla f(x(t_0)) = \beta'(0) \parallel \nabla f(x(t_0)) \parallel^2 = 0,$$

so that $\beta'(0) = 0$.

Taking into account (3.12), (3.10) applied to the vectors $z(\alpha)$ and $u = \beta'(\alpha)\nabla f(x(t_0)) + (x_1 - x_0)$ becomes

$$(\beta'(\alpha)\nabla f(x(t_0)) + (x_1 - x_0))^T \nabla^2 f(z(\alpha))(\beta'(\alpha)\nabla f(x(t_0)) + (x_1 - x_0)) \geq 0,$$

so that, from (3.13), we have $\beta''(\alpha)\nabla f(z(\alpha))^T \nabla f(x(t_0)) \leq 0$, $\forall \alpha \in I$.

Since $\nabla f(z(0))^T \nabla f(x(t_0)) > 0$, the continuity of the gradient map implies the existence of α^* such that $\nabla f(z(\alpha))^T \nabla f(x(t_0)) > 0$, $\forall \alpha \in (-\alpha^*, \alpha^*)$; consequently, we have $\beta''(\alpha) \leq 0$, $\forall \alpha \in (-\alpha^*, \alpha^*)$, i.e., the function $\beta(\alpha)$ is concave in $(-\alpha^*, \alpha^*)$. Since $\beta(0) = \beta'(0) = 0$, the concavity implies $\beta(\alpha) \leq 0$, $\forall \alpha \in (-\alpha^*, \alpha^*)$. Setting $\alpha = t - t_0$ the last inequality reduces to $\beta(t - t_0) \leq 0$, $\forall t \in (t_0 - \alpha^*, t_0 + \alpha^*)$. Since $f(x(t)) < f(x(t_0))$, $t > t_0$, and taking into account (3.11), where now $\alpha(x_1 - x_0) + x(t_0) = (t - t_0)(x_1 - x_0) + x(t_0) = x(t)$, we necessarily have $\beta(t - t_0) \neq 0$, $\forall t \in (t_0 - \alpha^*, t_0 + \alpha^*)$ so that $\beta(t - t_0) < 0$, $\forall t \in (t_0 - \alpha^*, t_0 + \alpha^*)$. By means of Taylor's expansion we have $f(x(t_0)) = f(\beta(t - t_0)\nabla f(x(t_0)) + x(t)) = f(x(t)) + \nabla f(x(t))^T \nabla f(x(t_0))\beta(t - t_0) + o(t - t_0)$ and thus, for $t > t_0$ sufficiently close to t_0, taking into account that $\nabla f(x(t))^T \nabla f(x(t_0)) > 0$ and $\beta(t - t_0) < 0$, we have $f(x(t)) > f(x(t_0))$ and this is a contradiction. □

The following theorem states another second order characterization for quasiconvexity without any assumption on the gradient map.

Theorem 3.4.4. *Let f be a twice continuously differentiable function defined on an open convex set $S \subseteq \Re^n$. Then, f is quasiconvex on S if and only if the following conditions hold:*
(i) $x_0 \in S$, $u \in \Re^n$, $u^T \nabla f(x_0) = 0$ imply $u^T \nabla^2 f(x_0)u \geq 0$;
(ii) $x_0 \in S$, $x_1 \in S$, $f(x_1) < f(x_0)$, $\nabla f(x_0) = 0$, $u^T \nabla^2 f(x_0)u = 0$ with $u = x_0 - x_1$, imply that for all $\epsilon > 0$ there exists $k \in (0, \epsilon)$ such that $x_0 + ku \in S$ and $f(x_0) \leq f(x_0 + ku)$.

Proof. Obviously, the quasiconvexity of f implies (i) and (ii). With respect to the converse statement, referring to the proof given in Theorem 3.4.3, the case $\nabla f(x(t_0)) = 0$ remains to be considered. By setting $u = x(t_0) - x_0$, we have $u^T \nabla f(x(t_0)) = 0$ so that, from (i) and taking into account that $x(t_0)$ is a maximum point on the segment $[x_0, x_1]$, it results that $u^T \nabla^2 f(x(t_0))u = 0$. Since $f(x_0) < f(x(t_0))$, (ii) implies the existence of $k > 0$ such that $f(x(t_0)) \leq f(x(t_0) + ku)$ and this is a contradiction since for $k > 0$, i.e., for $t > t_0$, function f is decreasing on the segment $[x(t_0), x_1]$. □

3.4.2 Pseudoconvex Functions

For a twice continuously differentiable (strictly) pseudoconvex function both Theorems 3.2.7 and 3.2.8 may be stated in terms of the first and second derivatives as follows.

Theorem 3.4.5. *Let φ be a twice continuously differentiable function defined on an open interval $I \subseteq \Re$. Then, φ is (strictly) pseudoconvex on I if and only if for every $t_0 \in I$ such that $\varphi'(t_0) = 0$ either $\varphi''(t_0) > 0$ or $\varphi''(t_0) = 0$ and t_0 is a (strict) local minimum for φ.*

Theorem 3.4.6. *Let f be a twice continuously differentiable function defined on an open convex set $S \subseteq \Re^n$. Then, f is (strictly) pseudoconvex on S if and only if for every $x_0 \in S$ and $u \in \Re^n$ such that $u^T \nabla f(x_0) = 0$, either $u^T \nabla^2 f(x_0)u > 0$ or $u^T \nabla^2 f(x_0)u = 0$ and function $\varphi(t) = f(x_0 + tu)$ attains a (strict) local minimum at $t = 0$.*

The following theorem specifies Theorem 3.2.9; the given characterization is more suitable for establishing the pseudoconvexity of a function.

Theorem 3.4.7. *Let f be a twice continuously differentiable function defined on an open convex set $S \subseteq \Re^n$.*
Then, f is (strictly) pseudoconvex on S if and only if the following conditions hold:
(i)
$$x \in S, \ u \in \Re^n, \ u^T \nabla f(x_0) = 0 \ \Rightarrow \ u^T \nabla^2 f(x_0)u \geq 0 \qquad (3.14)$$

(ii) If $x_0 \in S$ is a critical point for f, then x_0 is a (strict) local minimum for f on S.

Proof. If f is (strictly) pseudoconvex, then (i) and (ii) follow directly from Theorems 3.2.9 and 3.4.2.
For the converse statement, taking into account Theorem 3.2.8, we must prove that $u^T \nabla f(x_0) = 0$ implies that $\varphi(t) = f(x_0 + tu)$ attains a (strict) local minimum at $t = 0$.
If $\nabla f(x_0) = 0$, then the thesis follows from condition (ii). If $\nabla f(x_0) \neq 0$ the continuity of the gradient map implies that $\nabla f(x) \neq 0$ for all x belonging to a suitable neighbourhood $I(x_0)$ of x_0 so that, from Theorem 3.4.3, (i) implies that f is quasiconvex on $I(x_0)$. The thesis follows from Theorem 3.2.6 and Theorem 3.2.8. $\qquad \square$

Corollary 3.4.1. *Let f be a twice continuously differentiable function defined on an open convex set $S \subseteq \Re^n$. If $\nabla f(x) \neq 0$ for all $x \in S$, then f is pseudoconvex on S if and only if condition (3.14) holds.*

3.4.3 Characterizations in Terms of the Bordered Hessian

Let us note that condition (3.14) is equivalent to studying the positive semidefiniteness of a quadratic form on a linear subspace; this kind of study has been carried out for a long time (see for instance [91, 112]).

As regards our aim, let us recall the following notations and results [82].

Let $a \in \Re^n$, $a \neq 0$ and A be a real symmetric matrix of order n.

Let $B = \begin{bmatrix} 0 & a^T \\ a & A \end{bmatrix}$ the so-called bordered matrix. For all nonempty subset $R \subseteq \{1, 2, .., n\}$, denote with $|R|$ the cardinality of R and with $B_R = \begin{vmatrix} 0 & a_R^T \\ a_R & A_R \end{vmatrix}$ the bordered principal minor of order $|R|$ of B, where A_R is obtained from A by deleting rows and columns whose indices are not in R and a_R is obtained analogously from a.

Furthermore denote with $B_r = \begin{vmatrix} 0 & a_r^T \\ a_r & A_r \end{vmatrix}$ the bordered leading principal minor of order r, $r = 1, .., n$, of B, where A_r is obtained from A by keeping the first r rows and the first r columns and a_r is obtained analogously from a.

Theorem 3.4.8. *Let $a \in \Re^n$, $a \neq 0$ and let A be a real symmetric matrix of order n. The following conditions are equivalent:*
(i) $a^T h = 0$ implies $h^T A h \geq 0$;
(ii) For all nonempty subset $R \subseteq \{1, 2, .., n\}$ we have $B_R \leq 0$.

Theorem 3.4.9. *Let $a \in \Re^n$, $a \neq 0$ and let A be a real symmetric matrix of order n. The following conditions are equivalent:*
(i) $a^T h = 0$, $h \neq 0$ implies $h^T A h > 0$;
(ii) $B_r < 0$, $r = 1, .., n$.

Theorem 3.4.8 allows a restatement of Theorems 3.4.2, 3.4.7, and Corollary 3.4.1, in terms of the so-called bordered Hessian defined as

$$D(x) = \begin{bmatrix} 0 & \nabla^T f(x) \\ \nabla f(x) & \nabla^2 f(x) \end{bmatrix}$$

By denoting with $D_R(x)$, $R \subseteq \{1, 2, .., n\}$, the bordered principal minors of $D(x)$ and with $D_r(x)$, $r = 1, .., n$, the bordered leading principal minors of $D(x)$, we have the following theorems.

Theorem 3.4.10. *Let f be a twice continuously differentiable function defined on an open convex set $S \subseteq \Re^n$. If f is quasiconvex on S, then the following condition holds:*

$$D_R(x) \leq 0, \ \forall x \in S, \ \forall R \subseteq \{1, 2, .., n\}, \ R \neq \emptyset \tag{3.15}$$

The following example points out that condition (3.15) is not sufficient for quasiconvexity or pseudoconvexity.

Example 3.4.2. Consider the function $f(x_1, x_2) = -(x_1 - x_2)^2$.

The bordered Hessian is $D(x) = \begin{bmatrix} 0 & -2(x_1 - x_2) & 2(x_1 - x_2) \\ -2(x_1 - x_2) & -2 & 2 \\ 2(x_1 - x_2) & 2 & -2 \end{bmatrix}$.

For $R = \{1\}$ and $R = \{2\}$, we have respectively

$$\begin{vmatrix} 0 & -2(x_1 - x_2) \\ -2(x_1 - x_2) & -2 \end{vmatrix} = -4(x_1 - x_2)^2 \leq 0$$

$$\begin{vmatrix} 0 & 2(x_1 - x_2) \\ 2(x_1 - x_2) & -2 \end{vmatrix} = -4(x_1 - x_2)^2 \leq 0.$$

For $R = \{1, 2\}$, it results that $D_R(x_1, x_2) = |D(x_1, x_2)| = 0$.
Consequently, condition (3.15) is verified, but f is not pseudoconvex (in particular, not quasiconvex) since its critical points are global maximum points.

Theorem 3.4.11. *Let f be a twice continuously differentiable function defined on an open convex set $S \subseteq \Re^n$. Then, f is (strictly) pseudoconvex on S if and only if conditions (i) and (ii) hold:*
(i)

$$D_R(x) \leq 0, \ \forall x \in S, \ \forall R \subseteq \{1, 2, .., n\}, \ R \neq \emptyset \qquad (3.16)$$

(ii) If $x \in S$ is a critical point for f, then x is a (strict) local minimum for f on S.

Theorem 3.4.12. *Let f be a twice continuously differentiable function defined on an open convex set $S \subseteq \Re^n$, with $\nabla f(x) \neq 0$ for all $x \in S$. Then, f is pseudoconvex on S if and only if (3.16) holds.*

Theorem 3.4.9 allows us to state a sufficient condition for pseudoconvexity.

Theorem 3.4.13. *Let f be a twice continuously differentiable function defined on an open convex set $S \subseteq \Re^n$. Then, a sufficient condition for f to be pseudoconvex on S is*

$$D_r(x) < 0, \ \forall x \in S, \ \forall r = 1, 2, .., n \qquad (3.17)$$

Proof. (3.17) is equivalent to stating that for every $u \neq 0$ such that $u^T \nabla f(x) = 0$ we have $u^T \nabla^2 f(x) u > 0$. The thesis follows from Theorem 3.4.6. $\qquad \square$

The sufficient condition expressed in Theorem 3.4.13 still holds for a strictly quasiconvex function.

Theorem 3.4.14. *Let f be a twice differentiable function defined on an open convex set $S \subseteq \Re^n$. If condition (3.17) holds, then f is strictly quasiconvex on S.*

Proof. It is sufficient to note that f cannot have a constant restriction. In fact, if $\varphi(t) = f(x_0 + tu)$ is constant, then $u^T \nabla f(x_0) = 0$, $u^T \nabla^2 f(x_0) u = 0$, while (3.17) implies $u^T \nabla^2 f(x_0) u > 0$. $\qquad \square$

Remark 3.4.1. Let us note that the sufficient condition (3.17) is sometimes very restrictive. In fact it cannot be verified by any function that has level surfaces containing line segments (for instance, affine functions or, more generally, pseudolinear functions).

The following examples show some applications of the given results in terms of the bordered Hessian.

Example 3.4.3. Consider the function

$$f(x_1, x_2) = 2x_2 + \frac{x_1}{x_2 + 1}, \quad (x_1, x_2) \in S = \{(x_1, x_2) \in \Re^2 : x_2 + 1 > 0\}.$$

It results that $\nabla f(x_1, x_2) = (\frac{1}{x_2+1}, 2 - \frac{x_1}{(x_2+1)^2})^T \neq 0, \quad \forall (x_1, x_2) \in S.$

The bordered Hessian is $D(x_1, x_2) = \begin{bmatrix} 0 & \frac{1}{x_2+1} & 2 - \frac{x_1}{(x_2+1)^2} \\ \frac{1}{x_2+1} & 0 & \frac{-1}{(x_2+1)^2} \\ 2 - \frac{x_1}{(x_2+1)^2} & \frac{-1}{(x_2+1)^2} & \frac{2x_1}{(x_2+1)^3} \end{bmatrix}.$

In order to study the pseudoconvexity of f, we shall begin to calculate the bordered leading principal minors of $D(x_1, x_2)$.
It results that

$$\begin{vmatrix} 0 & \frac{1}{x_2+1} \\ \frac{1}{x_2+1} & 0 \end{vmatrix} = -\frac{1}{(x_2+1)^2} < 0, \quad |D(x_1, x_2)| = -\frac{4}{(x_2+1)^3} < 0.$$

From Theorem 3.4.13, f is pseudoconvex on S.

Example 3.4.4. Consider the function $f(x_1, x_2) = -x_1^2 - 2x_1 x_2$.
Since $\nabla f(x_1, x_2) = (-2x_1 - 2x_2, -2x_1)^T$, the function does not have critical points on $int\Re_+^2$ so that, in order to study the pseudoconvexity of the function, we can refer to Theorem 3.4.12.

The bordered Hessian is $D(x_1, x_2) = \begin{bmatrix} 0 & -2x_1 - 2x_2 & -2x_1 \\ -2x_1 - 2x_2 & -2 & -2 \\ -2x_1 & -2 & 0 \end{bmatrix}.$

We have

$$\begin{vmatrix} 0 & -2x_1 - 2x_2 \\ -2x_1 - 2x_2 & -2 \end{vmatrix} = -(-2x_1 - 2x_2)^2 \leq 0, \quad \begin{vmatrix} 0 & -2x_1 \\ -2x_1 & 0 \end{vmatrix} = -4x_1^2 \leq 0,$$

$|D(x_1, x_2)| = -8x_1^2 - 16x_1 x_2 \leq 0.$
It follows from Theorem 3.4.12 that f is pseudoconvex on $int\Re_+^2$. Since the gradient of f vanishes at the origin which is a global maximum point, f is not pseudoconvex on $\Re_+^2$.
Note that, from Theorem 2.2.12, f is quasiconvex on $\Re_+^2$.

3.5 Generalized Convexity at a Point

In this section we shall introduce the notion of convexity and generalized convexity at a point which will allow us to state, in the next chapter and in a more general form, local-global property, first-order sufficient optimality conditions and constraint qualifications.

The notion of convexity and generalized convexity at a point represents a significant relaxation of the concept of generalized convexity since it does not necessarily require the convexity of the domain of the function. For a better understanding, consider Definition 2.1; quasiconvexity is introduced by requiring that (2.2) holds for each point of the line segment $[x_1, x_2]$; such an assumption can be relaxed in different ways. The more general one is that for a fixed point x_0 and for any $x \in S$, (2.2) holds for each point of the intersection of the line segment $[x_0, x]$ with S. In what follows we consider the case of such an intersection (in general it may also be a finite set) which coincides with the line segment $[x_0, x]$, i.e., the case of the domain of the function is star-shaped at a point. More precisely, we have the following definition.

Definition 3.5.1. *Let S be a subset of $\Re^n$ and let x_0 be a point of S. S is said to be star-shaped at x_0 if $x \in S$ implies $x_0 + t(x - x_0) \in S$ for all $t \in [0, 1]$.*

From the given definition is follows that a convex set is star-shaped at each of its points. The following Fig. 3.2 shows some examples of non-convex star-shaped sets.

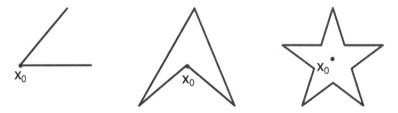

Fig. 3.2. Non-convex star-shaped sets

By considering a star-shaped set as a domain, we shall introduce the following definitions of generalized convex functions at a point.

Definition 3.5.2. *Let f be a function defined on a set $S \subseteq \Re^n$ which is star-shaped at $x_0 \in S$.*
(i) f is said to be quasiconvex at $x_0 \in S$ if

$$x \in S, \ f(x) \le f(x_0) \Rightarrow f(x_0 + t(x - x_0)) \le f(x_0) \tag{3.18}$$

for every $t \in [0, 1]$.
(ii) f is said to be strictly quasiconvex at $x_0 \in S$ if

$$x \in S, \ x \ne x_0, \ f(x) \le f(x_0) \Rightarrow f(x_0 + t(x - x_0)) < f(x_0) \tag{3.19}$$

for every $t \in (0, 1)$.

(iii) f is said to be semistrictly quasiconvex at $x_0 \in S$ if

$$x \in S, \ x \neq x_0, \ \ f(x) < f(x_0) \Rightarrow f(x_0 + t(x - x_0)) < f(x_0) \tag{3.20}$$

for every $t \in (0, 1]$.

In an analogous way the definitions of convexity and strict convexity at a point can be given.

Definition 3.5.3. *Let f be a function defined on a set $S \subseteq \Re^n$ which is star-shaped at $x_0 \in S$.*
(i) f is said to be convex at $x_0 \in S$ if

$$f(x_0 + t(x - x_0)) \leq f(x_0) + t(f(x) - f(x_0)) \tag{3.21}$$

for every $x \in S$ and for every $t \in [0, 1]$.
(ii) f is said to be strictly convex at $x_0 \in S$ if

$$f(x_0 + t(x - x_0)) < f(x_0) + t(f(x) - f(x_0)) \tag{3.22}$$

for every $x \in S$, $x \neq x_0$ and for every $t \in (0, 1)$.

Some graphs of convex and generalized convex functions at a point but not on the whole domain are depicted in Fig. 3.3. More precisely, in case a) the function is convex at $x_0 = (0, 0)$, in case b) the function is quasiconvex (neither semistrictly quasiconvex nor strictly quasiconvex) at x_0, and in case c) the function is strictly and semistrictly quasiconvex (not convex) at x_0.

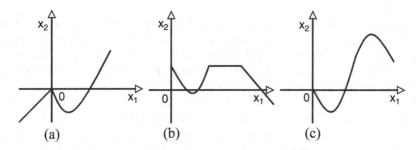

(a)   (b)   (c)

Fig. 3.3. Convex and generalized convex functions at a point

Convexity and generalized convexity at a point represent, as we have pointed out before, a significant relaxation of convexity. For this reason, as we shall see, not all the results developed so far throughout this chapter and the previous ones, hold true.

With respect to the relationships between the introduced classes of convex and generalized convex functions at a point, we have the following theorem whose proof follows directly from the given definitions.

Theorem 3.5.1. *Let f be a function defined on a set $S \subseteq \Re^n$ which is star-shaped at $x_0 \in S$.*
(i) If f is convex at $x_0 \in S$, then f is both quasiconvex and semistrictly quasiconvex at x_0;
(ii) If f is strictly convex at $x_0 \in S$, then f is strictly quasiconvex at x_0;
(iii) If f is strictly quasiconvex at $x_0 \in S$, then f is both quasiconvex and semistrictly quasiconvex at x_0.

The inclusion relationship between the class of quasiconvex functions and the class of lower semicontinuous semistrictly quasiconvex functions (see Theorem 2.3.2) does not hold for generalized convexity at a point even if S is a convex set and f is a continuous function, as is pointed out in the following examples.

Example 3.5.1. Consider the function

$$f(x) = \begin{cases} -x^2 & -1 \le x \le 1 \\ -x^2 + 6x - 6 & 1 < x \le 3 + \sqrt{3} \end{cases}.$$

It can be verified that f is continuous and semistrictly quasiconvex at $x_0 = 0$ but it is not quasiconvex at x_0 since $f(0) = f(3+\sqrt{3}) = 0$ with $f(3) = 3 > 0$.

Example 3.5.2. Consider the function

$$f(x) = \begin{cases} 0 & -1 \le x < 1 \\ -x + 1 & 1 \le x \le 2 \end{cases}.$$

It is easy to verify that f is quasiconvex at $x_0 = 0$ but it is not semistrictly quasiconvex at x_0 since, for instance, we have $f(x_0) = 0 > f(2) = -1$, but $f(1) = 0$.

The reason for which there is no any inclusion relationship between quasiconvexity and semistrictly quasiconvexity at a point, is related to the given definitions where nothing is said about the behaviour of the function when $f(x) > f(x_0)$. In Exercise 3.34 we shall present alternative definitions for which the inclusion relationship still holds.
Due to the generality of the given definitions, also the first-order characterizations of convexity and quasiconvexity do not hold. More precisely, (1.8) and (3.2) are necessary but not sufficient conditions for a function to be convex or quasiconvex, respectively, as is stated in Theorem 3.5.2 and in Example 3.5.3.

Theorem 3.5.2. *Let $S \subseteq \Re^n$ be a star-shaped set at $x_0 \in S$, and let f be a function defined on an open set containing S and differentiable at x_0.*
(i) If f is convex at x_0, then

$$f(x) \ge f(x_0) + (x - x_0)^T \nabla f(x_0), \quad \forall x \in S; \tag{3.23}$$

(ii) If f is quasiconvex at x_0, then

$$f(x) \le f(x_0) \Rightarrow (x - x_0)^T \nabla f(x_0) \le 0, \quad \forall x \in S. \tag{3.24}$$

Proof. (i) From (3.21) we have $\frac{f(x_0+t(x-x_0))-f(x_0)}{t} \leq f(x) - f(x_0)$, $\forall t \in (0,1]$.
By taking the limit when t approaches to 0^+, (3.23) is obtained.
(ii) From (3.18) we have $\frac{f(x_0+t(x-x_0))-f(x_0)}{t} \leq 0$, $\forall t \in (0,1]$. By taking the
limit for $t \to 0^+$, (3.24) is obtained. $\qquad\qquad\qquad\qquad\qquad$ $\square$

The following example shows that (3.23) and (3.24) are not sufficient conditions for convexity and quasiconvexity at x_0, respectively.

Example 3.5.3. Consider $S = \{x \in \Re : x \leq 3\}$, $x_0 = 0 \in S$, and the function
$f(x) = x(1-x)(x-2)$.
Since $f'(x_0)(x - x_0) = -2x$, it is easy to verify that (3.23) and (3.24) hold.
By applying (3.21) for $x = 2$, we have $f(2t) \leq 0$, $\forall t \in [0,1]$. Since $f(2t) > 0$,
$\forall t \in (\frac{1}{2}, 1)$, the function is not convex at x_0.
Furthermore, f is not quasiconvex at x_0 since we have $f(2) = 0 = f(0)$ with
$f(x) > 0$, $\forall x \in (1,2)$.

Peseudoconvexity at a point of a star-shaped set can be defined by relaxing
relations (3.5) and (3.6).

Definition 3.5.4. *Let f be a function defined on a set $S \subseteq \Re^n$ which is star-shaped at $x_0 \in S$.*
(i) f is said to be pseudoconvex at $x_0 \in S$ if f is differentiable at x_0 and

$$x \in S, \ f(x) < f(x_0) \Rightarrow (x - x_0)^T \nabla f(x_0) < 0; \qquad (3.25)$$

(ii) f is said to be strictly pseudoconvex at $x_0 \in S$ if f is differentiable at x_0 and

$$x \in S, \ f(x) \leq f(x_0) \Rightarrow (x - x_0)^T \nabla f(x_0) < 0. \qquad (3.26)$$

It easy to prove that the class of differentiable and convex functions at a
point x_0 is properly contained in the class of pseudoconvex functions at
x_0. Unfortunately, there are not any inclusion relationships among quasi-convexity, semistrictly quasiconvexity and pseudoconvexity at a point. For
instance, the function given in Example 3.5.3 is pseudoconvex at x_0 but it
is neither quasiconvex or semistrictly quasiconvex at x_0, while the function
$f(x) = \begin{cases} 0 & x = 0 \\ x^2 \log |x| & 0 < x \leq 1 \end{cases}$ is quasiconvex and semistrictly quasiconvex
at $x_0 = 0$ but not pseudoconvex at x_0.
As is shown in Theorem 3.2.6, quasiconvexity implies pseudoconvexity when
the gradient does not vanish on the feasible convex set. In order to extend
this property to a star-shaped set in addition to differentiability at a point we
need the continuity of the function on its domain, as is stated in the following
theorem.

Theorem 3.5.3. *Let f be a continuous function defined on an open set $S \subseteq \Re^n$ which is star-shaped at $x_0 \in S$. If f is differentiable and quasiconvex at $x_0 \in S$ with $\nabla f(x_0) \neq 0$, then f is pseudoconvex at x_0.*

Proof. The proof is analogous to the one given in Theorem 3.2.3 (it is sufficient to replace x_1 with x_0). □

Let us note that the converse statement of Theorem 3.5.3 is not true (see Example 3.5.3).
The following example points out that the assumption of continuity on the whole set is essential in the proof of Theorem 3.5.3.

Example 3.5.4. Consider $S = \{(x_1, x_2) \in \Re^2 : x_1 \geq -1\}$, $x_0 = (0,0) \in S$ and

$$f(x_1, x_2) = \begin{cases} x_1 x_2 + x_2 & (x_1, x_2) \neq (-1, 0) \\ -1 & (x_1, x_2) = (-1, 0) \end{cases}$$

Since $S^* = \{(x_1, x_2) \in S : f(x_1, x_2) \leq f(0,0)\} = \{(x_1, x_2) \in S : x_2 \leq 0\}$ is a convex set and x_0 is the maximum point for f on S^*, the function is quasiconvex at x_0. On the other hand, $f(-1, 0) = -1 < f(0,0)$ and $\nabla f(0,0)^T (-1, 0) = 0$, so that f is not pseudoconvex at x_0.

3.6 Exercises

3.1. Show that an upper semicontinuous function φ, defined on an open interval I, is quasiconvex if and only if it does not attain a semistrict local maximum point at any interior point $t \in I$ (see Remark 3.2.4).

3.2. Show that a strict local maximum is a semistrict local maximum and give an example which shows that the converse statement is not true.

3.3. Give an example which shows that the assumption of upper semicontinuity in Exercise 3.1 cannot be relaxed.

3.4. Let φ be a differentiable function on the open interval $I \subseteq \Re$. Show that φ is quasiconvex on I if and only if every point $t_0 \in I$ such that $\varphi'(t_0) = 0$ cannot be a semistrict local maximum point.

3.5. A differentiable convex function $f : \Re \to \Re$ that has a positive derivative at a point x_0 verifies the following limit: $\lim\limits_{x \to +\infty} f(x) = +\infty$. Is this result still valid for a quasiconvex function?

3.6. Let f be a differentiable function on the open convex set $S \subseteq \Re^n$. Show that f is quasiconvex on S if and only if for every $x_0 \in S$ and $u \in \Re^n$ such that $u^T \nabla f(x_0) = 0$, $\varphi(t) = f(x_0 + tu)$ does not attain a semistrict local maximum point at $t = 0$.

3.7. Which of the following functions is pseudoconvex?
(a) $f(x) = x \mid x \mid - x^2$;   b) $f(x) = x \mid x \mid + x^2$.

3.8. Let f be a differentiable function on an open convex set $S \subseteq \Re^n$. Show that f is pseudoconvex on S if and only if the following conditions hold:
1. f is semistrictly quasiconvex on S;
2. If $x_0 \in S$ is a critical point for f, then x_0 is a local minimum point for f.

3.9. Let f be a positive or negative pseudoconcave (strictly pseudoconcave) function on a convex set $S \subseteq \Re^n$. Show that the reciprocal function $z(x) = \frac{1}{f(x)}$ is pseudoconvex on S (strictly pseudoconvex).

3.10. Verify that the following functions are pseudoconvex:

(a) $f(x) = \log \sum_{i=1}^{n} x_i$, $x_i > 0$, $i = 1, .., n$;

(b)$f(x) = \log \sum_{i=1}^{n} x_i^3$, $x_i > 0$, $i = 1, .., n$;

(c) $f(x,y) = \log(x^4 + y^2) - \log(y - 1)$, $y > 1$.

3.11. Show that the function $f(x,y) = y + \frac{1}{x+1}$, $x > -1$ is pseudoconvex.

3.12. Show that the following functions are pseudoconvex:
$f(x,y) = \frac{x^2+x+y}{x+1}$, $x + 1 > 0$; $g(x,y) = \frac{xy+3y-5}{x+3}$, $x + 3 > 0$.

3.13. Show that the Cobb–Douglas function $f(x) = Ax_1^{\alpha_1} x_2^{\alpha_2} .... x_n^{\alpha_n}$, $A > 0$, $x_i > 0$, $\alpha_i > 0$, $i = 1, .., n$, is pseudoconcave.

3.14. Show that the C.E.S. function $f(x) = (a_1 x_1^\beta + a_2 x_2^\beta + .. + a_n x_n^\beta)^{\frac{1}{\beta}}$, $a_i > 0$, $x_i > 0$, $i = 1, .., n$, $\beta \neq 0$ is pseudoconcave if and only if $\beta \leq 1$.

3.15. Show that the generalized Cobb–Douglas function $z(x) = \prod_{i=1}^{k} (f_i(x))^{\alpha_i}$, $\alpha_i > 0$, $i = 1, .., k$, is pseudoconcave if $f_i(x)$, $i = 1, .., k$, are positive concave functions.

3.16. Let f, g be two differentiable functions on an open convex set $S \subseteq \Re^n$. Show that the function $z(x) = f(x) \cdot g(x)$ is:
(a) pseudoconcave if both functions are positive and concave;
(b) pseudoconvex if one function is negative and convex and the other one is positive and concave;
(c) strictly pseudoconvex if one function is negative and strictly convex and the other one is positive and concave.

3.17. Show that the function $z(x) = \frac{f(x)}{g(x)}$, where f and g are differentiable functions defined on an open convex set $S \subseteq \Re^n$, is strictly pseudoconvex if f is negative and strictly convex, g is positive and convex, or f is negative and convex, g is positive and strictly convex.

3.18. Let f be a continuous function defined on a convex set $S \subseteq \Re^n$. Show that f is quasilinear on S if and only if each of its level sets is convex.

3.19. Show that pseudolinearity is equivalent to requiring that the logical implication in the definition of pseudoconvexity and pseudoconcavity can be reversed, i.e., f is pseudolinear if and only if (i) and (ii) hold:
1. $x_1, x_2 \in S$, $f(x_2) < f(x_1) \Longleftrightarrow (x_2 - x_1)^T \nabla f(x_1) < 0$;
2. $x_1, x_2 \in S$, $f(x_2) > f(x_1) \Longleftrightarrow (x_2 - x_1)^T \nabla f(x_1) > 0$.

3.20. Verify that the function $f(x,y) = \dfrac{x+2-y\sqrt{(x+2)^2+y^2-1}}{(x+2)^2+y^2}$, suggested in [211], is pseudolinear on $int\Re_+^2$.

3.21. Verify that $f(x,y,z) = 2x+y-z-\frac{1}{2x+y-z+4}$ is pseudolinear on $S = \{(x,y,z): 2x+y-z+4>0\}$.

3.22. Find a pseudolinear function whose level sets are the following family of planes:
1. $k^2 - 2\alpha k - (ax+by+cz) = 0$; 2. $k^2 z - 2(x+y)k + z = 0$.

3.23. Show that the following functions are pseudolinear on $int\Re_+^2$:
(a) $f(x,y) = -x-1+\sqrt{x^2+2x+4y+1}$; (b) $f(x,y) = x-y+\sqrt{(x-y)^2+4y}$;
(c) $f(x,y) = \frac{1+\sqrt{xy+1}}{x}$; (d) $f(x,y) = \frac{y+x\sqrt{1-x^2+y^2}}{y^2-x^2}$, $y > x > 1$.

3.24. Show that the following functions are pseudolinear in their domain:
(a) $f(x) = \log(a^T x + a_0)$; (b) $f(x) = \log \sum_{i=1}^{n} x_i$, $x_i > 0$, $i = 1,..,n$;
(c) $f(x) = \log \frac{a^T x + a_0}{b^T x + b_0}$; (d) $f(x) = e^{x_1+...+x_n}$, $x_i > 0$, $i = 1,..,n$.

3.25. Let f be a twice continuously differentiable function defined on an open convex set $S \subseteq \Re^n$. Show that f is quasiconvex on S if and only if for every $x_0 \in S$, $u \in \Re^n$ such that $u^T \nabla f(x_0) = 0$, either $u^T \nabla^2 f(x_0)u > 0$ or $u^T \nabla^2 f(x_0)u = 0$ and $\varphi(t) = f(x_0 + tu)$ does not attain a semistrict local maximum point at $t = 0$.

3.26. By means of Theorems 3.4.6 and 2.2.12, verify that the function $f(x_1,x_2) = -x_1^2 - x_1 x_2$ is pseudoconvex on $int\Re_+^2$ and quasiconvex on $\Re_+^2$.

3.27. By means of the bordered Hessian, verify the pseudoconvexity of the function $f(x_1,x_2) = 3x_2 + \frac{x_2-x_1}{x_2+1}$, $x_2 > -1$.

3.28. Which of the following functions is pseudoconvex on $int\Re_+^2$?
(a) $f(x_1,x_2) = -x_1^2 - x_2^2 - 6x_1 x_2$;
(b) $f(x_1,x_2,x_3) = -x_1 x_2 - x_3^2 + x_3$.

3.29. Which of the following functions is pseudoconvex?
(a) $f(x_1,x_2) = -\frac{x_2^2}{x_1+1}$, $x_1 > -1$;
(b) $f(x_1,x_2) = 2x_1 + 3x_2 + \frac{x_2+5}{x_2+1}$, $x_2 > -1$.

3.30. Is $f(x_1,x_2) = -(x_1 - x_2)^2$ pseudoconvex on each of the two open half-planes associated with the line $x_1 = x_2$?

3.31. Find conditions on the parameters $a,b,c \in \Re$ for which the quadratic function $f(x_1,x_2) = \frac{1}{2}(ax_1^2 + 2bx_1 x_2 + cx_2^2)$ is pseudoconvex on $int\Re_+^2$ and quasiconvex on $\Re_+^2$.

3.32. Let f be a twice continuously differentiable function defined on an open convex set $S \subseteq \Re^n$ and assume that $\nabla f(x) \neq 0$, $\forall x \in S$. Show that f is pseudolinear on S if and only if (i) and (ii) hold:
1. $D_R(x) \leq 0$ if $|R|$ is odd;
2. $D_R(x) = 0$ if $|R|$ is even.

3.33. Verify that the following functions are pseudolinear on S:
(a) $f(x_1, x_2) = (x_1 + 2x_2)^3 + x_1 + 2x_2$, $S = \Re^2$;
(b) $f(x_1, x_2) = 2x_1 + x_2 + \frac{4x_1 + 2x_2 + 3}{2x_1 + x_2 + 6}$, $S = \{(x_1, x_2) \in \Re^2 : 2x_1 + x_2 + 6 > 0\}$.

3.34. The reason for which there are not any inclusion relationships between quasiconvexity and semistrictly quasiconvexity at a point, is related to the given definitions where nothing is said about the behaviour of the function when $f(x) > f(x_0)$. One way to avoid such a situation is to give a different definition of quasiconvexity and semistrictly quasiconvexity at a point. Consider a star-shaped set at a point x_0 and define quasiconvexity and semistrictly quasi convexity at a point requiring that (3.27) and (3.28) hold respectively:

$$f(x_0 + t(x - x_0)) \leq \max\{f(x), f(x_0)\}, \ t \in [0, 1] \tag{3.27}$$

$$f(x_0 + t(x - x_0)) < \max\{f(x), f(x_0)\}, \ t \in [0, 1] \tag{3.28}$$

Prove the following proposition: let f be a continuous function defined on a star-shaped $S \subset \Re^n$ at x_0. If f is semistrictly quasiconvex at x_0 according to (3.28) then f is quasiconvex at x_0 according to (3.27).

3.35. Give an example showing that the lower semicontinuity of the function is not enough to guarantee the proposition given in Exercise 3.34.

3.7 References

Arrow K. J., and Enthoven A. C. [7], Avriel M. [10], Avriel M., Diewert W. E., Schaible S., and Ziemba W. T. eds. [12], Avriel M., Diewert W. E., Schaible S., and Zang I. [13], Avriel M., and Schaible S. [11], Bazaraa M. S., Sherali H. D., and Shetty C. M. [18], Cambini A., and Martein L. [47, 45], Chew K. L., and Choo E. U. [62], Chabrillac Y., and Crouzeix J. P. [63], Cottle R. W., and Ferland J. A. [72, 73], Crouzeix J. P. [79, 80, 81], Crouzeix J. P., and Ferland J. A. [82], Carter M. [61], Debreu G. [91], Diewert W. E., Avriel M., and Zang I. [94], Ferland J. A. [109], Giorgi G., and Molho E. [118], Giorgi G., and Guerraggio A. [120], Giorgi G., and Thielfelder J. [122], Jeyakumar V., and Yang X. Q. [147], Karamardian S. [154], Katzner D. W. [160], Komlosi S. [168, 169, 170, 171], Mangasarian O. L. [192, 193, 194], Madden P. [191], Martos B. [209, 211], Ortega J. M., and Rheinboldt W. C. [217], Otani K. [219], Ponstein J. [223], Rapcsak T. [229], Schaible S., and Ziemba W. T. editors [248], Takayama A. [274], Thompson W. A., and Parke D. W. [275].

4

Optimality and Generalized Convexity

4.1 Introduction

In this chapter, the role of generalized convexity in Optimization is stressed. After presenting the Fritz John and Karush–Kuhn–Tucker necessary optimality conditions, which are proven by means of separation theorems, some constraint qualifications involving generalized convexity are illustrated.

One of the main reasons for introducing generalized convexity is the need to extend the fundamental properties of convex functions related to Optimization. In this regard, we shall see that semistrict quasiconvexity guarantees the local-global property and that this property, together with the minimality of a critical point and the sufficiency of the Karush–Kuhn–Tucker conditions, is guaranteed by pseudoconvexity.

In deriving sufficient optimality conditions and in investigating constraint qualifications, it will be clear that only generalized convexity at a point is needed, so that the given results are presented in a general form.

Under generalized convexity assumption, it will be proven that a maximum point is always located at the boundary of the domain; by requiring generalized convexity and generalized concavity, we have the important property that a minimum and a maximum point (if they exist) are attained at the boundary of the feasible set.

Finally, some classical applications in Economics will be presented.

4.2 Necessary Optimality Conditions Via Separation Theorems

In the differentiable case, when the domain of the function is described by constraint functions, the classical and well-known necessary optimality conditions are the Fritz John conditions and the Karush–Kuhn–Tucker conditions. Usually, the approach in deriving the Fritz John conditions is based on a separation theorem between a suitable subspace V and the non-positive orthant, while the Karush–Kuhn–Tucker conditions are derived by using a theorem of the alternative such as Farkas' Lemma. In this section, by studying the intersection between a subspace V and the non-positive orthant, we shall suggest a different proof of the Karush–Kuhn–Tucker conditions.

With this aim in mind, some preliminary results are needed.

Let us recall that a face $\mathcal{F}$ of $\Re^s_-$ is defined as

$$\mathcal{F} = \{z = \sum_{j \in J} \gamma_j (-e^j), \ \gamma_j \geq 0\}$$

where e^j is the unit vector having the j-th component equal to one and all others equal to zero and J is a proper subset of the set of indices $\{1, ..., s\}$. We shall use the convention $\mathcal{F} = \{0\}$ when $J = \emptyset$.

The following lemma holds.

Lemma 4.2.1. *Let V be a linear subspace of $\Re^s$ such that $V \cap int\Re^s_- = \emptyset$. Then, there exists a hyperplane which separates V and $int\Re^s_-$, i.e., there exists $\alpha \in \Re^s$ such that*

$$\alpha \geq 0, \ \alpha \neq 0, \ \alpha^T z = 0, \ \forall z \in V. \tag{4.1}$$

Proof. Since V and $int\Re^s_-$ are convex sets, by Theorem 1.2.14 there exists $\alpha \in \Re^s$, $\alpha \neq 0$ such that $\alpha^T z \geq 0$, $\forall z \in V$ and $\alpha^T z \leq 0$, $\forall z \in \Re^s_-$. This last inequality implies $\alpha^T (-e^i) = -\alpha_i \leq 0$, i.e., $\alpha_i \geq 0, i = 1, .., s$. By noting that $z \in V$ implies $-z \in V$, we have $\alpha^T z \geq 0$ and $\alpha^T (-z) \geq 0$, so that $\alpha^T z = 0$ for each $z \in V$. $\qquad\square$

Let us note that, in general, the vector α in (4.1) is not unique. It may happen that one or more components of α is zero whatever α may be. In order to look more closely at this aspect, we shall investigate the intersection between V and the boundary of the non-positive orthant by means of the so-called conical extension $V^* = V + \Re^s_+$ which is a convex and closed cone.

The following lemma points out that a subspace V and its conical extension have the same behaviour with respect to the intersection with $int\Re^s_-$.

Lemma 4.2.2. *The following properties hold:*
(i) $V \cap int\Re^s_- = \emptyset$ if and only if $V^* \cap int\Re^s_- = \emptyset$;
(ii) A hyperplane Γ separates V and $\Re^s_-$ if and only if Γ separates V^* and $\Re^s_-$.

Proof. (i) Since $V \subset V^*$, obviously $V^* \cap int\Re^s_- = \emptyset$ implies $V \cap int\Re^s_- = \emptyset$. Assume now that $V \cap int\Re^s_- = \emptyset$ and suppose, by contradiction, that there exists $z \in V^* \cap int\Re^s_-$. We have $z = v + w$, $v \in V$, $w \in \Re^s_+$, so that $v = z - w \in int\Re^s_-$ since $z \in int\Re^s_-$; consequently, $V \cap int\Re^s_- \neq \emptyset$, which contradicts the assumption.

(ii) Let Γ be a hyperplane which separates V and $\Re^s_-$ of equation $\alpha^T z = 0$, $\alpha \geq 0$. From Lemma 4.2.1 we have $\alpha^T z = 0$, $\forall z \in V$. Let $v^* = v + w \in V^*$, $v \in V$, $w \in \Re^s_+$. We have $\alpha^T v^* = \alpha^T v + \alpha^T w = \alpha^T w \geq 0$, i.e., Γ separates V^* and $\Re^s_-$. The converse statement is obvious. $\qquad\square$

When $V \cap int\Re_-^s = \emptyset$, we shall see that the intersection between the conical extension of V and the non-positive orthant is a face. The relevance of this result is due to the fact that it is possible to determine a set of indices J, which corresponds to multipliers which are zero in all separating hyperplanes, and also establish the existence of a separating hyperplane with positive multipliers associated with the indices which are not in J.

More precisely, we have the following theorem.

Theorem 4.2.1. *Let V be a linear subspace of $\Re^s$ such that $V \cap int\Re_-^s = \emptyset$. Then the following conditions hold:*

(i) $V^* \cap \Re_-^s$ is a face $\mathcal{F} = \{z = \sum_{j \in J} \gamma_j(-e^j),\ \gamma_j \geq 0\}$, where J is a proper subset of the set of indices $\{1, ..., s\}$;

(ii) If $J \neq \emptyset$, for each hyperplane of equation $\alpha^T z = 0$, $\alpha \geq 0$ which separates V and $\Re_-^s$, we have $\alpha_j = 0$, $\forall j \in J$. Furthermore, there exists a separating hyperplane such that $\alpha_i > 0$, $\forall i \notin J$;

(iii) If $J = \emptyset$, i.e., $V \cap \Re_-^s = \{0\}$, there exists a separating hyperplane such that $\alpha_i > 0$, $\forall i \in \{1, ..., s\}$.

Proof. (i) If $V \cap \Re_-^s = \{0\}$, the thesis follows by convention. Let $z \in V \cap \Re_-^s, z \neq 0$. Since $V \cap int\Re_-^s = \emptyset$, z is a boundary point of $\Re_-^s$ and thus there exists a proper subset of indices $J_z \subset \{1, .., s\}$ such that $z = \sum_{j \in J_z} \gamma_j(-e^j)$, $\gamma_j > 0$. Taking into account that $V^* = V + \Re_+^s$ is a convex cone, we have $\frac{1}{\gamma_k}(z + \sum_{j \in J_z, j \neq k} \gamma_j e^j) = -e^k \in V^*$, for every $k \in J_z$. Repeating this process for every element of $V \cap \Re_-^s$, we obtain a subset $J = \cup J_z$. Since $V \cap int\Re_-^s = \emptyset$, J is properly contained in $\{1, ..., s\}$. Consequently, the intersection $V^* \cap \Re_-^s$ is given by $\{z = \sum_{j \in J} \gamma_j(-e^j),\ \gamma_j \geq 0\}$, i.e., it is a face of $\Re_-^s$.

(ii) Let $\alpha^T z = 0$, $\alpha \geq 0$ be the equation of a hyperplane which separates V and $\Re_-^s$; from (ii) of Lemma 4.2.2, we have $\alpha^T z \geq 0$, $\forall v \in V^*$. Since $j \in J$ implies $-e^j \in V^*$, it results that $\alpha^T(-e^j) = -\alpha_j \geq 0$, i.e., $\alpha_j \leq 0$, so that necessarily we have $\alpha_j = 0$ for every $j \in J$. Consider now the case $i \notin J$, so that $-e^i \notin V^*$. Since V^* is the intersection of its supporting hyperplanes passing through the origin, there exists a hyperplane which separates V^* and $\Re_-^s$ and which does not contain $-e^i$; the equation of such a hyperplane is of the kind $(\alpha^i)^T z = 0$, $\alpha^i \geq 0$ where, necessarily, $\alpha_i^i > 0$. Let $\beta = \sum_{i \notin J} \alpha^i$. We have $\beta_j = 0$, $j \in J$, $\beta_i > 0$, $i \notin J$. Furthermore $\beta^T z = 0$, $\forall z \in V$, since $(\alpha^i)^T z = 0$, $\forall z \in V$, so that (ii) follows.

(iii) This follows by noting that $J = \emptyset$ implies $\beta_i > 0$, $i = 1, .., s$. $\qquad\square$

Corollary 4.2.1. *Let V be a linear subspace of $\Re^s$ such that $V \cap int\Re^s_- = \emptyset$. Then, there exists $\alpha \geq 0$ such that $\alpha^T v = 0$ for all $v \in V^*$ with $\alpha_i > 0$ if and only if $(-e^i) \notin V^*$.*

Proof. The thesis follows from (ii) of Theorem 4.2.1 by noting that $j \in J$ if and only if $(-e^i) \in V^*$.   □

Remark 4.2.1. By means of Corollary 4.2.1 it is possible to derive some theorems of the alternative. For instance, by setting $V = \{Bx,\ x \in \Re^n\}$, where B is an $s \times n$ matrix, we have that system $Bx \leq 0$ (i.e., $Bx \in \Re^s_- \setminus \{0\}$) has no solutions if and only if there exists $\alpha \in int\Re^s_+$ such that $\alpha^T B = 0$ (Stiemke's Theorem of the Alternative). In Exercises 4.1, 4.2, and 4.3, we shall see some extensions of Corollary 4.2.1 and their equivalence with some classical theorems of the alternatives.

Now we shall see how the previous results allow us to derive the Karush–Kuhn–Tucker conditions from the Fritz John conditions. To this end consider the following problem

$$P: \quad \min f(x),\ x \in S = \{x \in X : g_i(x) \leq 0,\ i = 1, ..., m\}$$

where $f, g_i, i = 1, ..., m$ are functions defined on an open set $X \subseteq \Re^n$. Corresponding to a feasible point x_0, let $I(x_0)$ be the set of indices associated with the constraints binding at x_0, i.e., $I(x_0) = \{i \in \{1, ..., m\} : g_i(x_0) = 0\}$, and let $k \in [1, m]$ be its cardinality. We can assume, without loss of generality, that $I(x_0) = \{1, ..., k\}$.
The following theorem holds.

Theorem 4.2.2. *Let x_0 be a feasible point for problem P. Suppose that f, g_i, $i \in I(x_0)$, are differentiable at $x_0 \in S$ and that g_i, $i \notin I(x_0)$, are continuous at x_0. If x_0 is a local minimum point, then (i) and (ii) hold:*
(i) There exist multipliers λ_0, λ_i, $i \in I(x_0)$, not all zero, such that:

$$\begin{cases} \lambda_0 \nabla f(x_0) + \displaystyle\sum_{i \in I(x_0)} \lambda_i \nabla g_i(x_0) = 0 \\ \lambda_0 \geq 0,\ \lambda_i \geq 0,\ i \in I(x_0) \end{cases} \qquad (4.2)$$

(ii) There exist multipliers that verify (4.2) with $\lambda_0 > 0$ if and only if

$$W \cap (int\Re_- \times \Re^k_-) = \emptyset. \qquad (4.3)$$

where $W = \{z = (\nabla f(x_0)^T d, \nabla g_1(x_0)^T d, ..., \nabla g_k(x_0)^T d)^T,\ d \in \Re^n\}$.

Proof. The continuity of the constraint functions not binding at x_0 implies that x_0 is also a local minimum point for the problem $\min f(x), x \in S^* = \{x \in X : g_i(x) \leq 0,\ i \in I(x_0)\}$.
(i) We have $W \cap int\Re^{k+1}_- = \emptyset$, otherwise there exists a direction $d \in \Re^n$

such that $d^T \nabla g_i(x_0) < 0$, $i \in I(x_0)$, and $d^T \nabla f(x_0) < 0$, i.e., d is a decreasing feasible direction[1], and this contradicts the local optimality of x_0. By applying Lemma 4.2.1 to the convex sets W and $int\Re_-^{k+1}$, there exist non-negative multipliers λ_0, λ_i, $i \in I(x_0)$, not all zero, such that

$$\lambda^T z = \lambda_0 z_0 + \lambda_1 z_1 + .. + \lambda_k z_k = 0, \ \forall z = (z_0, z_1, .., z_k)^T \in W \qquad (4.4)$$

i.e., $(\lambda_0 \nabla f(x_0) + \sum_{i \in I(x_0)} \lambda_i \nabla g_i(x_0))^T d = 0$, $\forall d \in \Re^n$.

By choosing $d = \lambda_0 \nabla f(x_0) + \sum_{i \in I(x_0)} \lambda_i \nabla g_i(x_0)$, the thesis follows.

(ii) (4.3) implies $-e^1 \notin W^* = W + int\Re_+^{k+1}$ so that, from (ii) of Theorem 4.2.1, it is possible to choose a separating hyperplane (4.4) with $\lambda_0 > 0$. Conversely, if $\lambda_0 > 0$ in (4.4), then (4.3) holds. In fact, if $z = (z_0, \bar{z}) \in W \cap (int\Re_- \times \Re_-^k)$, we have $\lambda^T z = 0 = \lambda_0 z_0 + \lambda^T \bar{z} < 0$ and this is absurd.   □

Remark 4.2.2. The necessary optimality conditions (4.2) are known as the Fritz John conditions, while (4.2) together with $\lambda_0 > 0$ are known as the Karush–Kuhn–Tucker conditions.

As regards to (4.2), it may happen that the multiplier associated with the objective function vanishes even if convexity is required, as is shown in the following example.

Example 4.2.1. Consider problem P where $f(x_1, x_2) = x_1$, $g_1(x_1, x_2) = x_1^2 - x_2$, $g_2(x_1, x_2) = x_2$. The feasible set is $S = \{(0,0)\}$ so that $x_0 = (0,0)$ is a global minimum point for the problem. On the other hand, $\nabla f(x_0) = (1,0)^T$, $\nabla g_1(x_0) = (0,-1)^T$, $\nabla g_2(x_0) = (0,1)^T$ and consequently, (4.2) is verified if and only if $\lambda_0 = 0$, $\lambda_1 = \lambda_2 > 0$, even if all the functions are convex.
Note that $W = \{z = (d_1, -d_2, d_1)^T, \ d_1, d_2 \in \Re\} \cap (int\Re_- \times \Re_-^2) \neq \emptyset$ according to (ii) of Theorem 4.2.2.

When $\lambda_0 = 0$ in (4.2), it is not possible to deduce the behaviour of the objective function at x_0; for this reason, the problem of finding conditions which imply $\lambda_0 \neq 0$ assumes a relevant aspect. Any condition which ensures $\lambda_0 \neq 0$ in (4.2) is called a constraint qualification.
In the next section we shall point out the role played by convexity and generalized convexity in establishing some constraint qualifications.
When a constraint qualification holds, the Karush–Kuhn–Tucker conditions may be stated as follows.

Theorem 4.2.3. *(The Karush–Kuhn–Tucker conditions)*
Consider problem P and let x_0 be a feasible point. Suppose that f, g_i, $i \in I(x_0)$, are differentiable at $x_0 \in S$ and that g_i, $i \notin I(x_0)$, are continuous at

---

[1] A vector $d \in \Re^n$, $d \neq 0$ is a feasible direction at $x_0 \in S$ if there exists $\epsilon > 0$ such that $x = x_0 + td \in S$, $\forall t \in [0, \epsilon]$.

x_0. If x_0 is a local minimum point and a constraint qualification holds, then there exist non-negative multipliers λ_i, $i \in I(x_0)$, such that:

$$\nabla f(x_0) + \sum_{i \in I(x_0)} \lambda_i \nabla g_i(x_0) = 0 \tag{4.5}$$

Remark 4.2.3. By assuming the differentiability at x_0 of all the constraint functions, the Fritz John conditions and the Karush–Kuhn–Tucker conditions may be restated, respectively, as follows.

$$\begin{cases} \lambda_0 \nabla f(x_0) + \sum_{i=1}^{m} \lambda_i \nabla g_i(x_0) = 0 \\ \lambda_0 \geq 0, \ \lambda_i \geq 0, \ i = 1, ..., m \\ \lambda_i g_i(x_0) = 0, \ i = 1, ..., m. \end{cases}$$

$$\begin{cases} \nabla f(x_0) + \sum_{i=1}^{m} \lambda_i \nabla g_i(x_0) = 0 \\ \lambda_i \geq 0, \ i = 1, ..., m \\ \lambda_i g_i(x_0) = 0, \ i = 1, ..., m. \end{cases}$$

In Sect. 4.4 we shall see that the necessary Karush–Kuhn–Tucker conditions become sufficient under suitable generalized convexity assumptions.

4.3 Generalized Convexity and Constraint Qualifications

As we have already remarked, a constraint qualification is a condition which ensures $\lambda_0 > 0$ in the Fritz John conditions. Since $W \cap (int\Re_- \times \Re^m_-) = \emptyset$ is a necessary and sufficient condition for the existence of such a positive multiplier, any condition which implies (4.3) is a constraint qualification.

Several constraint qualifications are suggested in the literature and their interrelationships are studied. In this section we shall limit ourselves to stressing the role of convexity and generalized convexity in establishing some constraint qualifications.

To this end, let x_0 be a local minimum point for problem P, where f and g_i, $i \in I(x_0) = \{i : g_i(x_0) = 0\}$, are differentiable at x_0 and X is star-shaped at x_0. Consider the following subsets of $I(x_0)$:

$J = \{i \in I(x_0) : g_i(x) \text{ is pseudoconcave at } x_0\}$;

$J_1 = \{i \in I(x_0) : g_i(x) \text{ is concave at } x_0\}$;

$I_L = \{i \in I(x_0) : g_i(x) \text{ is linear}\}$.

We shall prove that each of the following statements is a constraint qualification.

1. *The weak-reverse constraint qualification*
   Functions $g_i(x)$, $i \in I(x_0)$, are pseudoconcave at x_0.

2. *The reverse constraint qualification*
   Functions $g_i(x)$, $i \in I(x_0)$, are concave at x_0.

3. *The weak Arrow–Hurwicz–Uzawa constraint qualification*

$$\exists d \in \Re^n : d^T \nabla g_i(x_0) \leq 0, \ \forall i \in J, \ d^T \nabla g_i(x_0) < 0, \ \forall i \in I(x_0) \backslash J.$$

4. *The Arrow–Hurwicz–Uzawa constraint qualification*

$$\exists d \in \Re^n : d^T \nabla g_i(x_0) \leq 0, \ \forall i \in J_1, \ d^T \nabla g_i(x_0) < 0, \ \forall i \in I(x_0) \backslash J_1.$$

5. *Slater's weak constraint qualification*
   Functions $g_i(x)$, $i \in I(x_0)$, are pseudoconvex at x_0 and there exists $x^* \in S$ such that $g_i(x^*) < 0$, $\forall i \in I(x_0)$.

6. *Slater's constraint qualification*
   Functions $g_i(x)$, $i \in I(x_0)$, are convex at x_0 and there exists $x^* \in S$ such that $g_i(x^*) < 0$, $\forall i \in I(x_0)$.

7. *Slater's second constraint qualification*
   Functions $g_i(x)$, $i \in I(x_0)$, are convex on the convex set X and for each $i \in I(x_0)$ there exists $x^i \in S$ such that $g_i(x^i) < 0$.

8. *The modified Slater–Uzawa constraint qualification*
   Functions $g_i(x)$, $i \in I(x_0) \backslash I_L$, are pseudoconvex at x_0 and there exists $x^* \in S$ such that $g_i(x^*) < 0$, $\forall i \in I(x_0) \backslash I_L$ and $g_i(x^*) \leq 0$, $\forall i \in I_L$.

9. *Martos' constraint qualification*
   Functions $g_i(x)$, $i \in I(x_0)$, are pseudoconvex at x_0 and quasiconvex at x_0 for each $i \in J$. Furthermore, there exists $x^* \in S$ such that $g_i(x^*) < 0$, $\forall i \in I(x_0) \backslash J$.

10. *Arrow–Enthoven's constraint qualification*
    Functions $g_i(x)$, $i \in I(x_0)$, are continuous on X and quasiconvex at x_0 with $\nabla g_i(x_0) \neq 0$. Furthermore, there exists $x^* \in S$ such that, for all $i \in I(x_0)$, $g_i(x^*) < 0$.

The following theorem holds.

Theorem 4.3.1. $W \cap (int\Re_- \times \Re_-^m) = \emptyset$ *if one of the conditions (1–10) holds, i.e., every condition (1–10) is a constraint qualification.*

Proof. The proof is given by assuming, by contradiction, the existence of a direction $d^* \in \Re^n$ such that $\nabla f(x_0)^T d^* < 0, \nabla g_i(x_0)^T d^* \leq 0$, $i \in I(x_0)$.

1. Since g_i, $i \in I(x_0)$, is pseudoconcave at x_0, by setting $x = x_0 + td^*$, the inequality $\nabla g_i(x_0)^T (x - x_0) = t \nabla g_i(x_0)^T d^* \leq 0$ implies that $g_i(x_0 + td^*) \leq g_i(x_0)$ for every t such that $x_0 + td^* \in X$. Consequently, d^* is a feasible direction, so that $\nabla f(x_0)^T d^* < 0$ contradicts the optimality of x_0.

2. This is a particular case of 1.
3. By setting $\hat{d} = d^* + \frac{1}{n}d$, we have $\nabla g_i(x_0)^T \hat{d} = \nabla g_i(x_0)^T d^* + \frac{1}{n}\nabla g_i(x_0)^T d$. Consequently, $\nabla g_i(x_0)^T \hat{d} \leq 0$, $\forall i \in J$, and $\nabla g_i(x_0)^T \hat{d} < 0$, $\forall i \in I(x_0)\backslash J$. The pseudoconcavity of g_i at x_0, $i \in J$, together with the condition $\nabla g_i(x_0)^T \hat{d} < 0$, $i \in I(x_0)\backslash J$, implies that $\hat{d}$ is a feasible direction. Furthermore, for a large enough n, $\nabla f(x_0)^T \hat{d} = \nabla f(x_0)^T d^* + \frac{1}{n}\nabla f(x_0)^T d < 0$ and this contradicts the optimality of x_0.
4. This is a particular case of 3.
5. The pseudoconvexity of g_i at x_0, $i \in I(x_0)$, implies $\nabla g_i(x_0)^T(x^* - x_0) < 0$, so that $d = x^* - x_0$ is a feasible direction. By setting $\hat{d} = d^* + \frac{1}{n}d$, we have $\nabla g_i(x_0)^T \hat{d} < 0$, and, for a large enough n, $\nabla f(x_0)^T \hat{d} < 0$, so that $\hat{d}$ is a feasible decreasing direction which contradicts the optimality of x_0.
6. This is a particular case of 5.
7. Let x^* be a convex combination of the vectors x^i, i.e., $x^* = \sum_{i \in I(x_0)} \alpha_i x^i$,

   $\alpha_i > 0$, $\sum_{i \in I(x_0)} \alpha_i = 1$. By means of Jensen's Inequality, we have that

   $g_i(x^*) = \sum_{i \in I(x_0)} \alpha_i g_i(x^i) < 0$, and 6) holds.
8. The assumptions imply that $d = x^* - x_0$ is a feasible direction. By setting $\hat{d} = d^* + \frac{1}{n}d$, for a large enough n, $\hat{d}$ is a feasible decreasing direction which contradicts the optimality of x_0.
9. Since $g_i(x^*) \leq g_i(x_0)$, $\forall i \in I(x_0)$, the quasiconvexity of g_i implies $\nabla g_i(x_0)^T(x^* - x_0) \leq 0$, $\forall i \in I(x_0)$. In particular, from the pseudoconvexity of g_i we have $\nabla g_i(x_0)^T(x^* - x_0) < 0$, $\forall i \in I(x_0)\backslash J$. Set $d = x^* - x_0$ and consider $\hat{d} = d^* + \frac{1}{n}d$. We have $\nabla g_i(x_0)^T \hat{d} < 0$, $\forall i \in I(x_0)\backslash J$ and $\nabla g_i(x_0)^T \hat{d} \leq 0$, $\forall i \in J$. Taking into account the pseudoconcavity of g_i, $i \in J$, $\hat{d}$ is a feasible direction and since, for a large enough n, $\nabla f(x_0)^T \hat{d} < 0$, the optimality of x_0 is contradicted.
10. This is a particular case of 9.   □

Remark 4.3.1. Condition (9) is given in a different form with respect to the original constraint qualification suggested by Martos [211] since he introduced pseudoconvexity at a point in a more restrictive form than Definition 3.5.4. More precisely:

A function h is pseudoconvex at x_0 if (i) and (ii) hold:

(i) $h(x) < h(x_0) \Rightarrow \nabla h(x_0)^T(x - x_0) < 0$

(ii) $h(x) \leq h(x_0) \Rightarrow \nabla h(x_0)^T(x - x_0) \leq 0$.

4.4 Sufficiency of the Karush–Kuhn–Tucker Conditions

Consider problem P again. As we shall see, the validity of the Karush–Kuhn–Tucker conditions does not guarantee the optimality of x_0 (see Example 4.4.1 below); nevertheless, these conditions become sufficient under a suitable generalized convexity assumption on the objective function and on the constraints.

Example 4.4.1. Consider problem P where $f(x_1, x_2) = -x_1^2 - x_2^2 + 2x_1 + 4x_2$, $g(x_1, x_2) = -x_2$. With respect to the feasible point $x_0 = (1, 0)$, we have $\nabla f(x_0) + 4\nabla g(x_0) = 0$, so that the Karush–Kuhn–Tucker conditions hold but x_0 is not a local minimum point for the problem since $f(x_1, 0) < f(1, 0)$ for every $x_1 \neq 1$.

Theorem 4.4.1. *Consider problem P and let x_0 be a feasible point. Suppose that f is pseudoconvex at $x_0 \in S$ and that g_i, $i = 1, .., m$, are differentiable and quasiconvex at x_0. If there exist $\lambda_i \in \Re$, $i = 1, .., m$, such that*

$$
\begin{cases}
\nabla f(x_0) + \sum_{i=1}^{m} \lambda_i \nabla g_i(x_0) = 0 \\
\lambda_i \geq 0, \quad i = 1, ..., m \\
\lambda_i g_i(x_0) = 0, \quad i = 1, ..., m
\end{cases}
\tag{4.6}
$$

then x_0 is a global minimum point for P.

Proof. Assume, by contradiction, the existence of a feasible point $\bar{x}$ such that $f(\bar{x}) < f(x_0)$. From the pseudoconvexity of f we have $\nabla f(x_0)^T (\bar{x} - x_0) < 0$. Since $g_i(\bar{x}) \leq 0 = g_i(x_0)$, $i \in I(x_0)$, the quasiconvexity of g_i implies $\nabla g_i(x_0)^T (\bar{x} - x_0) \leq 0$, $i \in I(x_0)$.
From the complementarity condition $\lambda_i g_i(x_0) = 0$, $i = 1, ..., m$, we have $\lambda_i = 0$, $\forall i \notin I(x_0)$, so that $\nabla f(x_0)^T (\bar{x} - x_0) + \sum_{i=1}^{m} \lambda_i \nabla g_i(x_0)^T (\bar{x} - x_0) < 0$,
and this contradicts (4.6). $\qquad\square$

Remark 4.4.1. Let us note that the pseudoconvexity assumption in Theorem 4.4.1 cannot be substituted with quasiconvexity or semistrict quasiconvexity. Consider, for instance, the problem

$$min \ (x_1 + x_2)^3, \quad (x_1, x_2) \in S = \{(x_1, x_2) \in \Re^2 : -x_2 \leq 0\}.$$

The objective function is semistrictly quasiconvex (in particular quasiconvex) and $g(x_1, x_2) = -x_2$ is quasiconvex. It is easy to verify that the point $(0, 0)$ verifies conditions (4.6) with $\lambda = 0$, but it is not a local minimum point for the problem.

By requiring a generalized convexity assumption on the whole set X, we have the following corollaries.

Corollary 4.4.1. *Consider problem P where f is pseudoconvex on an open convex set X and $g_i, i = 1, 2, ..., m$, are differentiable and quasiconvex on X. If $x_0 \in S$ verifies (4.6), then x_0 is a global minimum point for P.*

Corollary 4.4.2. *Consider problem P where $f, g_i, i = 1, 2, ..., m$, are differentiable and quasiconvex on an open convex set X, with $\nabla f(x) \neq 0$, for all $x \in S$. If $x_0 \in S$ verifies (4.6), then x_0 is a global minimum point for P.*

Some applications of the sufficiency of the Karush–Kuhn–Tucker conditions in Economics shall be given in Sect. 4.8.

4.5 Local-Global Property

One of the most important questions in Optimization is knowing whether a local minimum is also global and a critical point is a global minimum. These important properties are not exclusive of convex functions but they hold under suitable generalized convexity assumptions. In this section, we shall point out the role of generalized convexity in establishing a local-global property for a minimum point. Since such a property involves only the behaviour of the function at a point, we can relax generalized convexity assumptions together with the convexity of the domain by requiring only generalized convexity at a point and the star-shapedness of the domain. By following this approach, all the results will be given in a more general form.

Theorem 4.5.1. *Let f be a function defined on a set $S \subseteq \Re^n$ which is star-shaped at $x_0 \in S$. Then, the following properties hold:*
(i) If x_0 is a strict local minimum point and f is quasiconvex at x_0, then x_0 is a strict global minimum point for f on S;
(ii) If x_0 is a local minimum point and f is semistrictly quasiconvex at x_0, then x_0 is a global minimum point for f on S;
(iii) If x_0 is a local minimum point and f is strictly quasiconvex at x_0, then x_0 is the unique global minimum point for f on S.

Proof. (i) Suppose that $x_0 \in S$ is a strict local minimum point, i.e., there exists a neighbourhood I of x_0 such that $f(x) > f(x_0)$, $\forall x \in I \cap S$. If x_0 is not a strict global minimum point, there exists $\bar{x} \in S$ such that $f(\bar{x}) \leq f(x_0)$. By the quasiconvexity of f at x_0 we have $f(x_0 + \lambda(\bar{x} - x_0)) \leq f(x_0)$, $\forall \lambda \in [0, 1]$, so that $x_0 + \lambda(\bar{x} - x_0) \in I \cap S$ for a small enough λ and thus x_0 cannot be a strict local minimum.
(ii) The proof is similar to the one given in (i).
(iii) Since a strictly quasiconvex function at x_0 is also semistrictly quasiconvex at x_0, then a local minimum point is also global.
Assume now that there exists another global minimum point $x_1 \in S$. The strict quasiconvexity at x_0 implies that $f(x_0 + \lambda(x_1 - x_0)) < f(x_0)$, $\forall \lambda \in (0, 1)$ and this contradicts the optimality of x_0.   □

The following examples point out that a non-strict local minimum point is not necessarily global for a quasiconvex function and that the semistrict quasiconvexity does not guarantee the uniqueness of a global minimum point.

Example 4.5.1. Consider $f(x) = \begin{cases} -x^2 & -1 \le x \le 0 \\ 0 & 0 < x \le 2 \end{cases}$

The function is quasiconvex at $x_0 = 1$ which is a local but not global minimum point.

Example 4.5.2. Consider $f(x) = \begin{cases} 0 & 0 \le x < 2 \\ (x-2)^2 & 2 \le x \le 3 \end{cases}$

The function is semistrictly quasiconvex at $x_0 = 0$; on the other hand, any point of the interval $[0, 2]$ is a global minimum.

In the differentiable case, unlike the pseudoconvex case (see Theorem 3.2.5), strict quasiconvexity is not sufficient for the optimality of a critical point. For instance, the strictly quasiconvex function $f(x) = x^3$ has a critical point at $x_0 = 0$, which is not a minimum point.

The following theorem shows that the class of pseudoconvex functions is the only one, among the classes of generalized convex functions, which maintains all the properties of convex functions related to minimum points.

Theorem 4.5.2. *Let f be a function defined on a set $S \subseteq \Re^n$ which is star-shaped at $x_0 \in S$. If x_0 is a local minimum point and f is pseudoconvex at x_0, then x_0 is a global minimum point.*

Proof. Assume the existence of $\bar{x} \in S$ such that $f(\bar{x}) < f(x_0)$. Since f is pseudoconvex at x_0, we have $\nabla f(x_0)^T (\bar{x} - x_0) < 0$ so that f is locally decreasing along the feasible direction $d = \bar{x} - x_0$, and the local optimality of x_0 is contradicted. □

Within the class of pseudoconvex functions, the strictly pseudoconvex functions guarantee the uniqueness of the global minimum point. More precisely, as a direct consequence of the previous Theorem and of Remark 3.2.3, we have the following result.

Theorem 4.5.3. *Let f be a function defined on a set $S \subseteq \Re^n$ which is star-shaped at $x_0 \in S$. If x_0 is a local minimum point and f is strictly pseudoconvex at x_0, then x_0 is the unique global minimum point.*

Let us note that a point is not necessarily a local minimum for a function f even if it is a local minimum for the restriction of f on every line segment. Consider, for instance, $f(x_1, x_2) = (x_2 - x_1^4)(x_2 - x_1^2)$, $x_1, x_2 \ge 0$. It can be verified that $x_0 = (0, 0)$ is a local minimum for f on every line segment starting from x_0. On the other hand, with respect to the restriction of f on the curve $x_2 = x_1^3$, $x_1 \ge 0$, we have $\varphi(x_1) = -x_1^5(x_1 - 1)^2$, so that $\varphi(x_1) < 0 = \varphi(0) = f(0, 0)$, $\forall x_1 > 0$. Consequently, x_0 is not a local minimum for f.

As we have remarked several times, for a generalized convex function f there is a strict connection between the behaviour of f and the behaviour of its restrictions on a line segment. This connection allows us to obtain the following result, as a direct consequence of Theorems 4.5.1 and 4.5.2.

Theorem 4.5.4. *Let f be a function defined on a set $S \subseteq \Re^n$ which is star-shaped at $x_0 \in S$.*
(i) If f is quasiconvex at $x_0 \in S$ and x_0 is a strict local minimum point with respect to every line segment $[x_0, x] \subset S$, then x_0 is a strict global minimum point for f on S.
(ii) If f is semistrictly quasiconvex at $x_0 \in S$ and x_0 is a local minimum point with respect to every line segment $[x_0, x] \subset S$, then x_0 is a global minimum point for f on S.
(iii) If f is differentiable and pseudoconvex at $x_0 \in S$ and x_0 is a local minimum point with respect to every line segment $[x_0, x] \subset S$, then x_0 is a global minimum point for f on S.

Let us note that all the previous results hold in the more restrictive assumption of generalized convexity on a convex set.

4.6 Maxima and Generalized Convexity

In this section we shall study the property of a maximum point, if one exists, related to generalized convexity. We shall see that a maximum point is always located on the boundary of the domain or, equivalently, is attained at an interior point if and only if the function is constant on the whole domain. This important property does not hold in general for a minimum point which can be located anywhere.
We shall begin to prove that if the maximum value of a semistrictly quasiconvex function is reached at a relative interior point of S, then the function is constant on S.

Lemma 4.6.1. *Let f be a continuous and semistrictly quasiconvex function on a convex set $S \subseteq \Re^n$. If $x_0 \in riS$ is such that $f(x_0) = \max\limits_{x \in S} f(x)$, then f is constant on S.*

Proof. Assume that there exists $\bar{x} \in S$ such that $f(\bar{x}) < f(x_0)$. From (v) of Theorem 1.2.7, there exists $x^* \in S$ such that x_0 is an interior point of the line segment $[x^*, \bar{x}]$. Since f is a continuous function, without loss of generality, we can assume that $f(x^*) > f(\bar{x})$. The semistrictly quasiconvexity of f implies $f(x) < f(x^*)$, $\forall x \in ri[x^*, \bar{x}]$ and this is absurd, since $x_0 \in ri[x^*, \bar{x}]$. $\qquad\square$

From Lemma 4.6.1, we directly have the following result.

Theorem 4.6.1. *Let f be a continuous and semistrictly quasiconvex function on a convex and closed set $S \subseteq \Re^n$. If f attains its maximum value on S, then it is reached at some boundary point.*

The previous theorem can be strengthened when the convex set S does not contain lines.

Theorem 4.6.2. *Let f be a continuous and semistrictly quasiconvex function on a convex and closed set $S \subseteq \Re^n$ containing no lines. If f attains its maximum value on S, then it is reached at an extreme point.*

Proof. If f is constant, then the thesis is obvious. Let x_0 be such that $f(x_0) = \max_{x \in S} f(x)$. From Theorem 4.6.1, x_0 belongs to the boundary of S. Let C be the minimal face of S containing x_0; if x_0 is not an extreme point, then $x_0 \in riC$. It follows from Lemma 4.6.1 that f is constant on C. On the other hand, C is a convex closed set containing no lines (see [234]), so that C has at least one extreme point $\bar{x}$ (see Theorem 1.2.10) which is also an extreme point of S. Consequently, f attains its maximum value at the extreme point $\bar{x}$. ☐

Let us note that a quasiconvex function may have a global maximum point which is not a boundary point. For instance, the function

$$f(x) = \begin{cases} -x^2 + 2x & 0 \leq x \leq 1 \\ 1 & x > 1 \end{cases}$$

is quasiconvex, since it is non-decreasing; on the other hand, f attains its maximum value at any point $x \geq 1$ which is not a boundary point of the domain $S = [0, +\infty)$.

In order to extend Theorem 4.6.2 to the class of quasiconvex functions, additional assumptions on the convex set S are required.

Theorem 4.6.3. *Let f be a continuous and quasiconvex function on a convex and compact set $S \subseteq \Re^n$. Then, there exists some extreme point on which f assumes its maximum value.*

Proof. From Weierstrass' Theorem, there exists $\bar{x} \in S$ with $f(\bar{x}) = \max_{x \in S} f(x)$. Since S is convex and compact, it is also the convex hull of its extreme points (see Theorem 1.2.8), so that there exists a finite number $x^1, ..., x^h$ of extreme points such that $\bar{x} = \sum_{i=1}^{h} \lambda_i x_i$, $\sum_{i=1}^{h} \lambda_i = 1$, $\lambda_i \geq 0$. Since f is quasiconvex, we have $f(\bar{x}) \leq \max\{f(x^1), ..., f(x^h)\}$ (see Theorem 2.2.5) and the thesis follows. ☐

Taking into account the inclusion relationships between the classes of generalized convex functions, we have the following results related to a pseudoconvex function.

Corollary 4.6.1. *Let f be a pseudoconvex function on a convex and closed set $S \subseteq \Re^n$. If f assumes maximum value on S, then it is reached at some boundary point.*

Corollary 4.6.2. *Let f be a pseudoconvex function on a convex and closed set $S \subseteq \Re^n$ containing no lines. If f assumes maximum value on S, then it is reached at an extreme point.*

From a computational point of view, the previous results are very important since they establish the need to investigate the boundary of the feasible set in order to find a global maximum. Nevertheless, for a generalized convex function, a local maximum is not necessarily global, so that the problem of maximizing a quasiconvex or a pseudoconvex function is a very difficult one. In the next section we shall see that this kind of difficulty vanishes if f is pseudolinear.

4.7 Minima, Maxima and Pseudolinearity

In this section we shall stress how suitable the behaviour of pseudolinear functions is, with respect to optimality.

As we have already remarked, a minimum point for a pseudoconvex function f can be located anywhere but, when f is also pseudoconcave, Corollary 4.6.1, applied to $-f$, allows us to state that the minimum value (if one exists) is attained at a boundary point. Consequently, for a pseudolinear function, both the minimum and the maximum are reached at a boundary point.

The following theorem gives a necessary and sufficient condition for a boundary point to be a global minimum (maximum) under the pseudoconvexity (pseudoconcavity) assumption.

Theorem 4.7.1. *Let f be a function defined on a convex set $S \subseteq \Re^n$ and let x_0 be a boundary point of S.*
(i) If f is pseudoconvex, then x_0 is a global minimum point for f if and only if the following inequality holds:

$$\nabla f(x_0)^T (x - x_0) \geq 0, \ \forall x \in S \qquad (4.7)$$

(ii) If f is pseudoconcave, then x_0 is a global maximum point for f if and only if the following inequality holds:

$$\nabla f(x_0)^T (x - x_0) \leq 0, \ \forall x \in S \qquad (4.8)$$

Proof. (i) Assume that x_0 is a minimum point for f and suppose, by contradiction, the existence of $x \in S$ such that $\nabla f(x_0)^T (x - x_0) < 0$. Taking into account the convexity of S, $d = x - x_0$ is a decreasing feasible direction and this contradicts the optimality of x_0. With respect to the converse statement,

assume the existence of x^* such that $f(x^*) < f(x_0)$. From the pseudoconvexity of f we have $\nabla f(x_0)^T(x^* - x_0) < 0$ and this contradicts (4.7).
(ii) It is sufficient to note that $-f$ is pseudoconvex. $\qquad\square$

With respect to an optimization problem which has a pseudolinear objective function defined on a polyhedral set S, we have the useful property that when the maximum and/or the minimum value exist, they are attained at a vertex of S (see Corollary 4.6.2).
By denoting with $d^1, ..., d^k$ the edges starting from a vertex $x_0 \in S$, the necessary and sufficient optimality conditions stated in Theorem 4.7.1 may be specified by means of the following theorem.

Theorem 4.7.2. Let f be a pseudolinear function defined on a polyhedral set $S \subseteq \Re^n$. Then:
(i) A vertex $x_0 \in S$ is a minimum point for f if and only if $\nabla f(x_0)^T d^i \geq 0$, $i = 1, \ldots, k$.
(ii) A vertex $x_0 \in S$ is a maximum point for f if and only if $\nabla f(x_0)^T d^i \leq 0$, $i = 1, \ldots, k$.

When S is a polyhedral compact set, the previous results may be extended to a quasilinear function.

Theorem 4.7.3. Let f be a quasilinear function defined on a polyhedral compact set $S \subseteq \Re^n$. Then:
(i) A vertex $x_0 \in S$ is a minimum point for f if and only if $\nabla f(x_0)^T d^i \geq 0$, $i = 1, \ldots, k$.
(ii) A vertex $x_0 \in S$ is a maximum point for f if and only if $\nabla f(x_0)^T d^i \leq 0$, $i = 1, \ldots, k$.

The optimality conditions stated in the previous theorems have suggested some simplex-like procedures for optimization problems having pseudolinear function as objective. These problems include linear programs and linear fractional programs which arise in many practical applications. Some sequential methods will be presented in Chapter 8.

4.8 Economic Applications

Assumptions of generalized convexity/concavity are found in several branches of Economics. For instance, a fundamental result in game theory regarding the existence of a Nash equilibrium involves generalized concavity:
"a game in strategic form has at least one Nash equilibrium if the strategy set is a compact and convex subset of an Euclidean space and if any payoff function is continuous and quasiconcave" (see for instance [99], [113], [115]).
Another useful application is the following generalization of the well-known Von Neumann min-max theorem:
"if $X \subset \Re^n$, $Y \subset \Re^n$ are two compact and convex sets, $F : X \times Y \to \Re$ is

a function which is upper semicontinuous and quasiconcave with respect to the first argument and lower semicontinuous and quasiconvex with respect to the second one, then F has a saddle point (x_0, y_0), i.e., $\min\limits_{y \in Y} \max\limits_{x \in X} F(x, y) = \max\limits_{x \in X} \min\limits_{y \in Y} F(x, y) = F(x_0, y_0)$" (see for instance [24]).

Applications of the Theory of Measurement and the Theory of Aggregation in Economics are found in [100] and characterizations of merely quasiconcave trader's utility functions are given in [278].

In this section we shall focus on some parametric constrained problems which are of great interest in Utility Theory and in the Theory of Firm.

In such problems, the optimal attainable value of the objective function depends, for a fixed vector of parameters (exogeneous variables), on the values of the choice variables. Consequently, the solutions' values and the optimal value of the objective function become functions of the parameters. A central part of the economic analysis is to show the properties of these functions.

In what follows we shall utilize two useful results regarding the following optimization problems.

$$P(\alpha): \ \max f(x), \ x \in S = \{x \in X \subseteq \Re^n : g(x, \alpha) \leq 0\}, \ \alpha \in A \subseteq \Re^s$$

$$P(\beta): \ \min f(x, \beta), \ x \in S = \{x \in X \subseteq \Re^n : g(x) \leq 0\}, \ \beta \in B \subseteq \Re^s$$

where X, A, B are open convex sets, and f, g are functions defined on X.

For the sake of simplicity, we shall assume the existence of optimal solutions for problem $P(\alpha)$ and $P(\beta)$ for every fixed α, β. By setting $z(\alpha), \psi(\beta)$ the optimal value of $P(\alpha)$ and $P(\beta)$, respectively, we have the following theorems.

Theorem 4.8.1. *Consider problem $P(\alpha)$. If $g(x, \alpha)$ is concave in the parameter α, then $z(\alpha)$ is quasiconvex.*

Proof. Let $\alpha_1, \alpha_2 \in A$ and let x_λ, $\lambda \in [0, 1]$ be an optimal solution of problem $P(\lambda\alpha_1 + (1 - \lambda)\alpha_2)$. By means of the concavity of g with respect to α, we have $0 \geq g(x_\lambda, \lambda\alpha_1 + (1-\lambda)\alpha_2) \geq \lambda g(x_\lambda, \alpha_1) + (1-\lambda)g(x_\lambda, \alpha_2)$. Consequently, since both λ and $1 - \lambda$ are non-negative, at least one of $g(x_\lambda, \alpha_1)$ and $g(x_\lambda, \alpha_2)$ is non-positive. Assume, without loss of generality, that $g(x_\lambda, \alpha_1) \leq 0$. Then, x_λ is feasible for problem $P(\alpha_1)$ so that $z(\alpha_1) \geq f(x_\lambda)$. It follows that, for every $\lambda \in [0, 1]$, $\max\{z(\alpha_1), z(\alpha_2)\} \geq z(\alpha_1) \geq f(x_\lambda) = z(\lambda\alpha_1 + (1 - \lambda)\alpha_2)$ and the thesis is achieved. $\qquad\square$

Theorem 4.8.2. *Consider problem $P(\beta)$. If $f(x, \beta)$ is concave in the parameter β, then $\psi(\beta)$ is concave.*

Proof. Let $\beta_1, \beta_2 \in B$ and let x_λ, $\lambda \in [0, 1]$ be an optimal solution of problem $P(\lambda\beta_1 + (1-\lambda)\beta_2)$. Since $g(x_\lambda) \leq 0$, x_λ is feasible for both problems $P(\beta_1)$ and $P(\beta_2)$ and this implies $f(x_\lambda, \beta_1) \geq \psi(\beta_1)$ and $f(x_\lambda, \beta_2) \geq \psi(\beta_2)$. By means of the concavity of f with respect to β, we have $f(x_\lambda, \lambda\beta_1 + (1 - \lambda)\beta_2) = \psi(\lambda\beta_1 + (1 - \lambda)\beta_2) \geq \lambda f(x_\lambda, \beta_1) + (1 - \lambda)f(x_\lambda, \beta_2) \geq \lambda\psi(\beta_1) + (1 - \lambda)\psi(\beta_2)$ and the thesis is achieved. $\qquad\square$

4.8.1 The Utility Maximization Problem

In this subsection we shall discuss the role of convexity/concavity and generalized convexity/concavity in consumer theory.

A consumer is an economic entity who gains satisfaction from the consumption of commodities and who uses the resources available (income) to buy commodities.

The consumer's problem consists of choosing the consumption bundle in order to maximize the satisfaction gained from its consumption, subject to the constraint that the total cost is not greater than the consumer's income.

To formalize this constraint let $p = (p_1, ..., p_n)^T$ the price vector, where $p_i > 0$ is the price of one unit of commodity i and let $x = (x_1, ..., x_n)^T$ the consumption bundle, where $x_i \geq 0$ is the amount of commodity i, $i = 1, ..., n$. If the total cost $p^T x$ is not greater than the consumer's income m, we must have the constraint $p^T x \leq m$ which is referred to as the *consumer's budget constraint*. The set of all feasible consumption bundles $S = \{x \in \Re^n_+ : p^T x \leq m\}$, $m > 0$, is called the *budget set*.

The consumer's behaviour is summarized in a preference relation which is described by means of a utility function $U : \Re^n_+ \to \Re_+$ which assigns a nonnegative numerical value to each consumption bundle $x \in \Re^n_+$.

$U(x) > U(y)$ means that x is preferred to y;

$U(x) = U(y)$ means that x is indifferent to y;

$U(x) \geq U(y)$ means that x is preferred or indifferent to y.

The consumer's problem may now be stated as the following utility maximization problem:

$$P_{UM} : \quad \max \; U(x), \; x \in S = \{x \in \Re^n_+ : p^T x \leq m\}, \; p > 0, \; m > 0.$$

Regarding the utility function U, we shall consider the following basic assumptions:

A_1 U is continuous on $\Re^n$ which means that the consumer's preferences cannot exhibit "jumps";

A_2 The upper level sets of U are convex or, equivalently, U is a quasiconcave function. The convexity of the upper level sets means that if the consumer is indifferent between x and y, then any weighted average of the bundles x and y cannot be worse than either x or y.

A_3 U is differentiable on an open set containing $\Re^n_+$ and $\nabla U(x) > 0$ for all $x \in int\Re^n_+$. This means that the consumer prefers more goods to fewer goods.

Assumption A_1 implies that the problem P_{UM} has at least one solution for all positive prices and non-negative levels of income since a continuous function on a compact set achieves a maximum (Weierstrass' Theorem). For given p

and m, the set $x(p, m)$ of solutions is known as the *consumer's demand correspondence*; when $x(p, m)$ is single-valued for all (p, m), it is known as the *consumer's demand function*.

Assumption A_2 implies that $x(p, m)$ is a convex set. Furthermore, if U is strictly quasiconvex, then $x(p, m)$ consists of a single element or, equivalently, $x(p, m)$ is a demand function.

Assumption A_3 implies Walras' law: $p^T x = m$, $\forall x \in x(p, m)$. In fact, an optimal solution for P_{UM} cannot be an interior point since the gradient does not vanish in $int\Re_+^n$ and furthermore, $p^T x > 0$ for all $x \in \Re_+^n$, $x \neq 0$.

Let us note that if $x(p, m)$ is a solution for P_{UM} for given (p, m), then it is also a solution for $(\lambda p, \lambda m)$, for any positive scalar λ, since the budget set is unchanged if prices and income are scaled up or down proportionally ($\lambda p^T x = \lambda m \Leftrightarrow p^T x = m$). It follows that $x(\lambda p, \lambda m) = x(p, m)$, $\lambda > 0$, i.e., the demand function is homogeneous of degree zero.

For each $p > 0$ and $m > 0$, the utility value $U(x(p, m))$ of the problem P_{UM} is denoted by $v(p, m)$ and is called the indirect utility function. The basic properties of this function are given in the following theorem (for details see [93]).

Theorem 4.8.3. *The indirect utility function $v(p, m)$ is:*
(i) Homogeneous of degree zero in (p, m);
(ii) Quasiconvex in (p, m);
(iii) Strictly increasing in m and non-increasing in p_j, for any $j = 1, ..., n$.

Proof. (i) This follows by noting that $x(\lambda p, \lambda m) = x(p, m)$ so that $v(\lambda p, \lambda m) = U(x(\lambda p, \lambda m)) = U(x(p, m)) = v(p, m)$.

(ii) This follows from Theorem 4.8.1, taking into account that function $g(p, m) = p^T x - m$ is linear (in particular, concave) in (p, m).

(iii) $m_1 < m_2$ implies that $S_1 = \{x \in \Re_+^n : p^T x \leq m_1\} \subset S_2 = \{x \in \Re_+^n : p^T x \leq m_2\}$ so that $v(p, m_1) \leq v(p, m_2)$. The strict inequality follows from Walras' law. Similarly, if $\bar{p}_j < \hat{p}_j$, the feasible set $\bar{S}_j$ associated with $\bar{p}_j$ contains the feasible set $\hat{S}_j$ associated with $\hat{p}_j$, so that v is non-increasing in p_j. $\square$

Another class of generalized concave functions which is useful in Economics is the class of *strongly pseudoconcave functions* which is a subclass of the one of the strictly pseudoconcave functions.

Definition 4.8.1. *A differentiable function f defined on an open convex set $S \subseteq \Re^n$ is said to be strongly pseudoconcave if it is strictly pseudoconcave and for every $x_0 \in S$ and $v \in \Re^n$ such that $\nabla f(x_0)^T v = 0$, there exist $\epsilon > 0$, $\alpha > 0$ such that $x_0 \pm \epsilon v \in S$ and $\varphi(t) = f(x_0 + tv) \leq \varphi(0) - \frac{1}{2}\alpha t^2$, for $0 \leq t < \epsilon$.*

It can be proven (see [93]) that a sufficient condition for the local continuous differentiability of a consumer's system of demand functions is obtained by assuming that the direct utility function is continuously twice differentiable and strongly pseudoconcave locally.

4.8.2 The Expenditure Minimization Problem

The expenditure minimization problem consists of choosing a consumption bundle in order to minimize the amount that the consumer must spend at prices p to get utility (expressed by a function U) not lower than a fixed level u. Formally, we have:

$$P_e : \ \min \ p^T x, \ x \in S = \{x \in \Re_+^n : U(x) \geq u\}, \ p > 0, \ u > 0.$$

The problem is well-posed since it has at least one solution. In fact, let $\bar{x}$ be a feasible solution and let $S_1 = \{x \in \Re_+^n : \ p^T x \leq p^T \bar{x}\}$. Obviously, problem P_e is equivalent to problem $\min p^T x$, $x \in \bar{S}$, where $\bar{S} = S \cap S_1$. Since $\bar{S}$ is compact (S_1 is compact and S is closed from the continuity of U), from Weierstrass' Theorem the minimum is achieved.

Property A_3 of the utility function implies that every optimal solution is binding to constraint $U(x) \geq u$, while property A_2 implies that the set of optimal solutions is convex and it consists of one single element if U is strictly quasiconvex.

For each $p > 0$ and $u > 0$, the optimal value of problem P_e is denoted by $e(p, u)$ and is called the consumer's expenditure function. The basic properties of this function are given in the following theorem (for details see [93]).

Theorem 4.8.4. *The expenditure function $e(p, u)$ is:*
(i) Homogeneous of degree one in p;
(ii) Concave in p;
(iii) Strictly increasing in u and non-increasing in p_j, for any $j = 1, ..., n$;
(iv) Convex in u if U is concave.

Proof. (i) For every $\lambda > 0$ we have $e(\lambda p, u) = \min_{x \in S}(\lambda p^T)x = \lambda \min_{x \in S} p^T x = \lambda e(p, u)$.

(ii) This follows from Theorem 4.8.2, taking into account that the function $f(x, p) = p^T x$ is linear (in particular, concave) in p.

(iii) The proof is analogous to the one given in (iii) of Theorem 4.8.3.

(iv) Let $e(p, u_1) = p^T x_1$ and $e(p, u_2) = p^T x_2$. Since $U(x_1) \geq u_1$, $U(x_2) \geq u_2$, from the concavity of U we have $U(\lambda x_1 + (1-\lambda)x_2) \geq \lambda U(x_1) + (1-\lambda)U(x_2) \geq \lambda u_1 + (1-\lambda)u_2$. Consequently, $\lambda x_1 + (1-\lambda)x_2$ is feasible for problem P_e with $u = \lambda u_1 + (1-\lambda)u_2$. It follows that $e(p, \lambda u_1 + (1-\lambda)u_2) \geq p^T(\lambda x_1 + (1-\lambda)x_2) = \lambda p^T x_1 + (1 - \lambda)p^T x_2 = \lambda e(p, u_1) + (1 - \lambda)e(p, u_2)$. $\square$

4.8.3 The Profit Maximization Problem and the Cost Minimization Problem

A firm is an economic entity which transforms inputs of goods into outputs of goods by some production process. We assume that only one output y

is produced by using n inputs $x_1, ..., x_n$ and that the producer's technology can be summarized by a production function $F(x)$, where $F(x)$ is the maximal amount of output that can be produced from the input vector $x = (x_1, ..., x_n)^T$.

A standard assumption in Economics is the convexity of the firm's *production set* $Y = \{(x, y) \in \Re_+^{n+1} : F(x) \geq y\}$, a set which represents the production plans that are technologically feasible for the firm. This assumption is equivalent to requiring that U is quasiconcave.

The producer's profit maximization problem consists of choosing the input levels x that the Firm buys and uses and the output level y that it produces and sells in order to maximize its profit, given prices p_0 of output and $p_1, ..., p_n$ of inputs.

Taking into account that the profit of the firm is $p_0 y - p^T x$, the profit maximization problem may be formulated as follows:

$$\Pi : \max(p_0 y - p^T x), \ (x, y) \in Y = \{(x, y) \in \Re_+^{n+1} : F(x) \geq y\}, \ p_0 > 0, \ p > 0.$$

Assume that problem Π is well-defined for given (p_0, p), and let $\pi(p_0, p)$ be its maximum value. The function $\pi(p_0, p)$ is called the profit function. The basic properties of this function are given in the following theorem.

Theorem 4.8.5. *The profit function $\pi(p_0, p)$ is:*
(i) Homogeneous of degree one;
(ii) Convex;
(iii) Non-decreasing in p_0 and non-increasing in p_j, for any $j = 1, ..., n$.

Proof. (i) For every $\lambda > 0$ we have

$$\pi(\lambda p_0, \lambda p) = \max_{(x,y) \in Y} \lambda(p_0 y - p^T x) = \lambda \max_{(x,y) \in Y} (p_0 y - p^T x) = \lambda \pi(p_0, p).$$

(ii) This follows from Theorem 4.8.2, taking into account that the function $-p_0 y + p^T x$ is linear (in particular concave) in (p_0, p) and that $\pi(p_0, p) = \max_{(x,y) \in Y} (p_0 y - p^T x) = - \min_{(x,y) \in Y} (-p_0 y + p^T x)$.

(iii) The proof is obvious. $\qquad\square$

The cost minimization problem consists of choosing a vector input x in order to minimize the amount that the producer must spend at prices p to get a pre-assigned output level y_0. Formally, we have:

$$C : \min \ p^T x, \ x \in S = \{x \in \Re_+^n : F(x) \geq y\}, \ p > 0, \ y > 0.$$

From a mathematical point of view, the cost minimization problem and the expenditure minimization problem are the same problem. It follows that C is a well-posed problem and every optimal solution is binding to the constraint $F(x) \geq y$. The value of the problem C for fixed (p, y) is denoted $C(p, y)$ and it is called the cost function for which we have the following basic properties.

Theorem 4.8.6. *The cost function $C(p, y)$ is:*
(i) Homogeneous of degree one in p and non-decreasing in y;
(ii) Concave in p;
(iii) Convex in y if F is concave.

4.9 Invex Functions

We shall conclude this chapter by presenting the main properties of a new class of generalized convex functions, the so-called invex functions, introduced by Hanson [135] with the aim of extending the validity of the sufficiency of the Karush–Kuhn–Tucker conditions. The term invex was created by Craven [75] and it stands for invariant convex.
Since the papers by Hanson and Craven, a great number of contributions related to invex functions and their generalizations, especially with regard to optimization problems, have been made (see for instance [21, 76, 77, 120, 137, 138, 146, 206, 221, 231, 235]; further references may be found at the end of the book).

Definition 4.9.1. *The differentiable function f defined on an open set $X \subseteq \Re^n$ is invex if there exists a vector function $\eta(x, y)$ defined on $X \times X$ such that*

$$f(x) - f(y) \geq \eta^T(x, y)\nabla f(y), \quad \forall x, y \in X \qquad (4.9)$$

Obviously, a differentiable convex function (on an open convex set X) is also invex (it is sufficient to choose $\eta(x, y) = (x - y)$).
A meaningful property characterizing invex functions is stated in the following theorem.

Theorem 4.9.1. *A function is invex (with respect to some η) if and only if every stationary point is a global minimum point.*

Proof. Let f be invex with respect to some $\eta(x, y)$. If x_0 is a stationary point for f, from (4.9), we have $f(x) - f(x_0) \geq \eta^T(x, x_0)\nabla f(x_0) = 0, \forall x \in X$, so that x_0 is a global minimum point. Now we shall prove that (4.9) holds for the function $\eta(x, y)$ defined as

$$\eta(x, y) = \begin{cases} 0 & if \ \nabla f(y) = 0 \\ \frac{(f(x) - f(y))\nabla f(y)}{\|\nabla f(y)\|^2} & if \ \nabla f(y) \neq 0 \end{cases}$$

If y is a stationary point and also a global minimum for f, we have $f(x) - f(y) \geq 0 = \eta^T(x, y)\nabla f(y)$; otherwise, $\eta^T(x, y)\nabla f(y) = f(x) - f(y)$, so that (4.9) holds. $\qquad \square$

It immediately follows from Theorem 4.9.1 that every function without stationary points is invex.

Moreover, since a stationary point is a global minimum point for a pseudo-convex function, the class of pseudoconvex functions is contained in the class of invex functions. Instead, there is not any inclusion relationships between the other classes of generalized convex functions and the class of invex functions; in fact, the function $f(x) = x^3$ is strictly quasiconvex but not invex, since $x = 0$ is a stationary point but it is not a minimum point; the following example shows that there exist invex functions which are not quasiconvex.

Example 4.9.1. Consider the function $f(x, y) = x^2 y^2$ on $\Re^2$.
All the stationary points of f, given by $(x, 0), (0, y)$, $x, y \in \Re$, are global minimum points, so that f is invex. On the other hand, by setting $A = (0, -4)$, $B = (3, -1)$, we have $f(A) = 0 < f(B) = 9$ and $(A - B)^T \nabla f(B) = (-3, -3)(6, -18)^T = 36 > 0$, so that f is not quasiconvex.

Some useful properties of generalized convex functions are lost in the invex case. For instance, the previous example shows that the set of all minimum points is not a convex set.
Now, we shall prove the sufficiency of the Karush–Kuhn–Tucker conditions under suitable invex assumptions.

Theorem 4.9.2. *Consider problem P and let x_0 be a feasible point. Suppose that f, $g_i, i \in I(x_0)$, are invex functions with respect to the same $\eta(x, y)$. If there exist $\lambda_i \in \Re$, $i = 1, .., m$, such that*

$$\begin{cases} \nabla f(x_0) + \sum_{i=1}^{m} \lambda_i \nabla g_i(x_0) = 0 \\ \lambda_i \geq 0, \quad i = 1, ..., m \\ \lambda_i g_i(x_0) = 0, \quad i = 1, ..., m \end{cases} \tag{4.10}$$

then x_0 is a global minimum point for P.

Proof. For any feasible point x, we have $f(x) - f(x_0) \geq \eta^T(x, x_0) \nabla f(x_0) = - \sum_{i \in I(x_0)} \lambda_i \eta^T(x, x_0) \nabla g_i(x_0) \geq - \sum_{i \in I(x_0)} \lambda_i (g_i(x) - g_i(x_0)) \geq 0.$
Consequently, x_0 is a global minimum point. □

The invex functions also play a role in stating constraint qualifications. We shall present two conditions. The first one can be viewed as a generalization of Slater's constraint qualification, while, unlike the first one, the second one cannot be embedded in the general separation approach suggested in Sect. 4.3.

Theorem 4.9.3. *Conditions (i) and (ii) are constraint qualifications.*
(i) The functions $g_i(x), i \in I(x_0)$, are invex at x_0 with respect to the same $\eta(x, x_0)$ and there exists $x^* \in S$ such that $g_i(x^*) < 0, i \in I(x_0)$.
(ii) Generalized Karlin constraint qualification:
The functions g_i, $i \in I(x_0)$, are invex with respect to the same η and there exists no vector $p \in \Re^s$, $p \geq 0$, $p \neq 0$, such that $\sum_{i \in I(x_0)} p_i g_i(x) \geq 0$, $\forall x \in X$.

Proof. (i) From Theorem 4.2.2, we must prove that $W \cap (int\Re_- \times \Re_-^m) = \emptyset$. Assume, by contradiction, the existence of a direction $d^* \in \Re^n$ such that $\nabla f(x_0)^T d^* < 0, \nabla g_i(x_0)^T d^* \leq 0, \ i \in I(x_0)$.
Since $g_i(x^*) < g_i(x_0) = 0, \ i \in I(x_0)$, from the invexity of $g_i, \ i \in I(x_0)$, we have $\eta(x^*, x_0)^T \nabla g_i(x_0) < 0, \ i \in I(x_0)$, so that $d = \eta(x^*, x_0)$ is a feasible direction. By setting $\hat{d} = d^* + \frac{1}{n}d$, we have $\nabla g_i(x_0)^T \hat{d} < 0$, and, for a large enough n, $\nabla f(x_0)^T \hat{d} < 0$, so that $\hat{d}$ is a feasible decreasing direction which contradicts the optimality of x_0.
(ii) The optimality of x_0 implies the following Fritz John conditions: there exist $\lambda_0 \geq 0, \ \lambda_i \geq 0, \ (\lambda_0, \lambda_1, ..., \lambda_s) \neq 0, \ i \in I(x_0)$, such that

$$\lambda_0 \nabla f(x_0) + \sum_{i \in I(x_0)} \lambda_i \nabla g_i(x_0) = 0$$

Assume that $\lambda_0 = 0$, i.e., $\sum\limits_{i \in I(x_0)} \lambda_i \nabla g_i(x_0) = 0$. For the invexity assumption, we have $g_i(x) - g_i(x_0) \geq \eta^T(x, x_0) \nabla g_i(x_0), \forall i \in I(x_0)$, so that $\sum\limits_{i \in I(x_0)} \lambda_i g_i(x) \geq$ $\eta^T(x, x_0) \sum\limits_{i \in I(x_0)} \lambda_i \nabla g_i(x_0) = 0$, and this contradicts the generalized Karlin constraint qualification. $\qquad\square$

Remark 4.9.1. Invexity also plays an important role in duality theory, since it is possible to establish duality results involving Wolfe dual or alternative duals by weakening the classical convexity requirements (see for instance [19, 20, 75, 77, 98, 135, 137, 161, 162, 163, 164, 165]).

4.10 Exercises

4.1. Show that Corollary 4.2.1 is equivalent to Farkas' Lemma which follows: Let A be an $m \times n$ matrix and $c \in \Re^n$. Then, exactly one of the following two systems has a solution:
system 1 $Ax \leq 0, \ c^T x > 0$ for some $x \in \Re^n$
system 2 $y^T A = c^T, \ y \geq 0$ for some $y \in \Re^m$
where $Ax \leq 0$ means $Ax \in \Re_-^m \setminus \{0\}$.

4.2. Let A and B be matrices of dimension $m \times n$, $s \times n$, respectively. Prove that exactly one of the following two systems has a solution:
system 1 $Ax < 0, \ Bx \leqq 0$
system 2 $\alpha^T A + \beta^T B = 0, \ \alpha \in \Re_+^m \setminus \{0\}, \ \beta \in \Re_+^s$
where $Ax < 0, \ Bx \leqq 0$ mean $Ax \in int\Re_-^m, \ Bx \in \Re_-^s$, respectively.

4.3. Let A and B be matrices of dimension $m \times n$, $s \times n$, respectively. Prove that exactly one of the following two systems has a solution:
system 1 $Ax < 0, \ Bx \leq 0$

system 2 $\alpha^T A + \beta^T B = 0$, with $\alpha \in \Re^m_+ \backslash \{0\}$, $\beta \in \Re^s_+$ or $\alpha \in \Re^m_+$, $\beta \in int\Re^s_+$ where $Ax < 0$, $Bx \leq 0$ mean $Ax \in int\Re^m_-$, $Bx \in \Re^s_- \backslash \{0\}$, respectively.

4.4. Referring to minimization problem P, consider the following cones:
$C^0 = \{d \in \Re^n : \nabla g_i(x^0)^T d < 0, \ i \in I(x_0)\}$;
$C = \{d \in \Re^n : \nabla g_i(x^0)^T d \leq 0, \ i \in I(x_0)\}$;
$F = \{d \in \Re^n \backslash \{0\} : \exists \ \epsilon > 0 : x_0 + td \in S, \ \forall \ t \in (0, \epsilon)\}$;
$T = \{d \in \Re^n : \exists \{x_n\} \subset S, \ x_n \to x_0, \ \exists \alpha_n \to +\infty : \alpha_n(x_n - x_0) \to d\}$.
Prove, by means of (4.3), that each of the following conditions is a constraint qualification:
(i) $C^0 \neq \emptyset$ (Cottle's constraint qualification);
(ii) $clF = C$ (Zangwill's constraint qualification);
(iii) $T = C$ (Abadie's constraint qualification).

4.5. Consider problem $P^* : \min_{x \in S} f(x)$, where $S = \{x \in X : g_i(x) \leq 0, \ h_j(x) = 0, \ i = 1, ..., m, \ j = 1, .., p\}$. Assume that f is pseudoconvex, g_i, $i = 1, .., m$, are quasiconvex, h_j, $j = 1, .., p$, are pseudolinear and that there exists a feasible point x_0 verifying the following conditions

$$
\begin{cases}
\nabla f(x_0) + \sum_{i=1}^m \lambda_i \nabla g_i(x_0) + \sum_{j=1}^p \mu_j \nabla h_j(x_0) = 0 \\
\lambda_i g_i(x_0) = 0, \ i = 1, .., m \\
\lambda_i \geq 0, \ i = 1, .., m, \ \mu_j \in \Re, \ j = 1, .., p
\end{cases}
$$

Prove that x_0 is a global minimum point for P^*.

4.6. Let S be a compact convex set and let f be a continuous convex function over S. By applying Jensen's Inequality, prove that a global maximum of f is attained at an extreme point of S.

4.7. Consider the utility maximization problem and the expenditure minimization problem corresponding to utility function $U(x) = \prod_{i=1}^n x_i^{\alpha_i}$, $x_i > 0$, $\alpha_i > 0$, $i = 1, .., n$, $\sum_{i=1}^n \alpha_i = 1$. Show that:
(a) the consumer's demand function of commodity i is $x_i(p, m) = m\frac{\alpha_i}{p_i}$;
(b) the consumer's expenditure function is $e(p, u) = u \prod_{i=1}^n (\frac{\alpha_i}{p_i})^{-\alpha_i}$.

4.8. Consider the profit maximization problem corresponding to the Cobb–Douglas production function $F(x) = \prod_{i=1}^n x_i^{\alpha_i}$, $x_i > 0, \alpha_i > 0, i = 1, .., n$, $\sum_{i=1}^n \alpha_i < 1$. Show that:

(a) The output supply function is $y(p_0, p) = p_0^\gamma \prod_{j=1}^{n} (\frac{\alpha_j}{p_j})^{\theta_j}$, where $\gamma =$

$$\frac{\sum\limits_{i=1}^{n} \alpha_i}{1 - \sum\limits_{i=1}^{n} \alpha_i} \text{ and } \theta_j = \frac{\alpha_j}{1 - \sum\limits_{i=1}^{n} \alpha_j}.$$

(b) The demand function for input i is $x_i(p_0, p) = \frac{\alpha_i}{p_i} p_0^\delta \prod_{j=1}^{n} (\frac{\alpha_j}{p_j})^{\theta_j}$, where

$$\delta = \frac{1}{1 - \sum\limits_{i=1}^{n} \alpha_i}.$$

4.9. Consider the following parametric problem

$$P(\alpha): \ \min f(x), \ x \in S = \{x \in X \subseteq \mathfrak{R}^n : g(x) \leq \alpha\}, \ \alpha \in A \subseteq \mathfrak{R}^s$$

where X, A are open convex sets and f, g are functions defined on X. By assuming the existence of optimal solutions for every fixed α, prove that the optimal value function $z(\alpha)$ of $P(\alpha)$ is a convex function.

4.10. Consider the standard linear parametric problem

$$P(\theta): \ \min c^T x, \ x \in S = \{x \in \mathfrak{R}^n : Ax = b + \theta u, x \geq 0\}.$$

Prove that the optimal value function $z(\theta)$ of $P(\theta)$ is a convex function.

4.11 References

Abadie J. M. [1], Arrow K. J., and Enthoven A. C. [7], Arrow K. J., Hurwicz L., and Uzawa H. [6, 8], Avriel M., and Zang I. [9], Avriel M. [10], Avriel M., Diewert W. E., Schaible S., and Ziemba W. T. eds. [12], Avriel M., Diewert W. E., Schaible S., and Zang I. [13], Bazaraa M. S., Sherali H. D., and Shetty C. M. [16, 18], Ben-Israel A., and Mond B. [21], Berge C. [24], Cambini A., and Martein L. [39, 45, 47], Carter M. [61], Chew K. L., and Choo E. U. [62], Craven B. D. [75, 76], Craven B. D., and Glover B. M. [77], Diewert W. E. [93], Eicherberger J. [99], Eichhorn W. [100], Forgo F. [113], Fudenberg D., and Tirole J. [115], Giorgi G. [117], Giorgi G., and Guerraggio A. [120], Giorgi G., and Thielfelder J. [122], Hanson M. A. [135], Hanson M. A., and Mond B. [137], Hanson M. A., and Rueda N. G. [138], Jeyakumar V. [146], Jeyakumar V., and Yang X. Q. [147], Karush W. [159], Kaul R. N., and Kaur S. [161, 162], Kaul R.N., Suneja S. K., and Lalitha C. S. [163], Khanh P. Q. [164, 165], Kuhn H. W., and Tucker A. W. [186], Madden P. [191], Mangasarian O. L. [193], Martin D. H. [206], Martos B. [211], Mas-Colell A., Whinston M. D., and

Green J. R. [212], Mititelu S. [213], Mond B., and Weir T. [215], Weber R. J. [278], Pini R. [221], Preda V. [225], Reiland T. W. [231], Rueda N. G., and Hanson M. A. [235], Whinston M. D., and Green J. R. [212], Takayama A. [274], Zangwill W. I. [285].

5

Generalized Convexity and Generalized Monotonicity

5.1 Introduction

As convexity plays an important role in solving mathematical programming problems, so, too, does monotonicity in solving variational inequality and nonlinear complementarity problems. Pioneering work was done by Cottle, Dantzig, Karamardian, Stampacchia, and many others (see for instance [71, 74, 134, 154, 155]).

Convexity and generalized convexity, which are central properties in many branches of Operational Research, gave rise to gradient maps with certain generalized monotonicity properties which are inherited from the generalized convexity of the underlying function. Subsequently, generalized monotonicity properties have been extended to general maps; various concepts were introduced in the literature (see for instance [87, 157, 158, 286]).

In this chapter we shall present the main concepts of generalized monotonicity and their relationships with the corresponding concepts of generalized convexity. The main references are Chaps. 2 and 9 of the recent Handbook of Generalized Convexity and Generalized Monotonicity [132] together with [49, 253, 254].

Economic Applications in Economics of generalized monotonicity can be found in [27, 148, 149, 151, 152, 184, 287].

5.2 Concepts of Generalized Monotonicity

In this section we shall introduce some classes of generalized monotone maps which are related, as we shall see in the next section, to the classes of generalized convex functions studied in the previous chapters.

Let S be a subset of $\Re^n$ and $F : S \to \Re^n$ be a map. The definitions of a monotone and a strictly monotone map are a natural generalization of the classical notions of non-decreasing and increasing functions of one variable.

Definition 5.2.1.
(i) F is monotone on S, if for all $x_1, x_2 \in S$,

$$(x_2 - x_1)^T (F(x_2) - F(x_1)) \geq 0. \tag{5.1}$$

(ii) F is strictly monotone on S, if for all $x_1, x_2 \in S$, $x_1 \neq x_2$,

$$(x_2 - x_1)^T (F(x_2) - F(x_1)) > 0. \tag{5.2}$$

Obviously, a strictly monotone map is monotone, too, but the converse statement is not true.

As regards the classes of generalized monotone maps that we are going to introduce, let us point out that pseudomonotonicity was introduced by Karamardian in [156], while quasimonotonicity was introduced by Hassouni in [139] and, independently, by Karamardian and Schaible in [157].

Definition 5.2.2. *F is pseudomonotone on S, if for all $x_1, x_2 \in S$,*

$$(x_2 - x_1)^T F(x_1) \geq 0 \Rightarrow (x_2 - x_1)^T F(x_2) \geq 0 \tag{5.3}$$

or, equivalently,

$$(x_2 - x_1)^T F(x_1) > 0 \Rightarrow (x_2 - x_1)^T F(x_2) > 0. \tag{5.4}$$

Definition 5.2.3. *F is quasimonotone on S, if for all $x_1, x_2 \in S$,*

$$(x_2 - x_1)^T F(x_1) > 0 \Rightarrow (x_2 - x_1)^T F(x_2) \geq 0. \tag{5.5}$$

In [130], strict quasimonotonicity and semistrict quasimonotonicity were introduced.

Definition 5.2.4. *Let S be convex.*
(i) F is strictly quasimonotone on S, if F is quasimonotone on S and for all $x_1, x_2 \in S$, there exists $\bar{x} \in ri[x_1, x_2]$ such that

$$(x_2 - x_1)^T F(\bar{x}) \neq 0. \tag{5.6}$$

(ii) F is semistrictly quasimonotone on S, if F is quasimonotone on S and for all $x_1, x_2 \in S$, with $x_1 \neq x_2$, the following implication holds:

$$(x_2 - x_1)^T F(x_1) > 0 \Rightarrow \exists\, \bar{x} \in ri\left[\frac{x_1 + x_2}{2}, x_2\right], \text{ such that } (x_2 - x_1)^T F(\bar{x}) > 0. \tag{5.7}$$

The inclusion relationships between the classes of generalized monotone functions that have been introduced are stated in the following theorem.

Theorem 5.2.1. *Let $S \subseteq \Re^n$ be a convex set and let $F : S \to \Re^n$ be a map.*
(i) If F is monotone on S, then F is pseudomonotone on S;
(ii) If F is pseudomonotone on S, then F is semistrictly quasimonotone on S;
(iii) If F is semistrictly quasimonotone on S, then F is quasimonotone on S;
(iv) If F is strictly monotone on S, then F is strictly quasimonotone on S;
(v) If F is strictly quasimonotone on S, then F is semistrictly quasimonotone on S.

Proof. (i) and (iii) follow directly from the definitions.
(ii) Since pseudomonotonicity implies quasimonotonicity, (5.7) remains to be proven. Let $x_1, x_2 \in S$ be such that $(x_2 - x_1)^T F(x_1) > 0$ and consider the point $\bar{x} = x_1 + \bar{t}(x_2 - x_1)$, $\bar{t} \in (\frac{1}{2}, 1)$. We have $(\bar{x} - x_1)^T F(x_1) = \bar{t}(x_2 - x_1)^T F(x_1) > 0$; the pseudomonotonicity of F implies $(\bar{x} - x_1)^T F(\bar{x}) > 0$ or, equivalently, $(x_2 - x_1)^T F(\bar{x}) > 0$. Consequently, (5.7) holds.
(iv) Since strict monotonicity implies quasimonotonicity, (5.6) remains to be proven. Assume, by contradiction, the existence of $x_1, x_2 \in S$ such that $(x_2 - x_1)^T F(\bar{x}) = 0$ for all $\bar{x} \in ri[x_1, x_2]$. Let $\bar{x}_1 = x_1 + t_1(x_2 - x_1)$, $\bar{x}_2 = x_1 + t_2(x_2 - x_1)$, with $0 < t_1 < t_2 < 1$. From the strict monotonicity we have $(\bar{x}_2 - \bar{x}_1)^T (F(\bar{x}_2) - F(\bar{x}_1)) > 0$, i.e., $(t_2 - t_1)(x_2 - x_1)^T (F(\bar{x}_2) - F(\bar{x}_1)) > 0$. Consequently, $0 = (x_2 - x_1)^T F(\bar{x}_2) > (x_2 - x_1)^T F(\bar{x}_1) = 0$, and this is a contradiction.
(v) Let $x_1, x_2 \in S$ such that $(x_2 - x_1)^T F(x_1) > 0$. The quasimonotonicity of F implies that $(x_2 - x_1)^T F(z) \geq 0$ for all $z \in ri[x_1, x_2]$. By applying the strict quasimonotonicity to points $\frac{x_1 + x_2}{2}$ and x_2, the thesis is achieved. □

The diagram of Fig. 5.1 summarizes the inclusion relationships between the various classes of monotone and generalized monotone maps.

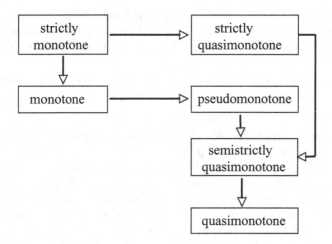

Fig. 5.1. Relationships between various types of monotonicity

All inclusions are proper as is shown in the following example.

Example 5.2.1.

(a) The map $F(x) = \begin{cases} 0, & 0 \le x \le 1 \\ x - 1, & 1 < x \le 2 \end{cases}$ is semistrictly quasimonotone and pseudomonotone but not strictly quasimonotone.

(b) The map $F(x) = \begin{cases} -x + 1, & 0 \le x \le 1 \\ 0, & 1 < x \le 2 \end{cases}$ is quasimonotone but not semistrictly quasimonotone.

(c) The map $F(x) = -|x|$ is strictly quasimonotone but not pseudomonotone.

(d) The map $F(x) = \begin{cases} 0, & 0 \le x < 1 \\ -1 + x, & 1 \le x < 2 \\ -x + 3, & 2 \le x \le 3 \end{cases}$ is semistrictly quasimonotone but not pseudomonotone.

An important case where quasimonotonicity reduces to pseudomonotonicity is stated in the following theorem.

Theorem 5.2.2. *Let $S \subseteq \Re^n$ be an open convex set and let $F : S \to \Re^n$ be a continuous map such that $F(x) \ne 0$ for all $x \in S$. Then, F is pseudomonotone on S if and only if F is quasimonotone on S.*

Proof. Since pseudomonotonicity implies quasimonotonicity, the converse statement remains to be proven. Assume, by contradiction, the existence of $x_1, x_2 \in S$ such that $(x_2 - x_1)^T F(x_1) \ge 0$ and $(x_2 - x_1)^T F(x_2) < 0$. From the quasimonotonicity assumption, we necessarily have $(x_2 - x_1)^T F(x_1) = 0$. Since $F(x_1) \ne 0$, there exists $u \in \Re^n$ such that $u^T F(x_1) > 0$. From the continuity of F and the continuity of the scalar product, there exists $\epsilon > 0$ such that $(x_2 + \epsilon u - x_1)^T F(x_2 + \epsilon u) < 0$. Since F is quasimonotone, we have $(x_2 + \epsilon u - x_1)^T F(x_1) \le 0$, i.e., $\epsilon u^T F(x_1) \le 0$, and this is a contradiction. $\square$

Let us note that a quasimonotone map is not necessarily continuous. Under a continuity assumption, we have the useful property that quasimonotonicity on an open convex set S is preserved on the closure of S, as is shown in the following theorem.

Theorem 5.2.3. *Let $S \subseteq \Re^n$ be a convex set with a nonempty interior, and let $F : clS \to \Re^n$ be a continuous map. If F is quasimonotone on $intS$, then F is quasimonotone on clS.*

Proof. We must prove that if $x, y \in clS$ are such that $(y - x)^T F(x) > 0$, then $(y - x)^T F(y) \ge 0$. If $x, y \in intS$, this is true by assumption. Let $\{x_n\} \subset intS$, $\{y_n\} \subset intS$, be sequences converging to x and y, respectively. For a large enough n, it results that $(y_n - x_n)^T F(x_n) > 0$ so that, for the quasimonotonicity of F on $intS$, we have $(y_n - x_n)^T F(y_n) \ge 0$. Consequently, the continuity of F and the continuity of the scalar product imply
$$\lim_{n \to +\infty} (y_n - x_n)^T F(y_n) = (y - x)^T F(y) \ge 0.$$
The proof is complete. $\square$

5.2.1 Differentiable Generalized Monotone Maps

The defining inequalities of various kinds of generalized monotonicity are, in general, hard to verify, so many studies have been devoted to deriving some characterization results.

Some necessary and/or sufficient conditions for quasimonotone and pseudomonotone differentiable maps are stated below. All proofs are omitted and can be found in [89].

Theorem 5.2.4. *Let $S \subseteq \Re^n$ be a convex set and let F be a differentiable map on S. If F is quasimonotone on S, then*

$$x \in intS, \ v \in \Re^n, \ v^T F(x) = 0 \Rightarrow v^T J_F(x)v \geq 0 \qquad (5.8)$$

where $J_F(x)$ denotes the Jacobian matrix of F evaluated at x.

Theorem 5.2.5. *Let $S \subseteq \Re^n$ be an open convex set and let F be a continuously differentiable map on S. Then:*
(i) F is quasimonotone on S if and only if (5.8) and (5.9) hold

$$\left. \begin{array}{c} x, \ x - v \in S \\ F(x) = 0, \ J_F(x)v = 0 \\ v^T F(x - v) > 0 \end{array} \right\} \Rightarrow \left\{ \begin{array}{c} \forall \bar{t} > 0, \ \exists t \in (0, \bar{t}] \text{ so that} \\ v^T F(x + tv) \geq 0 \end{array} \right. \qquad (5.9)$$

(ii) F is pseudomonotone on S if and only if (5.8) and (5.10) hold

$$\left. \begin{array}{c} x \in S, \ F(x) = 0, \\ J_F(x)v = 0 \end{array} \right\} \Rightarrow \left\{ \begin{array}{c} \forall \bar{t} > 0, \ \exists t \in (0, \bar{t}] \text{ so that} \\ v^T F(x + tv) \geq 0 \end{array} \right. \qquad (5.10)$$

5.3 Generalized Monotonicity of Maps of One Variable

The following theorem characterizes the generalized monotonicity of maps of one variable.

Theorem 5.3.1. *A function $f : \Re \to \Re$ is:*
(i) Monotone if and only if it is non-decreasing;
(ii) Strictly monotone if and only if it is increasing;
(iii) Pseudomonotone if and only if there exist disjoint consecutive intervals (possibly empty) I_1, I_2 and I_3 such that $I_1 \cup I_2 \cup I_3 = \Re$ and f is negative on I_1, zero on I_2 and positive on I_3;
(iv) Quasimonotone if and only if there exist disjoint intervals I_1 and I_2, one of which may be empty, such that $I_1 \cup I_2 = \Re$ and f is non-positive on I_1 and non-negative on I_2;

(v) Strictly quasimonotone if and only if there exist disjoint intervals I_1 and I_2 (one of which may be empty) with $I_1 \cup I_2 = \Re$, such that f is non-positive on I_1, non-negative on I_2 and there does not exist an open interval I in which $f(x) = 0$ for all $x \in I$;

(vi) Semistrictly quasimonotone if and only if there exist disjoint intervals I_1 and I_2 (one of which may be empty) with $I_1 \cup I_2 = \Re$, such that f is non-positive on I_1 and non-negative on I_2; furthermore, if $f(x) > 0$ $(f(x) < 0)$, then there does not exist an open interval $I \subset (x, +\infty)$ $(I \subset (-\infty, x))$ such that $f(z) = 0$ for all $z \in I$.

Proof. (i) and (ii) follow directly from the definitions.

(iii) Let $I_1 = \{x \in \Re : f(x) < 0\}$, $I_2 = \{x \in \Re : f(x) = 0\}$, and $I_3 = \{x \in \Re : f(x) > 0\}$. The pseudomonotonicity of f implies that if there exists $x_1 \in I_1$ $(x_3 \in I_3)$, then $y \in I_1$ $(y \in I_3)$ for all $y < x_1$ $(y > x_3)$. It follows that I_1, I_3, if nonempty, are intervals such that $inf I_1 = -\infty$, $sup I_3 = +\infty$. Consequently, $I_2 = \Re \backslash (I_1 \cup I_3)$, if nonempty, is an interval, too, of end points l_1, l_2 with $l_1 = sup I_1$ or $l_1 = -\infty$ if $I_1 = \emptyset$, and $l_2 = inf I_3$ or $l_2 = +\infty$ if $I_3 = \emptyset$.

As regards the converse statement, it is sufficient to note that the inequality $f(x)(y - x) > 0$ implies that $x, y \in I_3$ or $x, y \in I_1$.

(iv) This follows similarly to (iii).

(v) This follows directly from the definition, taking into account (iv).

(vi) Assume that f is semistrictly quasimonotone and let $x \in \Re$ such that $f(x) > 0$. By contradiction, let $I = [a, b] \subset (x, +\infty)$ such that $f(z) = 0$ for all $z \in I$. Set $\bar{a} = inf\{z : f(y) = 0, \forall y \in [z, b]\}$. Obviously, $x < \bar{a} \le a$ and, for a small enough $\epsilon > 0$, there exists $\bar{x} \in [\bar{a} - \epsilon, \bar{a}]$, such that $f(\bar{x}) > 0$. By applying (5.7) with $x_1 = \bar{x}$ and $x_2 = b$ we get a contradiction.

Conversely, the existence of I_1, I_2 which verify the assumptions, implies the quasimonotonicity of f so that (5.7) remains to be proven. If not, there exist $x_1, x_2 \in \Re$ such that $f(x_1)(x_2 - x_1) > 0$ and $f(\bar{x}) = 0$ for all $\bar{x} \in ri[\frac{x_1 + x_2}{2}, x_2]$, and this is a contradiction. □

The results given in the previous theorem allow us to recognize the generalized monotonicity of a function of one variable by means of its graph. In Fig. 5.2(a) a quasimonotone function is depicted which is not semistrictly, strictly quasimonotone, or pseudomonotone; Fig. 5.2(b) depicts the graph of a function which is strictly quasimonotone but neither strictly monotone, nor pseudomonotone.

As happens for generalized convex functions, characterizations of generalized monotonicity may be derived from the characterizations for maps of one variable. More precisely, let $S \subseteq \Re^n$ be a convex set and let F be a map defined on S. For all $z \in S$ and $u \in \Re^n$, consider the set:

$$I_{z,u} = \{t \in \Re : z + tu \in S\}$$

and the map of one variable

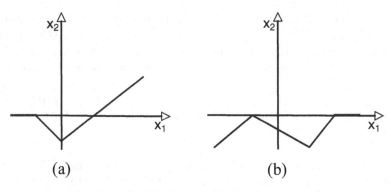

Fig. 5.2. Graphs of generalized monotone maps

$$F_{z,u}(t) = u^T F(z + tu).$$

The following theorem holds.

Theorem 5.3.2. *F is monotone, strictly monotone, pseudomonotone, quasi-monotone, strictly quasimonotone, and semistrictly quasimonotone on S if and only if, for all $z \in S$ and $u \in \Re^n$, $F_{z,u}$ is monotone, strictly monotone, pseudomonotone, quasimonotone, strictly quasimonotone, and semistrictly quasimonotone on $I_{z,u}$, respectively.*

Proof. It is sufficient to note that, by setting $x_1 = z + t_1 u$, $x_2 = z + t_2 u$, we have $(t_2 - t_1)F_{z,u}(t_1) = (x_2 - x_1)^T F(x_1)$ and $(t_2 - t_1)F_{z,u}(t_2) = (x_2 - x_1)^T F(x_2)$, so that to any assumption of generalized monotonicity on F there corresponds the same kind of generalized monotonicity on $F_{z,u}$.
For the converse statement, set $z = x_1$, $u = x_2 - x_1$.   □

5.4 Generalized Monotonicity of Affine Maps

One of the most important problems in studying generalized monotonicity is how to find characterizations of generalized monotone maps. In Sect. 5.2.1 we presented some results related to the differentiable case; now we shall present the main results related to the affine case.
The following theorem points out that, for an affine map, quasimonotonicity reduces to pseudomonotonicity even if the map assumes zero value in some points.

Theorem 5.4.1. *Let $S \subseteq \Re^n$ be an open convex set and let $F(x) = Mx + q$. Then, F is pseudomonotone on S if and only if F is quasimonotone on S.*

Proof. Since pseudomonotonicity implies quasimonotonicity, the converse statement remains to be proven. If $F(x) \neq 0$ for all $x \in S$, the thesis follows from Theorem 5.2.2. If $Mx + q = 0$, assume, by contradiction, the existence of $x, y \in S$ such that $(y - x)^T(Mx + q) = 0$ and $(y - x)^T(My + q) < 0$. Consider $x_1 = x + t_1(y - x)$ where $t_1 < 0$ is such that $x_1 \in S$. Since $y - x = \frac{y - x_1}{1 - t_1}$, we have $(y - x_1)^T F(y) < 0$ so that, from the quasimonotonicity of F, $(y - x_1)^T F(x_1) \leq 0$. On the other hand, $(y - x_1)^T F(x_1) = (1 - t_1)(y - x)^T(Mx + q + t_1 M(y - x)) = (1 - t_1)(y - x)^T((1 - t_1)(Mx + q) + t_1(My + q)) = t_1(1 - t_1)(y - x)^T(My + q) \leq 0$, i.e., $(y - x)^T(My + q) \geq 0$, and this is a contradiction. $\square$

For an affine map, the characterizations given in Theorem 5.2.5 reduce to condition (5.8). More precisely, we have the following theorem for which, for the sake of completeness, we shall propose a direct proof.

Theorem 5.4.2. *Let $S \subseteq \Re^n$ be an open convex set and let $F(x) = Mx + q$. Then, F is pseudomonotone on S if and only if*

$$x \in S, \ v \in \Re^n, \ v^T(Mx + q) = 0 \Rightarrow v^T Mv \geq 0. \tag{5.11}$$

Proof. Assume that F is pseudomonotone on S and let $x \in S$, $v \in \Re^n$, such that $v^T(Mx + q) = 0$. Consider the point $y = x + tv$ where $t \in \Re\backslash\{0\}$ is such that $y \in S$. We have $(y - x)^T(Mx + q) = 0$, so that from the pseudomonotonicity of the map, $(y - x)^T(My + q) \geq 0$, i.e., $tv^T(Mx + q + tMv) = t^2 v^T Mv \geq 0$ and (5.11) holds.

Conversely, assume by contradiction that F is not pseudomonotone, i.e., assume the existence of $x, y \in S$ such that $(y - x)^T(Mx + q) \geq 0$ and $(y - x)^T(My + q) < 0$. From the continuity of the affine map, there exists $\bar{x} = x + \bar{t}(y - x)$, $\bar{t} \in [0, 1)$, such that $(y - x)^T(M\bar{x} + q) = 0$. We have

$$(y - x)^T M(y - x) = (y - x)^T M\left(\frac{y - \bar{x}}{1 - \bar{t}}\right) = \frac{1}{1 - \bar{t}}(y - x)^T(My + q - M\bar{x} - q) =$$

$$\frac{1}{1 - \bar{t}}(y - x)^T(My + q) < 0. \text{ By setting } v = y - x, \text{ (5.11) is contradicted. } \square$$

Remark 5.4.1. Theorem 5.4.1 implies that (5.11) is also a characterization of quasimonotonicity of an affine map on an open convex set. Furthermore, note that the proof of pseudomonotonicity given in Theorem 5.4.1 does not require the openess of the domain S, so that (5.11) is a sufficient condition for pseudomonotonicity and quasimonotonicity even if S is not open.

The following example points out that a quasimonotone affine map on a convex set with a nonempty interior does not necessarily verify (5.11).

Example 5.4.1. Consider the map $F(x) = Mx$ on $\Re_+^2$, where $M = \begin{bmatrix} 0 & -1 \\ -2 & 0 \end{bmatrix}$. By setting $x = (x_1, x_2)^T$, $y = (y_1, y_2)^T$, we have that (5.12) implies (5.13),

$$(y - x)^T Mx = -x_2(y_1 - x_1) - 2x_1(y_2 - x_2) > 0 \tag{5.12}$$

$$(y - x)^T M y = -y_2(y_1 - x_1) - 2y_1(y_2 - x_2) \geq 0 \qquad (5.13)$$

This is true, if $y_1 - x_1$ and $y_2 - x_2$ are both negative. If $y_1 - x_1 > 0$ ($y_1 - x_1 \leq 0$) and $y_2 - x_2 \leq 0$ ($y_2 - x_2 > 0$), then $-y_2(y_1 - x_1) \geq -x_2(y_1 - x_1)$ and $-2y_1 > (y_2 - x_2) \geq -2x_1(y_2 - x_2)$, so that (5.12) implies (5.13). Since the case $y_1 - x_1 \geq 0$, $y_2 - x_2 \geq 0$ cannot occur, the map is quasimonotone on $\Re_+^2$. Now, we shall show that (5.11) does not hold at the boundary point $x = (0,0)^T$. In fact, we have $Mx = 0$, so that $v^T M x = 0$ for all $v \in \Re^2$, but $v^T M v = -3v_1 v_2 < 0$ if $v_1 v_2 > 0$.

Finally, note that the map is pseudomonotone on $int\Re_+^2$ but is not pseudomonotone on $\Re_+^2$ since, by choosing $x = (0,0)^T$, $y = (1,1)^T$, we have $(y - x)^T F(x) = 0$, and $(y - x)^T F(y) = -3 < 0$.

Example 5.4.1 shows that (5.11) does not necessarily hold at a boundary point x for which $Mx + q = 0$. In general, we have the following theorem whose proof can be found in [133].

Theorem 5.4.3. *Let $S \subseteq \Re^n$ be a convex set with a nonempty interior, and let $F(x) = Mx + q$ be quasimonotone on S.*
(i) If there exist $x \in S$, $v \in \Re^n$, such that (5.11) does not hold, then x belongs to the boundary of S and $F(x) = 0$.
(ii) If F has no zeros on the boundary of S, then F is pseudomonotone on S.

The link between the generalized monotonicity of $F(x) = Mx + q$ and the intrinsic properties of the matrix M has been studied by several authors (see, for instance, [84, 88]). Results related to the case $S = \Re_+^n$, which is of particular interest because of its relevance to complementarity problems, can be found in [50, 84, 126, 127, 128].

Now, we shall point out how generalized monotonicity is preserved under an affine transformation.

Consider the variable transformation $z = Ax + b$, where A is an $m \times n$ matrix and $b \in \Re^m$. Let $Z \subseteq \Re^m$ be a convex set and $S = \{x \in \Re^n : Ax + b \in Z\}$.

Theorem 5.4.4. *Let $G : Z \to \Re^m$ and let $F(x) = A^T G(Ax + b)$. If G is quasimonotone, pseudomonotone, strictly quasimonotone, and semistrictly quasimonotone on Z, then F is quasimonotone, pseudomonotone, strictly quasimonotone, and semistrictly quasimonotone on S, respectively.*

Proof. It is sufficient to note that, for every x_1, $x_2 \in S$, by setting $z_1 = Ax_1 + b$, $z_2 = Ax_2 + b$, we have $(x_2 - x_1)^T F(x_1) = (A(x_2 - x_1))^T G(Ax_1 + b) = (z_2 - z_1)^T G(z_1)$, $(x_2 - x_1)^T F(x_2) = (z_2 - z_1)^T G(z_2)$, so that to any assumption of generalized monotonicity on G there corresponds the same kind of generalized monotonicity on F. □

5.5 Relationships Between Generalized Monotonicity and Generalized Convexity

In this section we shall point out the connection between the generalized convexity of a function f and the generalized monotonicity of its gradient map ∇f.

We shall begin with the classical result related to convexity and monotonicity.

Theorem 5.5.1. *Let $S \subseteq \Re^n$ be a convex set and let f be a differentiable function on S.*
(i) f is convex on S if and only if ∇f is monotone on S;
(ii) f is strictly convex on S if and only if ∇f is strictly monotone on S.

Proof. (i) Assume that f is convex on S and let $x_1, x_2 \in S$. We have:

$$f(x_2) \geq f(x_1) + (x_2 - x_1)^T \nabla f(x_1) \tag{5.14}$$

$$f(x_1) \geq f(x_2) + (x_1 - x_2)^T \nabla f(x_2) \tag{5.15}$$

By adding (5.14) and (5.15), we obtain $(x_2 - x_1)^T (\nabla f(x_2) - \nabla f(x_1)) \geq 0$, i.e., ∇f is monotone on S.

Conversely, assume, by contradiction, the existence of $x_1, x_2 \in S$ such that $f(x_2) < f(x_1) + (x_2 - x_1)^T \nabla f(x_1)$. From the Mean Value Theorem, there exists $\bar{x} = x_1 + \bar{t}(x_2 - x_1)$, $\bar{t} \in (0, 1)$, such that $f(x_2) = f(x_1) + (x_2 - x_1)^T \nabla f(\bar{x})$, so that we have $(x_2 - x_1)^T \nabla f(\bar{x}) = f(x_2) - f(x_1) < (x_2 - x_1)^T \nabla f(x_1)$, i.e., $(x_2 - x_1)^T (\nabla f(\bar{x}) - \nabla f(x_1)) = \dfrac{1}{\bar{t}}(\bar{x} - x_1)^T (\nabla f(\bar{x}) - \nabla f(x_1)) < 0$, and this contradicts the monotonicity assumption.

Similarly, (ii) can be proven. □

The following lemma is useful in establishing the connection between the pseudoconvexity (quasiconvexity) of a function and the pseudomonotonicity (quasimonotonicity) of its gradient map.

Lemma 5.5.1. *Let $S \subseteq \Re^n$ be a convex set and let f be a differentiable function on S.*
(i) Assume that ∇f is pseudomonotone on S.
If $x_1, x_2 \in S$ are such that $(x_2 - x_1)^T \nabla f(x_1) \geq 0$, then the restriction of f on $[x_1, x_2]$ is non-decreasing.
If $x_1, x_2 \in S$ are such that $(x_2 - x_1)^T \nabla f(x_1) > 0$, then the restriction of f on $[x_1, x_2]$ is increasing.
(ii) Assume that ∇f is quasimonotone on S.
If $x_1, x_2 \in S$ are such that $(x_2 - x_1)^T \nabla f(x_1) > 0$, then the restriction of f on $[x_1, x_2]$ is non-decreasing and $f(x_1) < f(x_2)$.

Proof. (i) Let $\varphi(t) = f(x_1 + t(x_2 - x_1))$, $t \in [0, 1]$, and set $y = x_1 + t(x_2 - x_1)$. If $(x_2 - x_1)^T \nabla f(x_1) \geq 0$, then $(y - x_1)^T \nabla f(x_1) \geq 0$ for all $y \in [x_1, x_2]$, so that the pseudomonotonicity of $\nabla f(x)$ implies that $(y - x_1)^T \nabla f(y) \geq 0$ for all $y \in [x_1, x_2]$. It follows that $(y - x_1)^T \nabla f(y) = t(x_2 - x_1)^T \nabla f(y) = t\varphi'(t) \geq 0$, $\forall t \in [0, 1]$. Consequently, $\varphi(t)$ is non-decreasing on $[0, 1]$.

Similarly, the condition $(x_2 - x_1)^T \nabla f(x_1) > 0$, together with the pseudomonotonicity of $\nabla f(x)$, implies that $\varphi(t)$ is increasing on $[0, 1]$. Consequently, (i) holds.

(ii) The condition $(x_2 - x_1)^T \nabla f(x_1) > 0$, together with the quasimonotonicity of $\nabla f(x)$, implies that $\varphi(t)$ is non-decreasing on $[0, 1]$; furthermore, $\varphi'(0) = (x_2 - x_1)^T \nabla f(x_1) > 0$ implies that $\varphi(t)$ is locally increasing at $t = 0$. Consequently, (ii) holds. □

Theorem 5.5.2. *Let $S \subseteq \Re^n$ be a convex set and let f be a differentiable function on S.*

(i) f is pseudoconvex on S if and only if ∇f is pseudomonotone on S;

(ii) f is quasiconvex on S if and only if ∇f is quasimonotone on S.

Proof. (i) Assume that f is pseudoconvex on S and let x_1, $x_2 \in S$ such that $(x_2 - x_1)^T \nabla f(x_1) \geq 0$. Consequently, $f(x_1) \leq f(x_2)$. Since f is quasiconvex, too, we have $(x_1 - x_2)^T \nabla f(x_2) \leq 0$, i.e., $(x_2 - x_1)^T \nabla f(x_2) \geq 0$, so that $\nabla f(x)$ is pseudomonotone on S.

Conversely, assume by contradiction the existence of $x_1, x_2 \in S$ such that $f(x_1) > f(x_2)$ and $(x_2 - x_1)^T \nabla f(x_1) \geq 0$. This last inequality implies, from (i) of Lemma 5.5.1, that f is non-decreasing on $[x_1, x_2]$, so that $f(x_1) \leq f(x_2)$, and this is a contradiction.

(ii) Assume that f is quasiconvex on S and let x_1, $x_2 \in S$ such that $(x_2 - x_1)^T \nabla f(x_1) > 0$. Consequently, $f(x_1) < f(x_2)$ and $(x_1 - x_2)^T \nabla f(x_2) \leq 0$, i.e., $(x_2 - x_1)^T \nabla f(x_2) \geq 0$, so that $\nabla f(x)$ is quasimonotone on S.

Conversely, assume by contradiction the existence of $x_1, x_2 \in S$ such that $f(x_1) \geq f(x_2)$ and $(x_2 - x_1)^T \nabla f(x_1) > 0$. From (ii) of Lemma 5.5.1, we get a contradiction. □

Theorem 5.5.3. *Let $S \subseteq \Re^n$ be a convex set and let f be a differentiable function on S.*

(i) f is strictly quasiconvex on S if and only if ∇f is strictly quasimonotone on S;

(ii) f is semistrictly quasiconvex on S if and only if ∇f is semistrictly quasimonotone on S.

Proof. (i) If f is strictly quasiconvex on S, then f is quasiconvex, too, so that, from (ii) of Theorem 5.5.2, $\nabla f(x)$ is quasimonotone on S. If (5.6) does not hold, then there exist $x_1, x_2 \in S$ such that $(x_2 - x_1)^T \nabla f(\bar{x}) = 0$ for all $\bar{x} \in [x_1, x_2]$, and this implies the constancy of f on $[x_1, x_2]$, contradicting the strict quasiconvexity of f (see Theorem 2.2.11).

Conversely, if $\nabla f(x)$ is strictly quasimonotone, it is quasimonotone, too, so that f is quasiconvex. Furthermore, condition (5.6) with $F = \nabla f$, implies that f is not constant on each interval $[x_1, x_2] \subset S$, and thus f is strictly quasiconvex.

(ii) If f is semistrictly quasiconvex on S, then f is quasiconvex, too, so that, from (ii) of Theorem 5.5.2, $\nabla f(x)$ is quasimonotone on S. If $\nabla f(x)$ is not semistrictly quasimonotone, then there exist $x_1, x_2 \in S$ such that $(x_2 - x_1)^T \nabla f(x_1) > 0$ and $(x_2 - x_1)^T \nabla f(\bar{x}) \leq 0$ for all $\bar{x} \in ri[\frac{x_1+x_2}{2}, x_2]$. The quasimonotonicity of $\nabla f(x)$ implies that $(x_2 - x_1)^T \nabla f(\bar{x}) = 0$ for all $\bar{x} \in ri[\frac{x_1+x_2}{2}, x_2]$, so that f is constant on $[\frac{x_1+x_2}{2}, x_2]$. On the other hand, from (ii) of Lemma 5.5.1, $f(x_1) < f(x_2)$ and this contradicts the semistrict quasiconvexity of f on $[x_1, x_2]$.

Conversely, the semistrict quasimonotonicity of $\nabla f(x)$ implies the quasiconvexity of f.

Assume now, by contradiction, the existence of $x_1, x_2, \bar{x} \in ri[x_1, x_2] \subset S$ such that $f(x_1) > f(x_2)$ and $f(\bar{x}) = f(x_1)$. Let $\tilde{t} = max\{t \in [0,1] : \varphi(t) = f(x_1 + t(x_2-x_1)) = f(x_1) = \varphi(0)\}$; obviously, $\tilde{t} \in (0,1)$. Let $\epsilon > 0$ be such that $\epsilon < \frac{\tilde{t}}{2}$ and $\tilde{t} + \epsilon < 1$. Then, there exist $\bar{t} \in (\tilde{t}, \tilde{t} + \epsilon)$ such that $\varphi'(\bar{t}) < 0$. By setting $\bar{y} = x_1 + \bar{t}(x_2 - x_1)$, we have $\varphi'(\bar{t}) = (x_2 - x_1)^T \nabla f(\bar{y}) = \frac{1}{\bar{t}}(\bar{y} - x_1)^T \nabla f(\bar{y}) < 0$, i.e., $(x_1 - \bar{y})^T \nabla f(\bar{y}) > 0$. From (5.7), there exists $\hat{x} \in ri[x_1, \frac{x_1+\bar{y}}{2}]$ such that $(x_1 - \bar{y})^T \nabla f(\hat{x}) > 0$. Since $\hat{x} \in ri[x_1, \tilde{x}]$ with $\tilde{x} = x_1 + \tilde{t}(x_2 - x_1)$, we have $(x_1 - \bar{y})^T \nabla f(\hat{x}) = 0$, and this is a contradiction. $\qquad\square$

5.6 The Generalized Charnes–Cooper Transformation

As we have remarked several times, it is often difficult to test the pseudoconvexity of a function as well as the pseudomonotonicity of a map.

In this section we shall introduce a nonlinear variable transformation, a generalized Charnes–Cooper transformation. We shall see that this transformation preserves the pseudomonotonicity of the gradient map of a differentiable function f, or, equivalently, the pseudoconvexity of f. This property will be utilized in the next chapter in order to obtain pseudoconvexity results for particular classes of functions.

Consider the following transformation

$$y = \frac{Ax}{b^T x + b_0} \tag{5.16}$$

defined on the set $\Gamma = \{x \in \Re^n : b^T x + b_0 > 0\}$, where A is a nonsingular matrix of order n, $b \in \Re^n$ and $b_0 \neq 0$.

We refer to (5.16) as the generalized Charnes–Cooper transformation since, when $A = I$, it reduces to the variable transformation originally suggested by Charnes A. and Cooper W. W. in [64] (see also Sect. 7.4).

Theorem 5.6.1. *The inverse of the transformation (5.16) is given by*

$$x = \frac{b_0 A^{-1} y}{1 - b^T A^{-1} y} \tag{5.17}$$

defined on the set $\Gamma^* = \{y \in \Re^n : \frac{b_0}{1 - b^T A^{-1} y} > 0\}$.

Proof. From (5.16), we have $A^{-1}y = \frac{x}{b^T x + b_0}$, $b^T A^{-1} y = \frac{b^T x}{b^T x + b_0} = 1 - \frac{b_0}{b^T x + b_0}$,
so that $\frac{1}{b^T x + b_0} = \frac{1 - b^T A^{-1} y}{b_0}$.
The thesis is achieved, taking into account that $b^T x + b_0 > 0$ implies that $\frac{b_0}{1 - b^T A^{-1} y} > 0$ and that $x = (b^T x + b_0)A^{-1}y$. $\qquad\square$

Let f be a differentiable function defined on an open convex set $S \subseteq \Re^n$ and consider the function $\psi(y)$ obtained by applying the generalized Charnes–Cooper transformation to $f(x)$, hereby, assuming that $S \subseteq \Gamma$.
The following theorem, whose proof can be found in [49], shows that the generalized Charnes–Cooper transformation preserves the pseudomonotonicity of the gradient of a function.

Theorem 5.6.2. *Let f be a differentiable function defined on an open convex set $S \subseteq \Gamma$. Then, $\nabla f(x)$ is pseudomonotone if and only if $\nabla \psi(y)$ is pseudomonotone.*

Taking into account that a function is pseudoconvex if and only if its gradient map is pseudomonotone, we have the following result.

Corollary 5.6.1. *The generalized Charnes–Cooper transformation (5.16) preserves pseudoconvexity of an arbitrary differentiable function f, i.e., $f(x)$ is pseudoconvex if and only if $\psi(y)$ is pseudoconvex.*

Unfortunately, both the generalized Charnes–Cooper transformation and the classic Charnes–Cooper transformation do not preserve, in general, the pseudomonotonicity of a map, as is shown in the following example.

Example 5.6.1. The map $F(x) = x^3$, $x \in \Re$ is pseudomonotone (see (iii) of Theorem 5.3.1). Consider the transformation $y = -\frac{x}{x+1}$ $(x > -1)$. We have $x = -\frac{y}{y+1}$ $(y > -1)$. Then $F(x(y)) = (-\frac{y}{y+1})^3$ is not pseudomonotone for $y > -1$. To see this, let $y_1 = -\frac{1}{2}$, $y_2 = 1$. Then $(y_2 - y_1)F(x(y_1)) = \frac{3}{2} > 0$, but $(y_2 - y_1)F(x(y_2)) = -\frac{3}{16} < 0$.

The following results establish conditions under which the Charnes–Cooper transformation $(A = I)$ preserves the pseudomonotonicity of a map.

Theorem 5.6.3. *If the map $F(y)$ is homogeneous of degree one and pseudomonotone on a cone $C \subseteq \Re^n$, then the transformed map $\Phi(x) = F(\frac{x}{b^T x + b_0})$ is pseudomonotone on $\bar{C} = C \cap \{x \in \Re^n : b^T x + b_0 > 0\}$.*

Proof. We must prove that if $x, z \in \bar{C}$ are such that $(z - x)^T \Phi(x) > 0$, then $(z - x)^T \Phi(z) > 0$.
We have $(z - x)^T \Phi(x) = (z - x)^T F(\frac{x}{b^T x + b_0}) = \frac{1}{b^T x + b_0}(z - x)^T F(x) > 0$, i.e., $(z - x)^T F(x) > 0$. Since $z, x \in C$, taking into account the pseudomonotonicity and the homogeneity of F, we have $(z - x)^T F(z) > 0$, which implies that $(z - x)^T F(\frac{z}{b^T z + b_0}) > 0$. $\qquad\square$

Corollary 5.6.2. *If My is a pseudomonotone map on a cone $C \subseteq \Re^n$, then the map $\frac{Mx}{b^T x + b_0}$ is pseudomonotone on $\bar{C} = C \cap \{x \in \Re^n : b^T x + b_0 > 0\}$.*

Remark 5.6.1. Theorem 5.6.3 cannot be extended to the generalized Charnes–Cooper transformation when $A \neq I$. Consider for instance the map $\Phi(y) = My$, where $M = \begin{bmatrix} 0 & 2 \\ -1 & 0 \end{bmatrix}$, which is pseudomonotone on the interior of $\Re^2_+$. By applying the generalized Charnes–Cooper transformation $y = \frac{Ax}{b^T x + b_0}$, $A = \begin{bmatrix} 3 & 0 \\ 0 & 1 \end{bmatrix}$, $b = \begin{pmatrix} 1 \\ 1 \end{pmatrix}$, $b_0 = 1$, we obtain the map $\Phi(x) = M \frac{Ax}{b^T x + b_0}$ which is not pseudomonotone on $int\Re^2_+$ since for $z^T = (1, 2)$, $x^T = (\frac{1}{3}, 1)$ we have $(z - x)^T M \frac{Ax}{b^T x + b_0} = \frac{1}{7} > 0$, while $(z - x)^T M \frac{Az}{b^T z + b_0} = -\frac{1}{12} < 0$.

Finally, we shall present conditions under which the generalized Charnes–Cooper transformation taken as a map, not as a variable transformation, is pseudomonotone. With this aim in mind, consider, firstly, the following theorem.

Theorem 5.6.4. *Let $G : \Re^n \to \Re^n$, $g : \Re^n \to \Re$, and consider the map $F(x) = \frac{G(x)}{g(x)}$ defined on $H = \{x \in \Re^n : g(x) > 0\}$. If G is pseudomonotone on $S \subseteq \Re^n$, then F is pseudomonotone on $\bar{S} = S \cap H$.*

Proof. We must prove that if x, $z \in \bar{S}$ are such that $(z - x)^T F(x) > 0$, then $(z - x)^T F(z) > 0$. We have $(z - x)^T F(x) = (z - x)^T \frac{G(x)}{g(x)} > 0$, i.e., $(z - x)^T G(x) > 0$. The pseudomonotonicity of G implies that $(z - x)^T G(z) > 0$, i.e., $(z - x)^T \frac{G(z)}{g(z)} = (z - x)^T F(z) > 0$. □

Corollary 5.6.3. *If the affine map Ax is pseudomonotone on $S \subseteq \Re^n$, then the map $y = \frac{Ax}{b^T x + b_0}$ is pseudomonotone on $\bar{S} = S \cap \{x \in \Re^n : b^T x + b_0 > 0\}$.*

The following corollary gives a sufficient condition for the pseudomonotonicity of the generalized Charnes–Cooper transformation.

Corollary 5.6.4. *The map $y = \frac{Ax}{b^T x + b_0}$ is pseudomonotone on the half-space $\Gamma = \{x \in \Re^n : b^T x + b_0 > 0\}$ if the matrix A is positive semidefinite.*

Corollary 5.6.4 shows that the classic Charnes–Cooper transformation ($A = I$) is a pseudomonotone map.

5.7 References

Brighi L., and John R. [27], Cambini A., Dass B. K., and Martein L. [46], Cambini A., Martein L., and Schaible S. [49], Cambini A., and Martein L. [50], Cottle R. W. [70], Crouzeix J.-P., and Ferland J. A. [83], Crouzeix J.-P., and Schaible S. [84], Crouzeix J.-P. [85, 89], Crouzeix J. P., Martinez-Legaz

J. E., and Volle M. [86], Crouzeix J.-P., Marcotte P., and Zhu D. L. [87], Crouzeix J.-P., Hassouni A., Lahlou A., and Schaible S. [88], Dantzig G. B., and Cottle R. W. [90], Eberhard A., Hadjisavvas N., and Dinh The Luc. [97], Gowda M. S. [126, 127, 128], Hadjisavvas N., and Schaible S. [130, 133], Hadjisavvas N., Martinez-Legaz J. E., and Penot J.-P. [131], Hadjisavvas N., Komlósi S., and Schaible S. [132], Hartman P, and Stampacchia G. [134], John R. [148, 149, 150, 151, 152], Karamardian S. [156], Karamardian S., and Schaible S. [157], Karamardian S., Schaible S., and Crouzeix J.-P. [158], Konnov I. V. [183, 184], Konnov I. V., Dinh The Luc, and Rubinov A. M. [185], Marchi A., and Martein L. [197], Pini R., and Schaible S. [222], Schaible S. [254], Xu B., and Zhu D. L. [282], Jao J. C., and Chadli O. [284], Zhu D. L., and Marcotte P. [286], Zhu D. L. [287].

6

Generalized Convexity of Quadratic Functions

6.1 Introduction

Generalized convexity of quadratic functions has been widely studied; the main historical references are Martos [209, 210, 211], Ferland [108], Cottle and Ferland [73], Schaible [236, 243, 242, 248].

In this Chapter we shall put together some results related to generalized convex quadratic functions. After noting that quasiconvexity can differ from convexity only on a proper subset S of $\Re^n$ and that quasiconvexity reduces to pseudoconvexity on an open set, in Sect. 6.3 we shall characterize the maximal domains of quasiconvexity and pseudoconvexity of a quadratic function in the general form suggested by Schaible in [251]. All the results that we are going to develop are obtained by means of a different approach based on the second order characterization of pseudoconvexity given in Corollary 3.4.1 and on the properties established in Sect. 6.2.

The results will be specified in Sect. 6.4 in order to obtain the criteria established by Martos [209, 210, 211] related to generalized convexity over the non-negative orthant $\Re^n_+$.

6.2 Preliminary Results

Since the symmetric matrix Q associated with a quasiconvex (not convex) quadratic form has one and only one negative eigenvalue (see Theorem 3.4.1), in this section we shall establish, for these matrices, some properties which will play a fundamental role in characterizing the generalized convexity of a quadratic function.

To this end we shall introduce the following notations:

- $\lambda_1, ..., \lambda_n$ are the eigenvalues of the $n \times n$ symmetric matrix Q;
- $\{v^1, ..., v^n\}$ is an orthonormal basis of eigenvectors associated with $\lambda_1, ...,$ λ_n. In order to define each of the eigenvectors uniquely, we shall assume

that the first component of any eigenvector is positive (this can be obtained by multiplying it by (-1) if necessary).

- $kerQ$ is the kernel of Q, i.e., $kerQ = \{x \in \Re^n : Qx = 0\}$;
- $rankQ$ is the rank of Q, i.e., the maximum number of linearly independent columns (or rows) of Q;
- $\nu_-(Q)$ is the number of the negative eigenvalues of Q (according to their multiplicity).

Regarding the number of the negative eigenvalues of Q we have the following useful lemma.

Lemma 6.2.1. *Let Q be an $n \times n$ symmetric matrix and assume the existence of two vectors u, w such that*

$$u^T Q u < 0, \quad w^T Q w < 0, \quad u^T Q w = 0.$$

Then, Q has at least two negative eigenvalues.

Proof. Let $u = \sum_{i=1}^{n} \alpha_i v^i$, $w = \sum_{i=1}^{n} \beta_i v^i$, $\alpha_i, \beta_i \in \Re, i = 1, ..., n$. We have

$$u^T Q u = \sum_{i=1}^{n} \alpha_i^2 \lambda_i, \quad w^T Q w = \sum_{i=1}^{n} \beta_i^2 \lambda_i, \quad u^T Q w = \sum_{i=1}^{n} \alpha_i \beta_i \lambda_i.$$

The assumptions imply that $\sum_{i=1}^{n} \alpha_i^2 \lambda_i < 0$ and $\sum_{i=1}^{n} \beta_i^2 \lambda_i < 0$, so that at least one eigenvalue is negative. Without loss of generality assume $\lambda_1 < 0$. If $\alpha_1 = 0$ or $\beta_1 = 0$, then obviously we have a second negative eigenvalue. If $\alpha_1 \beta_1 \neq 0$, we have

$$\sum_{i=1}^{n} (\beta_1 \alpha_i - \alpha_1 \beta_i)^2 \lambda_i = \beta_1^2 \sum_{i=1}^{n} \alpha_i^2 \lambda_i + \alpha_1^2 \sum_{i=1}^{n} \beta_i^2 \lambda_i - 2\alpha_1 \beta_1 \sum_{i=1}^{n} \alpha_i \beta_i \lambda_i =$$

$$= \beta_1^2 \sum_{i=1}^{n} \alpha_i^2 \lambda_i + \alpha_1^2 \sum_{i=1}^{n} \beta_i^2 \lambda_i.$$

Consequently, $\sum_{i=1}^{n} (\beta_1 \alpha_i - \alpha_1 \beta_i)^2 \lambda_i < 0$ so that a second negative eigenvalue exists and the thesis is achieved. □

6.2.1 Some Properties of a Quadratic form Associated with a Symmetric Matrix Having One Simple Negative Eigenvalue

Now we shall establish some fundamental properties which will be used in the next sections in characterizing the quasiconvexity and pseudoconvexity of quadratic functions.

From now on we shall assume $\lambda_1 < 0$, $\lambda_i > 0, i = 2, .., p$, and $\lambda_i = 0$, $i = p+1, .., n$.
The following lemma holds.

Lemma 6.2.2. *Let Q be an $n \times n$ symmetric matrix and assume $\nu_-(Q) = 1$. Then, the following conditions hold:*
(i) If $u \in \Re^n$ is such that $u^T v^1 = 0$, then either $u \in kerQ$ or $u^T Qu > 0$;
(ii) $u \in kerQ$ if and only if $u^T Qu = 0$ and $u^T v^1 = 0$.

Proof. (i) Let $u = \sum_{i=1}^{n} \alpha_i v^i$. We have $0 = u^T v^1 = \alpha_1$, so that $u = \sum_{i=2}^{n} \alpha_i v^i$.
If $u \notin kerQ$, there exists $i \in \{2, .., p\}$ such that $\alpha_i \neq 0$. It follows that $u^T Qu = \sum_{i=2}^{p} (\alpha_i)^2 \lambda_i > 0$.
(ii) If $u \in kerQ$, obviously we have $u^T Qu = 0$ and $u^T v^1 = 0$. The converse statement follows directly from (i). $\qquad\square$

Now we shall introduce the following opposite cones associated with the matrix Q:

$$T = \{x : x^T Qx \leq 0, \ x^T v^1 \geq 0\}, \quad -T = \{x : x^T Qx \leq 0, \ x^T v^1 \leq 0\}$$

We shall see in the next section that cones T and $-T$ will play a fundamental role in characterizing the maximal domains of the quasiconvexity and pseudoconvexity of a quadratic function.
The following theorems hold, where ∂T denotes the boundary of T. Note that since T and $-T$ are opposite cones, the properties of $-T$ can be easily derived from the ones which will be established for T.

Theorem 6.2.1. *Let Q be an $n \times n$ symmetric matrix and assume $\nu_-(Q) = 1$. Then, the following conditions hold:*
(i) $kerQ = T \cap (-T)$;
(ii) T is a pointed cone if and only if $kerQ = \{0\}$.

Proof. (i) From (ii) of Lemma 6.2.2 we have $kerQ \subseteq T \cap (-T)$. If $x \in T \cap (-T)$ we necessarily have $x^T Qx \leq 0$, $x^T v^1 = 0$; consequently, (i) of Lemma 6.2.2 implies that $x \in kerQ$.
(ii) Since T is pointed if and only if $T \cap (-T) = \{0\}$, the thesis follows from (i). $\qquad\square$

Theorem 6.2.2. *Let Q be an $n \times n$ symmetric matrix and assume $\nu_-(Q) = 1$. Then, the following conditions hold:*
(i) $x_0 \in intT$ if and only if $x_0^T Qx_0 < 0$ and $x_0^T v^1 > 0$;
(ii) $x_0 \in \partial T \backslash kerQ$ if and only if $x_0^T Qx_0 = 0$ and $x_0^T v^1 > 0$;
(iii) $intT \cap int(-T) = \emptyset$;
(iv) $T \cup (-T) = \{x \in \Re^n : x^T Qx \leq 0\}$;
(v) $int(T \cup (-T)) = intT \cup int(-T)$.

Proof. (i) This is obvious.

(ii) This follows by noting that $x_0^T Q x_0 = 0$ if and only if $x_0 \in \partial T \cup \partial(-T)$ and that $x_0 \notin kerQ$ if and only if $x_0^T v^1 \neq 0$.

(iii) It follows from (i) and from its analogous result for cone $-T$.

(iv) This follows directly from the definitions of T and $-T$.

(v) Since $int(T \cup (-T)) = \{x \in \Re^n : x^T Q x < 0\} \supseteq intT \cup int(-T)$, we must prove that $intT \cup int(-T) \supseteq \{x \in \Re^n : x^T Q x < 0\}$. Let x such that $x^T Q x < 0$. From Lemma 6.2.2 we necessarily have $x^T v^1 \neq 0$ and the thesis follows. □

The following theorem points out the convexity of cones T and $-T$.

Theorem 6.2.3. *Let Q be an $n \times n$ symmetric matrix. If $\nu_-(Q) = 1$, then T is a closed convex cone.*

Proof. Let P be the orthonormal matrix which has the eigenvectors $v^1, ..., v^n$ as columns, and let H be the diagonal matrix with the first p diagonal entries given by $(-\lambda_1)^{-\frac{1}{2}}, (\lambda_2)^{-\frac{1}{2}}, ..., (\lambda_p)^{-\frac{1}{2}}$, and all the others equal to 1. It is well known that the linear transformation $x = PHy$ reduces the quadratic form $x^T Q x$ to the canonical form $\sum_{i=2}^{p} y_i^2 - y_1^2 = \parallel \bar{y} \parallel^2 - y_1^2$, where $\bar{y} = (y_2, .., y_p)^T$.

Let $C = \{(y_1, \bar{y}) : \parallel \bar{y} \parallel^2 - y_1^2 \leq 0, y_1 \geq 0\} = \{(y_1, \bar{y}) : \parallel \bar{y} \parallel \leq y_1, y_1 \geq 0\}$.

It is easy to verify that C is a closed cone; we shall prove that C is convex. Let $z = (z_1, \bar{z}) \in C, w = (w_1, \bar{w}) \in C$. Since $\parallel \bar{z} \parallel \leq z_1, \parallel \bar{w} \parallel \leq w_1$, we have $\parallel t\bar{z} + (1-t)\bar{w} \parallel \leq t \parallel \bar{z} \parallel + (1-t) \parallel \bar{w} \parallel \leq tz_1 + (1-t)w_1$ for all $t \in [0, 1]$. Consequently, $tz + (1-t)w \in C$ for all $t \in [0, 1]$ so that C is convex.

Taking into account that $x^T v^1 = y^T H^T P^T v^1$ and that $v^1 = Pe^1$, where e^1 is the unit vector $e^1 = (1, 0, .., 0)^T$, we have $x^T v^1 = y^T He^1 = (-\lambda_1)^{-\frac{1}{2}} y^T e^1 = (-\lambda_1)^{-\frac{1}{2}} y_1$. Consequently, $y_1 \geq 0$ if and only if $x^T v^1 \geq 0$ and this implies $PH(C) = T$. The thesis follows from the linearity of the transformation PH. □

Remark 6.2.1. Given a convex set C and a linear map A, one has $A(riC) = ri(AC)$, but, in general, the image of a closed convex set is not closed. When C is a closed convex cone such that $C \cap (-C) = kerA$, then $A(clC) = cl(AC)$. Consequently, from (i) of Theorem 6.2.1 and from Theorem 6.2.3, we have the following corollary.

Corollary 6.2.1. *Let Q be an $n \times n$ symmetric matrix and assume $\nu_-(Q) = 1$. Then, the following conditions hold:*

(i) $Q(intT) = ri(Q(T)), Q(int(-T)) = ri(Q(-T))$;

(ii) $Q(T)$ and $Q(-T)$ are closed convex cones.

Consider now the set

$$Z = \{z \in \Re^n \backslash \{0\} : \exists w \in intT \text{ such that } z^T w = 0\}$$

and denote with T^+ and T^- the positive polar and the negative polar of T, respectively. The following theorem characterizes Z in terms of the two polar cones.

Theorem 6.2.4. *Let Q be an $n \times n$ symmetric matrix. If $\nu_-(Q) = 1$, then $Z = (T^+ \cup T^-)^c$.*

Proof. Since $w \in int T$ if and only if either $z^T w > 0$ for all $z \in T^+$ or $z^T w < 0$ for all $z \in T^-$, we necessarily have $Z \cap T^+ = \emptyset$ and $Z \cap T^- = \emptyset$. Consequently, $Z \subseteq (T^+ \cup T^-)^c$. Consider now an element $z \in (T^+ \cup T^-)^c$ and assume by contradiction that $z^T w \neq 0$, $\forall w \in int T$. The convexity of $int T$ implies that $z^T w > 0 \ \forall w \in int T$ or $z^T w < 0$, $\forall w \in int T$, i.e., $z \in (T^+ \cup T^-)$, and this is a contradiction. It follows that $(T^+ \cup T^-)^c \subseteq Z$ and the thesis is achieved. $\square$

The following lemma characterizes the image of the cones T and $-T$ under the linear transformation $z = Qx$. The obtained results will play a fundamental role in characterizing the maximal domains of quasiconvexity of a quadratic function.

Lemma 6.2.3. *Let Q be an $n \times n$ symmetric matrix and assume $\nu_-(Q) = 1$. Then: $Q(int T) = ri T^-$, $Q(T) = T^-$, $Q(int(-T)) = ri T^+$, $Q(-T) = T^+$, $Q((T \cup (-T))^c) = Z \cap (ker Q)^\perp$.*

Proof. First of all we shall prove that $Q(int T) \subseteq ri T^-$, $Q(int(-T)) \subseteq ri T^+$, $Q((T \cup (-T))^c) \subseteq (ri T^- \cup ri T^+)^c$.
Let $x_0 \in int T$. Since $x_0^T Q x_0 < 0$, $Q x_0 \notin T^+$ and, taking into account Lemma 6.2.1, $Q x_0 \notin Z$. Consequently, $Q x_0 \in T^-$ and, from Corollary 6.2.1, $Q(int T) \subseteq ri T^-$. Similarly we have $Q(int(-T)) \subseteq ri T^+$.
Now we shall prove that $Q((T \cup (-T))^c) \cap ri T^- = \emptyset$.
Let $z_0 \in (T \cup (-T))^c$, i.e., $z_0^T Q z_0 > 0$, and let $x_0 \in int T$, so that $Q x_0 \in ri T^-$. If $Q z_0 \in ri T^-$, then $Q([z_0, x_0]) \subseteq ri T^-$ because of the convexity of $ri T^-$. On the other hand, $x_0 \in int T$, $z_0 \notin T$ imply the existence of $\bar{x} \in [z_0, x_0] \cap \partial T$ for which $Q \bar{x} \in \partial T^-$. Since $Q x_0 \in ri T^-$, we get a contradiction.
In a similar way it can be proven that $Q((T \cup (-T))^c) \cap ri T^+ = \emptyset$, so that $Q((T \cup (-T))^c) \subseteq (ri T^- \cup ri T^+)^c$.
Since $ker Q = T \cup (-T)$ implies that $T^+ \cap T^- \subseteq (ker Q)^\perp = Im Q = \{w = Q x, \ x \in \Re^n\}$, from Corollary 6.2.1 the thesis is achieved. $\square$

6.3 Quadratic Functions

Consider the quadratic function

$$Q(x) = \frac{1}{2} x^T Q x + q^T x \tag{6.1}$$

where Q is an $n \times n$ symmetric matrix, $q \in \Re^n$ and let

$$Q_0(x) = \frac{1}{2}\, x^T Q x \tag{6.2}$$

be the quadratic form associated with (6.1).

In this section we shall characterize quadratic functions which are generalized convex.

The following theorem shows that quasiconvexity reduces to convexity if the domain is the whole space $\Re^n$.

Theorem 6.3.1. *The quadratic function $Q(x)$ is quasiconvex on $\Re^n$ if and only if $Q(x)$ is convex on $\Re^n$.*

Proof. Since convexity implies quasiconvexity, the converse statement remains to be proven. By contradiction, assume that $Q(x)$ is not convex. Then, Q has at least one negative eigenvalue λ_1. Let w be a normalized eigenvector associated with λ_1 and let $\varphi(t) = Q(tw) = \frac{1}{2}\lambda_1 t^2 + t\, q^T w, t \in \Re$. The restriction $\varphi(t)$ has a strict local maximum point at $\bar t = -\frac{q^T w}{\lambda_1}$ and consequently, $\varphi(t)$, and in turn $Q(x)$, is not quasiconvex which contradicts the assumption. $\quad\square$

Theorem 6.3.1 implies that quasiconvexity can differ from convexity only on a proper subset S of $\Re^n$. From now on, following Martos [211], we shall insert the word "merely" to distinguish quadratic quasiconvex (pseudoconvex, etc.) functions that are not convex.

Remark 6.3.1. From Theorem 3.4.1, a necessary condition for a quadratic function to be merely quasiconvex is that the matrix Q *has one simple negative eigenvalue*, i.e., $\nu_-(Q) = 1$.

Remark 6.3.2. Note that for quadratic functions, quasiconvexity reduces to semistrict quasiconvexity since $Q(x_1) > Q(x_2)$, x_1, $x_2 \in S$, implies that the restriction of $Q(x)$ on the line through x_1 and x_2 is a decreasing or a strictly convex quadratic function.

The following theorem shows that quasiconvexity reduces to pseudoconvexity on every open convex set of $\Re^n$.

Theorem 6.3.2. *The quadratic function $Q(x)$ is quasiconvex on an open convex set $S \subseteq \Re^n$ if and only if it is pseudoconvex on S.*

Proof. Since pseudoconvexity implies quasiconvexity, the converse statement remains to be proven. The thesis follows from Theorem 6.3.1 if $Q(x)$ is convex, otherwise Q has one simple negative eigenvalue λ_1. Let w be a normalized eigenvector associated with λ_1. From Theorem 3.2.6, it is sufficient to prove that the gradient of $Q(x)$ cannot vanish. By contradiction, assume the existence of $x_0 \in S$ such that $\nabla Q(x_0) = Q x_0 + q = 0$, and consider the restriction $\varphi(t) = Q(x_0 + tw)$. Since $\varphi(t) = \frac{1}{2}\lambda_1 t^2 + Q(x_0)$, this restriction has a strict local maximum point at $t = 0$, so that $\varphi(t)$ and in turn $Q(x)$, is not quasiconvex, which contradicts the assumption. $\quad\square$

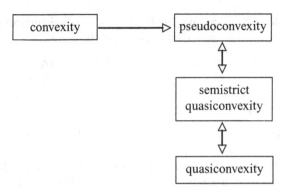

Fig. 6.1. Relationships between various types of convexity for quadratic functions

For quadratic functions on open convex sets, the diagram in Sect. 3.2.3 is simplified as is shown in Fig. 6.1.
Note that Theorem 6.3.2 implies:

- A quadratic function which is merely pseudoconvex on an open convex set S has no critical points;
- A quadratic function which is merely quasiconvex on a convex set S is merely pseudoconvex at least on $int S$;
- A quadratic function which is merely pseudoconvex on an open convex set S is merely quasiconvex (not necessarily pseudoconvex) on the closure of S (see Theorem 2.2.12).

Remark 6.3.3. It is important to point out that any characterization of pseudoconvexity of a quadratic function $Q(x)$ on an open convex set S allows us to simultaneously obtain criteria for the quasiconvexity of $Q(x)$ on the closure of S. This fact simplifies the analysis in the sense that, in order to characterize the quasiconvexity of $Q(x)$ on S, it is sufficient to study the pseudoconvexity on the interior of S.

The following corollary points out that the second order characterization of pseudoconvexity (see Corollary 3.4.1) becomes easy to handle for quadratic functions.

Corollary 6.3.1.
(i) The quadratic function $Q(x) = \frac{1}{2}\,x^T Q x + q^T x$ is pseudoconvex on an open convex set S if and only if (6.3) holds

$$x \in S, \ w \in \Re^n, w^T(Qx + q) = 0 \Rightarrow w^T Q w \geq 0. \tag{6.3}$$

(ii) The quadratic form $Q_0(x) = \frac{1}{2}\,x^T Q x$ is pseudoconvex on an open convex set S if and only if (6.4) holds

$$x \in S, \ w \in \Re^n, w^T Q x = 0 \Rightarrow w^T Q w \geq 0. \tag{6.4}$$

Now we are able to find the maximal domains of quasiconvexity (pseudoconvexity) of a quadratic form and of a quadratic function.

Theorem 6.3.3. *Let Q be an $n \times n$ symmetric matrix. If $\nu_-(Q) = 1$, then the quadratic form $Q_0(x) = \frac{1}{2} x^T Q x$ is merely quasiconvex on the closed convex cones $T, -T$. Furthermore, T and $-T$ are the maximal domains of quasiconvexity of $Q_0(x)$.*

Proof. Taking into account Remark 6.3.3, we shall prove that $Q_0(x)$ is pseudoconvex on $int T$. If not, from (ii) of Corollary 6.3.1, there exist $x_0 \in int T$, $w \in \Re^n$ such that $w^T Q x_0 = 0$ and $w^T Q w < 0$. Since $x_0^T Q x_0 < 0$, from Lemma 6.2.1 Q has at least two negative eigenvalues, which contradicts the assumptions. Similarly, we obtain that $Q_0(x)$ is pseudoconvex on $int(-T)$.

By means of Theorem 2.2.12, $Q_0(x)$ is quasiconvex on T and on $-T$.

The maximality of the domains remains to be proven. To see this, assume that $Q_0(x)$ is pseudoconvex on an open set S such that $S \cap (T \cup (-T)) \neq \emptyset$ and let $y \in S$, $y \notin T \cup (-T)$. Then $y^T Q y > 0$ and, from Lemma 6.2.3, $Q y \in Z$. Consequently, there exists $x_0 \in int T$ such that $x_0^T Q y = 0$, $x_0^T Q x_0 < 0$, which contradicts (6.4). The thesis is achieved. □

Taking into account Remark 6.3.1, Theorem 6.3.3 may be re-stated as follows.

Theorem 6.3.4. *A quadratic form $Q_0(x)$ is merely quasiconvex on a convex set S, with $int S \neq \emptyset$, if and only if*
(i) $\nu_-(Q) = 1$;
(ii) $S \subseteq T$, or $S \subseteq -T$.

We shall prove that the maximal domains of quasiconvexity of a quadratic function are obtained by the ones $(\pm T)$ associated with the quadratic form by means of a suitable translation. To this end, firstly we shall state the following theorem which gives a necessary condition for a quadratic function to be quasiconvex and which points out that, unlike the convex case, the sum of a quasiconvex function with a linear function is not, in general, quasiconvex.

Theorem 6.3.5. *Assume that the quadratic function $Q(x) = \frac{1}{2} x^T Q x + q^T x$ is merely quasiconvex on an open set $S \subset \Re^n$. Then, $rank Q = rank[Q, q]$.*

Proof. The thesis is trivial if $q = 0$. Consider $Im Q = \{Q x, x \in \Re^n\}$ and assume by contradiction that $q \notin Im Q$. Then, for every fixed $x \in \Re^n$, $Q x + q \notin Im Q$ and, in particular, $Q x + q \notin T^+ \cup T^- \subseteq Im Q$. From Lemma 6.2.3, we have $Q x + q \in Z$ so that, from Theorem 6.2.4, there exists $w \in int T$ such that $w^T (Q x + q) = 0$. Let $x_0 \in S$ and consider the restriction $\varphi(t) = Q(x_0 + tw)$. By means of simple calculations we have $\varphi'(0) = w^T (Q x_0 + q) = 0$, $\varphi''(0) = w^T Q w < 0$, so that $t = 0$ is a strict local maximum for $\varphi(t)$ and this implies that $Q(x)$ is not quasiconvex on S, which contradicts the assumption. It follows that $q \in Im Q$ or, equivalently, $rank Q = rank[Q, q]$. □

Remark 6.3.4. Note that $w \in ImQ$ if and only if $w \in (kerQ)^{\perp}$. In particular, $rankQ = rank[Q, q]$ implies that $q \in (kerQ)^{\perp}$.

Theorem 6.3.6. *Consider the quadratic function* $Q(x) = \frac{1}{2} x^T Q x + q^T x$, *and assume the existence of* $s \in \Re^n$ *such that* $Qs + q = 0$.
(i) $Q(x)$ *is merely quasiconvex on the closed convex cones* $s + T, s - T$ *if and only if* $Q_0(x) = \frac{1}{2} x^T Q x$ *is merely quasiconvex on* T, $-T$, *respectively.*
(ii) If $\nu_-(Q) = 1$, *then* $s + T$ *and* $s - T$ *are the maximal domains of quasiconvexity of* $Q(x)$ *and we have*

$$s + T = \{x \in \Re^n : (x - s)^T Q(x - s) \le 0, \ (v^1)^T(x - s) \ge 0\} \qquad (6.5)$$

$$s - T = \{x \in \Re^n : (x - s)^T Q(x - s) \le 0, \ (v^1)^T(x - s) \le 0\} \qquad (6.6)$$

Proof. (i) From (i) of Corollary 6.3.1, $Q(x)$ is pseudoconvex on $s \pm intT$ if and only if (6.3) holds with $S = s \pm intT$. Since $x \in s \pm intT$ if and only if $x - s = u \in \pm intT$, we have $Qx + q = Q(x - s) = Qu$ so that (6.3) is equivalent to (6.4) with $S = \pm intT$. Consequently, $Q(x)$ is merely pseudoconvex on $s \pm intT$ if and only if $Q_0(x)$ is merely pseudoconvex on $\pm intT$ and the thesis follows.
(ii) Theorem 6.3.3 implies that $\pm T$ are the maximal domains of quasiconvexity of $Q_0(x)$ so that, taking into account (i), $s \pm T$ are the maximal domains of quasiconvexity of $Q(x)$.
Finally, $x \in s \pm T$ if and only if $x - s \in \pm T$, so that (6.5), (6.6) hold.
The proof is complete. $\qquad\square$

Remark 6.3.5. If Q is a singular matrix, then a stationary point of $Q(x)$ is not unique. However, the characterization of the maximal domains of quasiconvexity is independent of the particular stationary point used. To see this, let s_1, s_2 be two distinct stationary points, i.e., $Qs_1 + q = Qs_2 + q = 0$. We have $s_1 = s_2 + u$, $u \in kerQ \subset T \cup (-T)$. It follows that $s_1 \pm T = s_2 + u \pm T = s_2 \pm T$.

The previous results allow us to characterize the merely quasiconvexity of a quadratic function.

Theorem 6.3.7. *The quadratic function* $Q(x) = \frac{1}{2} x^T Q x + q^T x$ *is merely quasiconvex on a convex set* S *with nonempty interior if and only if the following conditions hold:*
(i) $\nu_-(Q) = 1$;
(ii) There exists $s \in \Re^n$ *such that* $Qs + q = 0$;
(iii) $S \subseteq s \pm T$.

Proof. Assume that $Q(x)$ is merely quasiconvex on S. We necessarily have $\nu_-(Q) = 1$ and, from Theorem 6.3.5, (ii) follows; (iii) is a direct consequence of Theorem 6.3.6.
Conversely, the thesis follows from Theorem 6.3.3 and from Theorem 6.3.6. $\quad\square$

Corollary 6.3.2. *If $Q(x) = \frac{1}{2} x^T Q x + q^T x$ is merely quasiconvex on a convex set S with nonempty interior, then $Q_0(x)$ is merely quasiconvex on $S - s$, where s is such that $Qs + q = 0$.*

Proof. The thesis follows from Theorem 6.3.7, taking into account Theorem 6.3.3. □

From Theorem 6.3.2, all the characterizations obtained so far for a quadratic quasiconvex function hold for a quadratic pseudoconvex function if the convex domain S is contained in $s \pm intT$.

The following theorem characterizes the strict pseudoconvexity of a quadratic function.

Theorem 6.3.8. *The quadratic function $Q(x) = \frac{1}{2} x^T Q x + q^T x$ is strictly pseudoconvex on a convex set $S \subseteq s \pm intT$ with nonempty interior if and only if the following conditions hold:*
(i) $Q(x)$ is pseudoconvex;
(ii) Q is nonsingular.

Proof. Assume that $Q(x)$ is strictly pseudoconvex. Since $Q(x)$ is pseudoconvex, too, we must prove that Q is nonsingular. If not, let $x_0 \in S$ and $u \neq 0$ such that $Qu = 0$. Taking into account that $q = -Qs$, we have $\varphi(t) = Q(x_0 + tu) = \varphi(0)$ for all t and this contradicts the strict pseudoconvexity of φ and, in turn, the strictly pseudoconvexity of $Q(x)$.

Assume now that (i) and (ii) hold. If $Q(x)$ is not strictly pseudoconvex, then there exist x_1, $x_2 \in S$, $x_1 \neq x_2$, such that $Q(x_1) = Q(x_2)$ and $\nabla Q(x_1)^T (x_2 - x_1) = (Qx_1 + q)^T (x_2 - x_1) = 0$. Set $u = x_2 - x_1$ and consider the restriction $\varphi(t) = Q(x_1 + tu)$. We have $\varphi'(t) = u^T Q u\, t + (x_1 - s)^T Q u$. Since φ is pseudoconvex and $\varphi'(0) = 0$, $t = 0$ is a minimum point and thus we necessarily have $u^T Q u = 0$. This last equality implies that $u \in \pm T$ so that $Qu \in T^+$ or $Qu \in T^-$. The nonsingularity of Q implies that $Qu \neq 0$ and, since $x_1 - s \in \pm intT$, we have $(x_1 - s)^T Q u \neq 0$, and this is a contradiction. □

The following examples clarify the results given in Theorem 6.3.4 and in Theorem 6.3.7.

Example 6.3.1. Consider the quadratic form $Q_0(x) = 2x_1^2 - x_2^2 - x_1 x_2$.
We have $Q = \begin{bmatrix} 4 & -1 \\ -1 & -2 \end{bmatrix}$, $\lambda_1 = 1 - \sqrt{10} < 0$, $\lambda_2 = 1 + \sqrt{10} > 0$, $v^1 = \frac{v}{\|v\|}$ with $v = (1, 3 + \sqrt{10})^T$.
Theorem 6.3.4 implies that $Q_0(x)$ is quasiconvex on the maximal domains T, $-T$. It is easy to verify that $T = \{x = \alpha(1,1)^T + \beta(-1,2)^T, \ \alpha, \ \beta \geq 0\}$, so that the positive and negative polars of T are respectively,
$T^+ = \{x = \alpha_1(-1,1)^T + \beta_1(2,1)^T, \ \alpha_1, \ \beta_1 \geq 0\}$,
$T^- = \{x = \alpha_1(-1,1)^T + \beta_1(2,1)^T, \ \alpha_1, \ \beta_1 \leq 0\}$.
Now we shall verify that the image of T under the linear transformation Q is T^-.

In fact, $Q(T) = \{y = \alpha Q(1,1)^T + \beta Q(-1,2)^T, \ \alpha, \ \beta \geq 0\} = \{y = -3\alpha(-1,1)^T - 3\beta(2,1)^T, \ \alpha, \ \beta \geq 0\} = \{y = \alpha_1(-1,1)^T + \beta_1(2,1)^T, \ \alpha_1, \ \beta_1 \leq 0\} = T^-$.

By means of similar calculations we can verify that $Q(-T) = T^+$.

The cones T, $-T$ and the supporting hyperplane $(v^1)^T x = 0$ are drawn in Fig. 6.2.

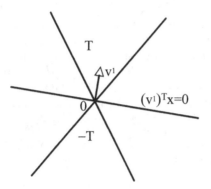

Fig. 6.2. Maximal cones

Let us note that the nonsingularity of Q implies that the quadratic function $Q(x) = Q_0(x) + q^T x$ is quasiconvex for every $q \in \Re^n$ on the maximal domains $s + T$ and $s - T$, where $s = -Q^{-1}q$.

Example 6.3.2. Consider the quadratic function $Q(x) = -x_1^2 - x_2^2 - 2x_1x_2 + 2x_1 + 2x_2$. We have $Q = \begin{bmatrix} -2 & -2 \\ -2 & -2 \end{bmatrix}$, $q = (2,2)^T$, $\lambda_1 = -4 < 0$, $\lambda_2 = 0$, $v^1 = \frac{v}{\|v\|}$ with $v = (1,1)^T$.

Note that $Q(x)$ is a concave function; nevertheless, since $rank\ Q = rank\ [Q,q]$, $Q(x)$ is also merely quasiconvex on $s+T$ and on $s-T$, where s is any vector of the kind $s = (s_1, -s_1+1)^T$, $s_1 \in \Re$. Taking into account that $x^T Q x \leq 0$, $\forall x \in \Re^2$, we have $s+T = \{x \in \Re^2 : (v^1)^T(x-s) \geq 0\} = \{x \in \Re^2 : x_1 + x_2 - 1 \geq 0\}$.

The following theorem characterizes the merely quasiconvexity of a quadratic function over a half-space.

Theorem 6.3.9. *The quadratic function* $Q(x) = \frac{1}{2}x^T Q x + q^T x$ *is merely quasiconvex on the half-space* $H = \{x \in \Re^n : h^T x + h_0 \geq 0\}$ *if and only if (6.7) holds:*

$$\nu_-(Q) = 1, \ kerQ = h^\perp, \ \exists \ \beta \in \Re : \ q = \beta h, \ h_0 \leq \beta \frac{\|h\|^4}{h^T Q h}. \tag{6.7}$$

Proof. Assume that $Q(x)$ is merely quasiconvex on H. From Theorem 6.3.7, we have $H \subseteq s+T \subseteq \Gamma = \{x \in \Re^n : (v^1)^T(x-s) \geq 0\}$, or $H \subseteq s-T \subseteq \Gamma_1 =$

$\{x \in \Re^n : (v^1)^T(x - s) \leq 0\}$. Since $H \subseteq \Gamma$ or $H \subseteq \Gamma_1$, ∂H and $\partial \Gamma$ are necessarily parallel hyperplanes so that $h = kv^1$, $k \neq 0$, i.e., h is an eigenvector associated with the negative eigenvalue λ_1. Obviously, $k > 0$ implies $H \subseteq \Gamma$, while $k < 0$ implies $H \subseteq \Gamma_1$. We shall limit ourselves to considering the case $k > 0$ since the other one is perfectly analogous.

Note that $H \subseteq \Gamma$ if and only if $h_0 \leq -h^T s$; when $h_0 = -h^T s$ we have $H = s + T = \Gamma$ and $T = \{x \in \Re^n : x^T Q x \leq 0, h^T x \geq 0\}$.

Since the quadratic form $\frac{1}{2}x^T Q x$ is merely quasiconvex on T and on $-T$, by means of continuity, we have $\frac{1}{2}x^T Q x \leq 0$, $\forall x \in \Re^n$ and, because $\nu_-(Q) = 1$, this implies that $ker Q = h^\perp$ and $Im Q = \{kh, k \in \Re\}$. Since $rank Q = rank\ [Q, q]$, there exists $\beta \in \Re$ such that $q = \beta h$. If $\beta = 0$, we have $s \in ker Q = h^\perp$ so that $h_0 \leq -h^T s = 0$ and (6.7) holds. If $\beta \neq 0$, we can choose any $s \in s_0 + h^\perp$, where $Q s_0 = -q$. In particular, $s = (s_0 + h^\perp) \cap Im Q$ is an eigenvector of Q and thus $Q s = \lambda_1 s$. It follows that $h_0 \leq -h^T s = -\frac{h^T Q s}{\lambda_1} = \frac{h^T \beta h}{\lambda_1} = \beta \frac{\|h\|^2}{\lambda_1} = \beta \frac{\|h\|^4}{h^T Q h}$.

Conversely, by choosing $s = -\frac{\beta}{\lambda_1} h$, it is easy to verify that (6.7) implies (i), (ii), and (iii) of Theorem 6.3.7. $\qquad\square$

Corollary 6.3.3. *The quadratic function $Q(x) = \frac{1}{2}x^T Q x + q^T x$ is merely quasiconvex on the half-space $H = \{x \in \Re^n : h^T x + h_0 \geq 0\}$ if and only if $Q = \mu h h^T$, $q = \beta h$, with $\mu < 0$ and $h_0 \leq \frac{\beta}{\mu}$.*

6.4 Quadratic Functions of Non-negative Variables

By specifying the results given in the previous section, it is possible to establish criteria for generalized convex quadratic functions on $\Re_+^n$. These results were obtained for the first time by Martos in [209, 210], by introducing of the concept of positive subdefinite matrices.

Now we shall characterize the quasiconvexity of a quadratic form on the non-negative orthant. The first result points out the relationships between the non-negative orthant and maximal cone T.

Theorem 6.4.1. *The quadratic form $Q_0(x)$ is merely quasiconvex on $\Re_+^n$ if and only if the following conditions hold:*
(i) $\nu_-(Q) = 1$;
(ii) $\Re_+^n \subseteq T$.

Proof. From Theorem 6.3.4, $Q_0(x)$ is merely quasiconvex on $\Re_+^n$ if and only if (i) holds and either $\Re_+^n \subseteq T$ or $\Re_+^n \subseteq -T$. This last inclusion cannot hold; in fact $(v^1)^T x \leq 0$, $\forall x \in \Re_+^n$ implies that $v^1 \in \Re_-^n$ and this is a contradiction since the first non-zero component of v^1 is positive. The thesis follows. $\qquad\square$

The following theorem characterizes a quadratic form on the non-negative orthant in terms of the sign of the elements of matrix Q.

Theorem 6.4.2. *The quadratic form $Q_0(x)$ is merely quasiconvex on $\Re^n_+$ if and only if*

(i) $\nu_-(Q) = 1$;

(ii) $Q \le 0$ [1].

Proof. Assume that (i) and (ii) hold and let Γ be the subspace spanned by the normalized eigenvectors associated with the non-negative eigenvalues of Q; we have $\Gamma = \{x \in \Re^n : (v^1)^T x = 0\}$. Since $x^T Q x \ge 0$ for all $x \in \Gamma$, and $x^T Q x \le 0$ for all $x \in \Re^n_+$, Γ is a supporting hyperplane to $\Re^n_+$ at the origin, so that $v^1 \in \Re^n_+$. Consequently, the elements of $\Re^n_+$ satisfy the inequalities $x^T Q x \le 0$, $(v^1)^T x \ge 0$ so that $\Re^n_+ \subseteq T$ and the thesis follows from Theorem 6.4.1.

Assume now that $Q_0(x)$ is merely quasiconvex on $\Re^n_+$.
From Theorem 6.4.1, (i) holds and furthermore $\Re^n_+ \subseteq T$ so that $x^T Q x \le 0$ for all $x \in \Re^n_+$; in particular $(e^i)^T Q e^i = q_{ii} \le 0$, $i = 1, ..., n$. Consider now the submatrix of Q, $Q_{ij} = \begin{bmatrix} q_{ii} & q_{ij} \\ q_{ij} & q_{jj} \end{bmatrix}$ and the restriction $\varphi(x_i, x_j) = \frac{1}{2}(q_{ii} x_i^2 + 2 q_{ij} x_i x_j + q_{jj} x_j^2)$.

Since $\varphi(x_i, x_j) \le 0$, $\forall (x_i, x_j) \in \Re^2_+$, we have $q_{ij} \le 0$ when $q_{ii} q_{jj} = 0$. Consider the case $q_{ii} < 0$, $q_{jj} < 0$. The quasiconvexity of φ implies that Q_{ij} has at most one negative eigenvalue, so that $q_{ii} q_{jj} - q_{ij}^2 \le 0$. If $q_{ii} q_{jj} - q_{ij}^2 < 0$, the equation $q_{ii} x_i^2 + 2 q_{ij} x_i x_j + q_{jj} x_j^2 = 0$ has, for every fixed x_j (or x_i), two roots which cannot be positive since $\varphi(x_i, x_j) \le 0$, $\forall (x_i, x_j) \in \Re^2_+$; consequently, $q_{ij} \le 0$.

If $q_{ii} q_{jj} - q_{ij}^2 = 0$, we have $\varphi(x_i, x_j) = \frac{1}{2 q_{ii}}(q_{ii} x_i + q_{ij} x_j)^2$. This function has a line r of critical points which are global maximum points; since the quasiconvexity of φ implies that $r \cap int\Re^2_+ = \emptyset$, we necessarily have $q_{ij} \le 0$.
It follows that $q_{ij} \le 0$, $\forall i, j = 1, .., n$, i.e., $Q \le 0$ and the proof is complete. $\square$

Theorem 6.4.3. *Let $Q(x)$ be merely quasiconvex on $\Re^n_+$. Then, $Q_0(x)$ is merely quasiconvex on $\Re^n_+$.*

Proof. From (ii) of Theorem 6.3.6, either $\Re^n_+ \subseteq s + T$ or $\Re^n_+ \subseteq s - T$. Let $v_j^1 > 0$ be the first non-zero component of v^1; since $t e^j \in \Re^n_+$, $\forall t > 0$, we have $(v^1)^T (t e^j - s) = t v_j^1 - (v^1)^T s > 0$, for a large enough t. It follows that $\Re^n_+ \subseteq s + T$. Consequently, we must prove that $\Re^n_+ \subseteq T$, i.e., $x^T Q x \le 0$, $(v^1)^T x \ge 0$, $\forall x \in \Re^n_+$. Assume the existence of $\bar{x}$ such that $\bar{x}^T Q \bar{x} > 0$ $((v^1)^T \bar{x} < 0)$. Since $t\bar{x} \in \Re^n_+$, $\forall t > 0$, for a large enough t we have $(t\bar{x} - s)^T Q(t\bar{x} - s) > 0$ $((v^1)^T (t\bar{x} - s) < 0)$ and this is a contradiction. Consequently, $\Re^n_+ \subseteq T$, so that Q_0 is merely quasiconvex on $\Re^n_+$. $\square$

The following example shows that the converse statement of Theorem 6.4.3 does not hold; we need some additional assumptions on the vector q which will be given in Theorem 6.4.4.

---

[1] $Q \le 0$ means $q_{ij} \le 0$, $\forall i, j$.

Example 6.4.1. Consider the quadratic function $Q(x_1, x_2) = -x_1 x_2 + x_1 - x_2$.
We have $Q = \frac{1}{2} \begin{bmatrix} 0 & -1 \\ -1 & 0 \end{bmatrix}$, $q = \begin{pmatrix} 1 \\ -1 \end{pmatrix}$.

$Q_0(x_1, x_2) = -x_1 x_2$ is merely quasiconvex on $\Re_+^2$ according to Theorem 6.4.2. On the other hand, $Q(x_1, x_2)$ is not quasiconvex on $\Re_+^2$, since its restriction on the line $x_2 = x_1 + 2$ has a strict local maximum point at $(1, 3)$.

Theorem 6.4.4. *The quadratic function* $Q(x) = \frac{1}{2} x^T Q x + q^T x$ *is merely quasiconvex on* $\Re_+^n$ *if and only if the following conditions hold:*
(i) $\nu_-(Q) = 1$;
(ii) $Q \leq 0$;
(iii) There exists $s \in \Re^n$ *such that* $Qs + q = 0$, $q^T s \geq 0$;
(iv) $q \leq 0$.

Proof. Assume that (i–iv) hold. From Theorem 6.3.6 we must prove that
$\Re_+^n \subseteq s + T = \{x \in \Re^n : (x - s)^T Q(x - s) \leq 0, \ (v^1)^T (x - s) \geq 0\}$.
We have $(x - s)^T Q(x - s) = x^T Q x + 2q^T x - q^T s$, so that (ii), (iii) and (iv) imply that $(x - s)^T Q(x - s) \leq 0, \ \forall \ x \in \Re_+^n$. From Theorem 6.4.2 and Theorem 6.4.1 we have $\Re_+^n \subseteq T$ and this implies that $v^1 \in \Re_+^n$.
On the other hand, $s^T v^1 = \frac{1}{\lambda_1} s^T Q v^1 = -\frac{1}{\lambda_1} q^T v^1 \leq 0$, and, consequently, $(v^1)^T (x - s) \geq 0, \ \forall \ x \in \Re_+^n$, so that $\Re_+^n \subseteq T$.
Assume now that $Q(x)$ is merely quasiconvex on $\Re_+^n$. From Theorem 6.3.7 we have $\nu_-(Q) = 1$, $Qs + q = 0$ for some $s \in \Re^n$, and $\Re_+^n \subseteq s + T$, while from Theorem 6.4.3 and from Theorem 6.4.2, we have $Q \leq 0$ and $\Re_+^n \subseteq T$. It remains to be proven that $q \leq 0$ and $q^T s \geq 0$. The inclusion $\Re_+^n \subseteq s + T$ implies that $0 \in s + T$, i.e., $s \in -T$. Consequently, $s^T Q s \leq 0$ and since $q^T s = -s^T Q s$, we have $q^T s \geq 0$. Furthermore, from Lemma 6.2.3, $Qs \in T^+$, i.e., $q \in T^-$; it follows that $q^T x \leq 0, \ \forall \ x \in T$ and, in particular, $q^T x \leq 0, \ \forall \ x \in \Re_+^n$ so that $q \leq 0$.
The proof is complete. $\qquad\qquad\qquad\qquad\qquad\qquad\qquad\qquad\qquad\qquad\qquad$ $\square$

6.5 Pseudoconvexity on a Closed Set

The aim of this section is to characterize the maximal domains of pseudo-convexity of a non-convex quadratic function. In particular, we are interested in analyzing pseudoconvexity on the non-negative orthant $\Re_+^n$ since many extremum quadratic problems have a feasible region contained in $\Re_+^n$ and not just in $int\Re_+^n$. Since $\Re_+^n$ is a closed set, we shall refer to the notion of pseudoconvexity at a point given in Sect. 3.5.
Since a non-convex quadratic function is quasiconvex on a convex set S if only if it is pseudoconvex on $int S$, we must further investigate the study of the pseudoconvexity on the boundary of S, starting from the maximal domains of quasiconvexity of a quadratic form.
Regarding this, the following lemma holds.

Lemma 6.5.1. *Consider the quadratic form $Q_0(x) = \frac{1}{2}x^T Q x$, and assume $\nu_-(Q) = 1$.*
Then, $Q_0(x)$ is pseudoconvex at $x_0 \in \pm T$ if and only if $\nabla Q_0(x_0) \neq 0$.

Proof. Since $Q_0(x)$ is merely pseudoconvex on $\pm int T$ (see Theorem 6.3.3 and Theorem 6.3.2), we must investigate the boundary of cones T and $-T$. We shall consider ∂T since the other case is analogous.
$Q_0(x)$ is pseudoconvex at $x_0 \in \partial T$ if and only if (6.8) holds:

$$x_0 \in \partial T, \ x \in T, \ Q_0(x) < Q_0(x_0) \Rightarrow (x - x_0)^T Q x_0 < 0. \qquad (6.8)$$

Obviously, (6.8) implies that $\nabla Q_0(x_0) = Q x_0 \neq 0$.
Conversely, let $x_0 \in \partial T$; we necessarily have $Q_0(x_0) = 0$, so that $Q_0(x) < Q_0(x_0) = 0$ implies that $x \in int T$. On the other hand, $Q x_0 \neq 0$ implies that $Q x_0 \in T^- \backslash \{0\}$ so that $x^T Q x_0 < 0, \ \forall x \in int T$. Since $(x - x_0)^T Q x_0 = x^T Q x_0$, the thesis is achieved. $\qquad \square$

The following theorems, which are a direct consequence of Lemma 6.5.1, characterize the maximal domains of pseudoconvexity of a non-convex quadratic form.

Theorem 6.5.1. *Consider the quadratic form $Q_0(x)$ and assume that $\nu_-(Q) = 1$. Then the following properties hold:*
(i) $Q_0(x)$ is merely pseudoconvex on the maximal domains $T \backslash ker Q, -T \backslash ker Q$;
(ii) $Q_0(x)$ is merely pseudoconvex on $T \backslash \{0\}, -T \backslash \{0\}$ if and only if Q is non-singular.

Theorem 6.5.1 may be re-stated as follows.

Theorem 6.5.2. *A quadratic form $Q_0(x)$ is merely pseudoconvex on a convex set S with nonempty interior if and only if*
(i) $\nu_-(Q) = 1$;
(ii) $S \subseteq T \backslash ker Q$, or $S \subseteq -T \backslash ker Q$.

Taking into account (i) of Theorem 6.2.1, the maximal domains $T \backslash ker Q$ and $-T \backslash ker Q$ can be characterized by means of the inequalities $(v^1)^T x > 0$, $(v^1)^T x < 0$, respectively. More exactly, we have the following theorem.

Theorem 6.5.3. *Consider the quadratic form $Q_0(x) = \frac{1}{2} x^T Q x$ and assume $\nu_-(Q) = 1$. Then, the maximal domains of pseudoconvexity of $Q_0(x)$ are given by*

$$T \backslash ker Q = \{x \in \Re^n : x^T Q x \leq 0, \ (v^1)^T x > 0\}$$

$$-T \backslash ker Q = \{x \in \Re^n : x^T Q x \leq 0, \ (v^1)^T x < 0\}.$$

The relation between the pseudoconvexity of a quadratic function and the pseudoconvexity of the corresponding quadratic form is specified in the following theorem.

Theorem 6.5.4. *Consider the quadratic function* $Q(x) = \frac{1}{2}x^T Q x + q^T x$, *and assume the existence of* $s \in \Re^n$ *such that* $Qs + q = 0$. *Then,* $Q(x)$ *is pseudoconvex on* $s \pm T$ *if and only if* $Q_0(x) = \frac{1}{2}x^T Q x$ *is pseudoconvex on* $\pm T$.

Proof. $Q(x)$ is pseudoconvex at $x_0 \in s + T$ if and only if

$$x \in s + T, \ Q(x) < Q(x_0) \Rightarrow \nabla Q(x_0)^T (x - x_0) < 0. \tag{6.9}$$

Set $y_0 = x_0 - s \in T$, $y = x - s \in T$. We have $Q(x) = \frac{1}{2}(x - s)^T Q(x - s) - \frac{1}{2}s^T Q s = Q_0(y) - \frac{1}{2}s^T Q s$, $Q(x_0) = Q_0(y_0) - \frac{1}{2}s^T Q s$. It follows that $Q(x) < Q(x_0)$ if and only if $Q_0(y) < Q_0(y_0)$. Furthermore, $\nabla Q(x_0) = Q x_0 + q = Q(x_0 - s) = Q y_0 = \nabla Q_0(y_0)$, so that $\nabla Q(x_0)^T (x - x_0) < 0$ if and only if $\nabla Q_0(y_0)^T (y - y_0) < 0$. Consequently, (6.9) is equivalent to

$$y_0 \in T, \ y \in T, \ Q_0(y) < Q_0(y_0) \Rightarrow \nabla Q_0(y_0)^T (y - y_0) < 0$$

i.e., the pseudoconvexity of $Q(x)$ on $s + T$ is equivalent to the pseudoconvexity of $Q_0(y)$ on T.

Analogously, the pseudoconvexity of $Q(x)$ on $s - T$ is equivalent to the pseudoconvexity of $Q_0(y)$ on $-T$. □

The following lemma extends Lemma 6.5.1 to a quadratic function.

Lemma 6.5.2. *Consider the quadratic function* $Q(x) = \frac{1}{2}x^T Q x + q^T x$ *with* $\nu_-(Q) = 1$, *and assume the existence of* $s \in \Re^n$ *such that* $Qs + q = 0$. *Then,* $Q(x)$ *is merely pseudoconvex at* $x_0 \in s \pm T$ *if and only if* $\nabla Q(x_0) \neq 0$.

Proof. From Theorem 6.5.4, $Q(x)$ is merely pseudoconvex at $x_0 \in s \pm T$ if and only if Q_0 is merely pseudoconvex at $y_0 = x_0 - s \in \pm T$.

Since $Q y_0 = Q(x_0 - s) = Q x_0 + q = \nabla Q(x_0)$, the thesis follows from Lemma 6.5.1. □

As a direct consequence of the previous results, we have the following characterization of the maximal domains of pseudoconvexity of a quadratic function.

Theorem 6.5.5. *The quadratic function* $Q(x) = \frac{1}{2}x^T Q x + q^T x$ *is merely pseudoconvex on a convex set* S *with nonempty interior if and only if the following conditions hold:*
(i) $\nu_-(Q) = 1$;
(ii) there exists $s \in \Re^n$ *such that* $Qs + q = 0$;
(iii) $S \subseteq s + (T \backslash \ker Q) = \{x \in \Re^n : (x - s)^T Q(x - s) \leq 0, \ (v^1)^T (x - s) > 0\}$, *or* $S \subseteq s - (T \backslash \ker Q) = \{x \in \Re^n : (x - s)^T Q(x - s) \leq 0, \ (v^1)^T (x - s) < 0\}$.

6.5.1 Pseudoconvexity on the Non-negative Orthant

The above criteria can be specified to the case where S is the non-negative orthant.

Theorem 6.5.6. *Let* $Q_0(x) = \frac{1}{2}x^T Q x$ *be merely quasiconvex on* $\Re_+^n$. *Then,* $Q_0(x)$ *is merely pseudoconvex on* $\Re_+^n \setminus \{0\}$ *if and only if* Q *does not contain a column (or a row) of zeros.*

Proof. From Lemma 6.5.1, $Q_0(x)$ is pseudoconvex on $\Re_+^n \setminus \{0\}$ if and only if $Qx \neq 0$, $\forall x \in \Re_+^n \setminus \{0\}$. By denoting with q^j the j-th column of Q, $j = 1, ..., n$, we have $Qx = \sum_{j=1}^n x_j q^j, x_j \geq 0$. Since $Q \leq 0$ (see Theorem 6.4.2), $Qx = 0$ if and only if for some j we have $q^j = 0$. $\qquad\square$

Theorem 6.5.7. *Let* $Q(x) = \frac{1}{2}x^T Q x + q^T x$ *be merely quasiconvex on* $\Re_+^n$. *Then,* $Q(x)$ *is merely pseudoconvex on* $\Re_+^n$ *if and only if* $q \neq 0$.

Proof. From Lemma 6.5.2, $Q(x)$ is pseudoconvex on $\Re_+^n$ if and only if $Qx + q \neq 0$, $\forall x \in \Re_+^n$. Since $Q \leq 0$, $q \leq 0$ (see Theorem 6.4.4), the thesis follows. $\qquad\square$

Remark 6.5.1. Further developments, such as the criteria of quasiconvexity and pseudoconvexity in terms of the bordered Hessian, can be found in the References at the end of this Chapter.

6.5.2 Generalized Convexity of a Quadratic form on $\Re_+^2$

By applying Theorems 6.4.2 and 6.5.6 to a 2×2 matrix we obtain the following criteria.

Theorem 6.5.8. *Consider the matrix* $Q = \begin{bmatrix} \alpha & \beta \\ \beta & \gamma \end{bmatrix}$. *Then, the quadratic form* $Q_0(x) = \frac{1}{2}x^T Q x$ *is merely quasiconvex on* $\Re_+^2$ *if and only if the following conditions hold:*
(i) $\alpha \leq 0$, $\beta \leq 0$, $\gamma \leq 0$, $(\alpha, \beta, \gamma) \neq (0, 0, 0)$;
(ii) $\alpha\gamma - \beta^2 \leq 0$.
Furthermore, $Q_0(x)$ *is pseudoconvex on* $\Re_+^2 \setminus \{(0, 0)\}$ *if and only if in addition to (i) and (ii) we have* $(\alpha, \beta) \neq (0, 0)$ *and* $(\beta, \gamma) \neq (0, 0)$.

Example 6.5.1. Consider the matrices

$$A = \begin{bmatrix} \alpha & 0 \\ 0 & 0 \end{bmatrix}, \alpha < 0; \quad B = \begin{bmatrix} 0 & \beta \\ \beta & 0 \end{bmatrix}, \beta < 0;$$

$$C = \begin{bmatrix} 0 & 0 \\ 0 & \gamma \end{bmatrix}, \gamma < 0; \quad D = \begin{bmatrix} \alpha & \beta \\ \beta & 0 \end{bmatrix}, \alpha < 0, \beta < 0.$$

The quadratic forms associated with all the matrices are quasiconvex on $\Re_+^2$ but only the quadratic forms associated with the matrices B and D are pseudoconvex on $\Re_+^2 \setminus \{(0, 0)\}$.

6.6 A Special Case

The necessary and sufficient conditions stated in the previous sections are, in general, not easy to use for testing the quasiconvexity (pseudoconvexity) of a quadratic function. Nevertheless, when $Q(x)$ has a particular structure, it is possible to obtain a characterization that is easy to test. In this section we shall consider the following class of functions:

$$f(x) = (a^T x + a_0)(b^T x + b_0) + c^T x. \tag{6.10}$$

Theorem 6.6.1. *Consider the function f in (6.10) and assume that the vectors a and b are linearly independent. Then, f is merely quasiconvex on a convex set $S \subset \Re^n$ with nonempty interior if and only if*
(i) there exist $\alpha, \beta \in \Re$ such that $c = \alpha a + \beta b$;
(ii) $S \subseteq \{x \in \Re^n : a^T x + a_0 + \beta \geq 0, \ b^T x + b_0 + \alpha \leq 0\}$ or
$S \subseteq \{x \in \Re^n : a^T x + a_0 + \beta \leq 0, \ b^T x + b_0 + \alpha \geq 0\}$.

Proof. We have $f(x) = \frac{1}{2}x^T Q x + q^T x + q_0$, where

$$Q = ab^T + ba^T, \ q = b_0 a + a_0 b + c, \ q_0 = a_0 b_0.$$

The linear independence of a and b implies that $dim(ImQ) = dim(\{z = \mu_1 a + \mu_2 b, \mu_1, \mu_2 \in \Re\}) = 2$, so that $dim(kerQ) = n - 2$. Consequently, taking into account that the quadratic form $x^T Q x = 2a^T x b^T x$ is not constant in sign, we necessarily have a unique negative eigenvalue, i.e., $\nu_-(Q) = 1$. From Theorem 6.3.7, f is quasiconvex on S if and only if there exists $s \in \Re^n$ such that $Qs + q = 0$ and $S \subseteq s + T$ or $S \subseteq s - T$. We have $Qs + q = 0$ if and only if $q \in ImQ$ or, equivalently, if and only if there exist $\alpha, \beta \in \Re$ such that $c = \alpha a + \beta b$, i.e., if and only if (i) holds. Furthermore, $Qs + q = (b^T s + b_0 + \alpha)a + (a^T s + a_0 + \beta)b$ so that $Qs + q = 0$ if and only if $b^T s = -(b_0 + \alpha)$ and $a^T s = -(a_0 + \beta)$. By means of simple calculations we have $(x - s)^T Q(x - s) = 2(a^T x + a_0 + \beta)(b^T x + b_0 + \alpha)$ and thus $(x - s)^T Q(x - s) \leq 0$ if and only if (ii) holds.
The proof is complete. $\qquad\square$

Remark 6.6.1. When a and b are linearly dependent, f is convex on $\Re^n$ or it is concave on $\Re^n$. In this last case f turns out to be quasiconvex on a convex set S if and only if $c = \alpha a$ and S is contained in one of the two half-spaces associated with the hyperplane given by the set of critical points of the function (see also Exercise 6.13).

Corollary 6.6.1. *Consider the function f in (6.10) and assume that a and b are linearly independent. Then, f is merely pseudoconvex on a convex set $S \subset \Re^n$ with nonempty interior if and only if*
(i) there exist $\alpha, \beta \in \Re$ such that $c = \alpha a + \beta b$;
(ii) $S \subseteq \{x \in \Re^n : a^T x + a_0 + \beta > 0, \ b^T x + b_0 + \alpha \leq 0\} \cup \{x \in \Re^n :$

$a^T x + a_0 + \beta \geq 0, \ b^T x + b_0 + \alpha < 0 \}$ *or*
$S \subseteq \{x \in \Re^n : a^T x + a_0 + \beta < 0, \ b^T x + b_0 + \alpha \geq 0\} \cup \{x \in \Re^n : a^T x + a_0 + \beta \leq 0, \ b^T x + b_0 + \alpha > 0\}.$

Proof. Taking into account that $\nabla f(x_0) = 0$ if and only if $x_0 \in \{x \in \Re^n : a^T x + a_0 + \beta = 0, b^T x + b_0 + \alpha = 0\}$, the thesis follows from Lemma 6.5.2. $\Box$

In order to characterize the quasiconvexity of f on $\Re_+^n$, we shall state, firstly, the following lemma.

Lemma 6.6.1. *Consider the matrix* $Q = ab^T + ba^T$. *Then,* $Q \leq 0$ *if and only if* $a \geq 0, b \leq 0$ *or* $a \leq 0, b \geq 0$.

Proof. Obviously, if $a \geq 0, b \leq 0$ or $a \leq 0, b \geq 0$, then $Q \leq 0$. Conversely, since the thesis is trivial if $a = 0$ or $b = 0$, we shall consider the case $a \neq 0, b \neq 0$. Assume, by contradiction, the existence of i, j such that $a_i > 0, a_j < 0$ and consider the submatrix $Q_{ij} = \begin{bmatrix} 2a_i b_i & a_i b_j + a_j b_i \\ a_i b_j + a_j b_i & 2a_j b_j \end{bmatrix}$. If $b_i b_j \neq 0$, $a_i b_i \leq 0$, $a_j b_j \leq 0$ imply that $b_i < 0, b_j > 0$, respectively, so that $a_i b_j + a_j b_i > 0$ and this is absurd. If $b_i = 0$ and $b_j \neq 0$, we have $Q_{ij} = \begin{bmatrix} 0 & a_i b_j \\ a_i b_j & 2a_j b_j \end{bmatrix}$, so that $a_j b_j \leq 0$ implies that $b_j > 0$ while $a_i b_j \leq 0$ implies that $b_j < 0$ and, once again, we get a contradiction. The case $b_j = 0$, $b_i \neq 0$ is analogous, so that the case $b_i = 0$, $b_j = 0$ remains to be considered. Let k be such that $b_k \neq 0$ and consider the submatrix $\begin{bmatrix} a_i b_k + a_k b_i & a_i b_j + a_j b_i \\ 2a_k b_k & a_k b_j + a_j b_k \end{bmatrix} = \begin{bmatrix} a_i b_k & 0 \\ 2a_k b_k & a_j b_k \end{bmatrix}$; $a_i b_k < 0$ implies that $b_k < 0$ while $a_j b_k < 0$ implies that $b_k > 0$ and this is absurd. Consequently, we have $a \geq 0$ or $a \leq 0$. For symmetric reasons, the components of b also have the same sign so that, necessarily, $a \geq 0$, $b \leq 0$ or $a \leq 0$, $b \geq 0$. The thesis is achieved. $\Box$

Theorem 6.6.2. *Consider the function* f *in (6.10) and assume that* a *and* b *are linearly independent. Then,* f *is merely quasiconvex on* $\Re_+^n$ *if and only if there exist* $\alpha, \beta \in \Re$ *such that* $c = \alpha a + \beta b$ *and one of the following conditions holds:*
(i) $a \geq 0, \ b \leq 0, \ \alpha \leq -b_0, \ \beta \geq -a_0$;
(ii) $a \leq 0, \ b \geq 0, \ \alpha \geq -b_0, \ \beta \leq -a_0$.

Proof. From Theorem 6.4.4 we have $Q = ab^T + ba^T \leq 0$, while from Lemma 6.6.1 we have $a \geq 0, \ b \leq 0$ or $a \leq 0, \ b \geq 0$. The thesis follows from Theorem 6.6.1. $\Box$

Corollary 6.6.2. *Consider the function* f *in (6.10) and assume that* a *and* b *are linearly independent. Then,* f *is merely pseudoconvex on* $\Re_+^n$ *if and only if there exist* $\alpha, \beta \in \Re$ *such that* $c = \alpha a + \beta b$ *and one of the following conditions holds:*

(i) $a \geq 0$, $b \leq 0$ and $\alpha < -b_0$, $\beta \geq -a_0$ or $\alpha \leq -b_0$, $\beta > -a_0$;
(ii) $a \leq 0$, $b \geq 0$ and $\alpha > -b_0$, $\beta \leq -a_0$ or $\alpha \geq -b_0$, $\beta < -a_0$.

Proof. Referring to Theorem 6.5.7, it is sufficient to note that $b_0 a + a_0 b + c \neq 0$ if and only if $a_0 + \beta \neq 0$ or $b_0 + \alpha \neq 0$. □

6.7 Exercises

6.1. Let $Q(x) = \frac{1}{2} x^T Q x + q^T x$ be merely quasiconvex on a convex set $S \subset \Re^n$ with $\text{int} S \neq \emptyset$. Prove that $Q(x)$ is bounded from above on S.

6.2. Consider the quadratic program $\sup\{Q(x) : x \in S\}$, where Q is merely pseudoconvex on the closed convex polyhedron $S \subset \Re^n$. Prove that the supremum is attained at an extreme point of S provided S has one.

6.3. Consider the function $Q(x) = 2x_1^2 - x_2^2 - x_1 x_2 - 4x_1 + x_2$.
(a) Find the maximal domain $s + T$ of quasiconvexity of $Q(x)$;
(b) Prove that $Q(x) \leq -2$, $\forall x \in s + T$;
(c) Verify that $f(x) = \sqrt{-2 - Q(x)}$ is concave on $s + T$.

6.4. Find a quadratic form $Q_0(x)$ whose maximal domain of quasiconvexity is the cone $T = \{x \in \Re^2 : x = \alpha(-1,1)^T + \beta(1,-2)^T, \alpha \geq 0, \beta \geq 0\}$.

6.5. Find the maximal domains of quasiconvexity and the maximal domains of pseudoconvexity of $Q(x_1, x_2) = 2x_1^2 - x_1 x_2 - x_2$.

6.6. Consider $Q(x_1, x_2) = -4(x_1 - 3x_2)^2 + 4(x_1 - 3x_2)$. Find a closed half-space H such that H is a maximal domain of quasiconvexity of $Q(x_1, x_2)$.

6.7. Consider the closed half-space $H = \{(x_1, x_2) \in \Re^2 : 3x_1 + 2x_2 + 6 \geq 0\}$. Give an example of a quadratic function having H as a maximal domain of quasiconvexity.

6.8. Let $Q(x)$ be a merely quasiconvex quadratic function on a closed half-space H. Can $Q(x)$ be pseudoconvex on H?

6.9. Give an example of a merely quasiconvex quadratic form such that:
(a) T contains $\Re_+^2$ properly;
(b) $\Re_+^2$ contains T properly;
(c) $T \cap \Re_+^2 \neq T$, $T \cap \Re_+^2 \neq \Re_+^2$.

6.10. Prove that the maximal domains of quasiconvexity of a quadratic form $Q_0(x_1, x_2)$ are $\Re_+^2$ and $\Re_-^2$ if and only if $Q_0(x_1, x_2) = k x_1 x_2$ with $k < 0$.

6.11. Let $Q_0(x) = \frac{1}{2} x^T Q x$ be merely pseudoconvex on $\Re_+^2$. Prove that $Q(x) = Q_0(x) + q^T x$ is merely pseudoconvex on $\Re_+^2$ for all $q \leq 0$ if and only if
$$Q = \begin{bmatrix} 0 & \beta \\ \beta & 0 \end{bmatrix}, \beta < 0.$$

6.12. Consider the function $f(x) = (a^T x + a_0)(b^T x + b_0)$ and assume that a and b are linearly independent. Show that f is quasiconvex on a convex set $S \subset \Re^n$ with $int S \neq \emptyset$ if and only if $S \subseteq \{x \in \Re^n : a^T x + a_0 \geq 0, b^T x + b_0 \leq 0\}$ or $S \subseteq \{x \in \Re^n : a^T x + a_0 \leq 0, b^T x + b_0 \geq 0\}$.

6.13. Consider the function $f(x) = (a^T x + a_0)(b^T x + b_0) + c^T x$ and assume that a and b are linearly dependent. Prove that f is merely quasiconvex on a suitable convex set $S \subset \Re^n$ with $int S \neq \emptyset$ if and only if there exist $k < 0$, $\alpha \in \Re$, such that $b = ka$, $c = \alpha a$.

6.8 References

Avriel M., Diewert W. E., Schaible S., and Ziemba W. T., eds. [12], Avriel M., Diewert W. E., Schaible S., and Zang I. [13], Avriel M., and Schaible S. [11], Cottle R. W., and Ferland J. A. [72, 73], Ferland J. A. [107, 108, 109], Giorgi G., and Thielfelder J. [122], Martos B. [209, 211], Schaible S. [236, 237, 238, 239, 243, 247, 242, 251], Schaible S., and Ziemba W. T. eds. [248].

7
Generalized Convexity of Some Classes of Fractional Functions

7.1 Introduction

Economic applications are often characterized by maximizing the efficiency of an economic system. This leads to optimization problems whose objective function is a ratio. Examples include maximization of productivity, maximization of return on investment, maximization of return/risk, minimization of cost/time. Linear fractional and generalized fractional problems may be found in different fields such as data envelopment analysis, tax programming, risk and portfolio theory, logistics and location theory (see for instance [14, 15, 66, 67, 166, 214]). The interest in studying fractional problems is confirmed in the extensive survey (with twelve hundred entries) which appeared in [256]; another updated survey may be found in [114].

Since the early sixties, the close relationship between generalized convexity and fractional programming has been highlighted and from the beginning, fractional programming has benefited from advances in generalized convexity, and vice versa (see, for instance, [114, 256]).

In this chapter we shall characterize the pseudoconvexity of some of the most important classes of fractional functions such as the ratio between a quadratic and a linear function and the sum of a linear and a linear fractional function. We shall also point out how Charnes–Cooper's variable transformation turns out to be a useful tool in studying the pseudoconvexity of some other classes of functions.

7.2 The Ratio of a Quadratic and an Affine Function

Consider the following quadratic fractional function

$$f(x) = \frac{n(x)}{d(x)} = \frac{\frac{1}{2}x^T Q x + q^T x + q_0}{d^T x + d_0} \tag{7.1}$$

on the set $D = \{x : d^T x + d_0 > 0\}$, where Q is an $n \times n$ symmetric matrix, $q, d \in \Re^n$, $d \neq 0$ and $q_0, d_0 \in \Re$.

We have seen in Theorem 3.2.10 that when $n(x)$ is convex on D, then f is pseudoconvex on D; for this reason in what follows we shall assume that Q is not positive semidefinite.

The following examples show that the pseudoconvexity of the numerator on D does not guarantee the pseudoconvexity of f on D and viceversa.

Example 7.2.1. Consider the function

$$f(x_1, x_2) = \frac{n(x_1, x_2)}{d(x_1, x_2)} = \frac{-\frac{1}{2}(x_1 - x_2)^2 - 4x_1 + 4x_2}{x_1 - x_2 - 1}$$

Taking into account Corollary 6.3.3, $n(x_1, x_2)$ is pseudoconvex on $D = \{(x_1, x_2) : x_1 - x_2 - 1 > 0\}$ since we have $h^T = (1, -1)$, $\beta = -4$, $\mu = -1$, $h_0 = -1 < \frac{\beta}{\mu} = 4$. On the other hand, $f(x_1, x_2)$ is not pseudoconvex on D since the restriction $\varphi(t) = f(x_0 + tu)$, $x_0^T = (3, 1)$, $u^T = (1, 0)$, is not pseudoconvex. In fact, we have $\varphi(t) = \dfrac{-\frac{1}{2}(2 + t)^2 - 4(2 + t)}{t + 1}$, $\varphi'(t) = \dfrac{1}{2(t + 1)^2}\left(9 - (t + 1)^2\right)$, so that $t_0 = 2$ is a feasible strict local maximum point for $\varphi(t)$ and, consequently, $\varphi(t)$ is not pseudoconvex.

Example 7.2.2. Consider the function

$$f(x_1, x_2) = \frac{n(x_1, x_2)}{d(x_1, x_2)} = \frac{\frac{1}{2}(x_1^2 - x_2^2) + 1}{1 - x_2}$$

on the convex set $D = \{(x_1, x_2) : 1 - x_2 > 0\}$. The Hessian matrix of f is

$$\nabla^2 f(x_1, x_2) = \begin{bmatrix} \frac{1}{1-x_2} & \frac{x_1}{(1-x_2)^2} \\ \frac{x_1}{(1-x_2)^2} & \frac{x_1^2+1}{(1-x_2)^3} \end{bmatrix}.$$

Since $\nabla^2 f(x_1, x_2)$ is positive definite on D, the function is convex and, in particular, pseudoconvex on D. On the other hand, $n(x_1, x_2)$ is pseudoconvex on the maximal domains $T, -T$, where $T = \{(x_1, x_2) : x_1^2 - x_2^2 \leq 0, x_2 \geq 0\}$. Since $D \not\subseteq T$ and $D \not\subseteq -T$, $n(x_1, x_2)$ is not pseudoconvex on D.

The previous examples show that the study of the pseudoconvexity of the ratio between a quadratic function and an affine function is not easy to carry on. A complete characterization of the pseudoconvexity of f can be found in [44]. Before presenting this result, we shall point out some properties of a quadratic function which are mantained in the ratio.

The following theorem shows that for function f quasiconvexity reduces to pseudoconvexity.

Theorem 7.2.1. *Consider the function f, where the matrix Q is not positive semidefinite. Then, f is quasiconvex on D if and only if f is pseudoconvex on D.*

Proof. We have

$$\nabla f(x) = \frac{1}{d^T x + d_0}[Qx + q - f(x)d],$$

$$(d^T x + d_0)\nabla^2 f(x) = Q + \frac{1}{d^T x + d_0}[2f(x)dd^T - (Qx + q)d^T - d(Qx + q)^T].$$

Since $h^T \nabla f(x) = 0$ if and only if $h^T(Qx + q) = f(x)h^T d$, we have

$$(d^T x + d_0)h^T \nabla^2 f(x)h = h^T Qh + \frac{2d^T h}{d^T x + d_0}[f(x)(d^T h) - h^T(Qx + q)] = h^T Qh.$$

From the quasiconvexity of f we have (see Theorem 3.4.4)

$$x \in D, \quad h^T \nabla f(x) = 0 \Rightarrow h^T \nabla^2 f(x)h \geq 0.$$

It follows that $\nabla f(x) \neq 0$, $\forall x \in D$, since $\nabla f(x) = 0$ implies that $(d^T x + d_0)h^T \nabla^2 f(x)h = h^T Qh \geq 0$ for all $h \in \Re^n$ in contradiction with the assumption on matrix Q. The thesis follows from Theorem 3.2.6. □

The following theorem states a necessary condition for f to be pseudoconvex. In particular, the theorem shows that the pseudoconvexity of f implies the pseudoconvexity of the quadratic function $n(x)$ on its maximal domains.

Theorem 7.2.2. *Consider the function f, where the matrix Q is not positive semidefinite. If f is pseudo convex, then the following conditions hold:*
(i) $\nu_-(Q) = 1$;
(ii) $rank Q = rank[Q, q] = rank[Q, d]$.

Proof. (i) Since Q is not positive semidefinite we have $\nu_-(Q) \geq 1$. Suppose by contradiction $\nu_-(Q) > 1$ and let v^1 and v^2 be two linearly independent eigenvectors associated with two negative eigenvalues of Q. Let W be the linear subspace generated by v^1 and v^2. Since $dim \, W = 2$ and $dim \, d^\perp = n - 1$, there exists $v \in W \cap d^\perp, v \neq 0$; $v \in W$ implies that v is a linear combination of v^1, v^2 so that $v^T Qv < 0$. Consider the line $x = x_0 + tv$, $x_0 \in D$, $t \in \Re$ which is contained in D since $d^T x + d_0 = d^T x_0 + d_0 > 0$. It is easy to verify that the restriction $\varphi(t) = f(x_0 + tv)$ is of the kind $\varphi(t) = at^2 + \beta t + \gamma$ with $\alpha < 0$ and this contradicts the pseudoconvexity of f.
(ii) The thesis is obvious if Q is nonsingular. Otherwise, let v^i, $i = 1, ..., p$, be eigenvectors associated with the non-null eigenvalues of Q and let v^j, $j = p + 1, ..., n$, be eigenvectors associated with the null eigenvalues. Since $\nu_-(Q) = 1$, let v^1 be such that $Qv^1 = \lambda_1 v^1, \lambda_1 < 0$. In correspondence to a feasible point x_0 assume the existence of v^j, $j = p + 1, ..., n$, such that $\nabla f(x_0)^T v^j \neq 0$. Consider the restriction $\varphi(t) = f(x_0 + tv)$, where $v = v^1 + kv^j$, $k \in \Re$. We have $\varphi'(0) = \nabla f(x_0)^T v = \nabla f(x_0)^T v^1 + k\nabla f(x_0)^T v^j$ so that, for $k^* = \frac{-\nabla f(x_0)^T v^1}{\nabla f(x_0)^T v^j}$, it results that $\varphi'(0) = 0$ and $\varphi''(0) = \frac{v^T Qv}{d^T x_0 + d_0} = \frac{\lambda_1 \| v^1 \|^2}{d^T x_0 + d_0}$ < 0. It follows that $\varphi(t)$ is not pseudoconvex and this is a contradiction.

Consequently, $\nabla f(x_0)^T v^j = 0$, $\forall j \in \{p+1, ..., n\}$, i.e., $\nabla f(x_0) \in ImQ$, i.e., there exists $u \in \Re^n, u \neq 0$ such that $Qu = \dfrac{Qx_0 + q - f(x_0)d}{d^T x_0 + d_0}$ or, equivalently, there exists u^0 such that $Qu^0 = q - f(x_0)d$. Since f is not constant on D, let $x^1 \in D$ with $f(x^1) \neq f(x_0)$. Analogously, there exists $u^1 \in \Re^n$ such that $Qu^1 = q - f(x^1)d$. We have $Q\dfrac{u^0 - u^1}{f(x^1) - f(x_0)} = d$, so that $rankQ = rank[Q, d]$ and, consequently, $rankQ = rank[Q, q]$. $\square$

The fundamental result regarding the characterization of the pseudoconvexity of the ratio between a quadratic function and an affine function is given in the following theorem whose proof can be found in [44].

Theorem 7.2.3. *The function*

$$f(x) = \frac{n(x)}{d(x)} = \frac{\frac{1}{2}x^T Qx + q^T x + q_0}{d^T x + d_0}$$

is pseudoconvex on $D = \{x \in \Re^n : d^T x + d_0 > 0\}$ *if and only if one of the following conditions holds:*
(i) $\nu_-(Q) = 0$ *(i.e., Q is positive semidefinite);*
(ii) $\nu_-(Q) = 1$, $\bar{x}$ *and* $\bar{y}$ *exist so that* $Q\bar{x} = -q$ *and* $Q\bar{y} = d$, $d^T \bar{y} = 0$, $d^T \bar{x} + d_0 = 0$ *and* $n(\bar{x}) \geq 0$;
(iii) $\nu_-(Q) = 1$, $\bar{x}$ *and* $\bar{y}$ *exist so that* $Q\bar{x} = -q$, $Q\bar{y} = d$, $d^T \bar{y} < 0$ *and* $(d^T \bar{x} + d_0)^2 + 2n(\bar{x})d^T \bar{y} \leq 0$.

Remark 7.2.1. It can be shown [55] that the study of the pseudolinearity of f is equivalent to the study of the pseudolinearity of the sum between a linear and a linear fractional function which will be carried out in the next section.

When the quadratic function $n(x)$ assumes a particular form, it is possible to specify Theorem 7.2.3. Consider, for instance, the following function studied in [52]:

$$g(x) = \frac{(a^T x + a_0)(b^T x + b_0) + c^T x + c_0}{d^T x + d_0}$$

$a, b, c, d \in \Re^n$, $d \neq 0$, $a_0, b_0, c_0, d_0 \in \Re$, $x \in D = \{x \in \Re^n : d^T x + d_0 > 0\}$. Regarding the pseudoconvexity of g, we have the following characterization related to the case where the vectors a, b are linearly independent (the case where a, b are linearly dependent is suggested in Exercise 7.3).

Theorem 7.2.4. *Consider the function g where a, b are linearly independent. Then, g is pseudoconvex on D if and only if there exist $\gamma_1, \gamma_2 \in \Re$ such that $c = \gamma_1 a + \gamma_2 b$, there exist $\delta_1, \delta_2 \in \Re$ such that $d = \delta_1 a + \delta_2 b$, and one of the following conditions holds: (i)* $d_0 = \delta_1(\gamma_2 + a_0) + \delta_2(\gamma_1 + b_0)$, $\delta_1\delta_2 = 0$, $c_0 + a_0 b_0 \geq (\gamma_1 + b_0)(\gamma_2 + a_0)$;
(ii) $[d_0 - \delta_1(\gamma_2 + a_0) - \delta_2(\gamma_1 + b_0)]^2 + 4\delta_1\delta_2[c_0 + a_0 b_0 - (\gamma_1 + b_0)(\gamma_2 + a_0)] \leq 0$, $\delta_1\delta_2 < 0$.

Proof. We have $g(x) = \dfrac{\frac{1}{2}x^T Q x + q^T x + q_0}{d^T x + d_0}$ where $Q = ab^T + ba^T$, and $q = b_0 a + a_0 b + c$, $q_0 = a_0 b_0 + c_0$. With reference to Theorem 7.2.3, taking into account that $\nu_-(Q) = 1$ and that ImQ is the subspace generated by a and b, the existence of $\bar{x}, \bar{y}$ such that $Q\bar{x} = -q$, $Q\bar{y} = d$ is equivalent to stating that c and d are linear combinations of a and b. More precisely, we have

$c = \gamma_1 a + \gamma_2 b$, $\gamma_1 = -(b^T \bar{x} + b_0)$, $\gamma_2 = -(a^T \bar{x} + a_0)$,
$d = \delta_1 a + \delta_2 b$, $\delta_1 = b^T \bar{y}$, $\delta_2 = a^T \bar{y}$.

By means of simple calculations we have

$d^T \bar{y} = 2\delta_1 \delta_2$, $d^T \bar{x} + d_0 = -\delta_1(\gamma_2 + a_0) - \delta_2(\gamma_1 + b_0) + d_0$,
$n(\bar{x}) = c_0 + a_0 b_0 - (\gamma_1 + b_0)(\gamma_2 + a_0)$.

It follows that (ii) and (iii) of Theorem 7.2.3 are equivalent to (i) and (ii) together with $c, d \in ImQ$. $\qquad\square$

7.3 The Sum of a Linear and a Linear Fractional Function

Consider the function

$$f(x) = a^T x + \frac{c^T x + c_0}{d^T x + d_0}$$

where $a, c, d \in \Re^n$, $d \neq 0$, $c_0, d_0 \in \Re$, $x \in D = \{x \in \Re^n : d^T x + d_0 > 0\}$. Such a class of functions has been studied by several authors in the framework of optimization problems (see Chapter 8).

In general, f may have local minimum points which are not global as is shown in the following example.

Example 7.3.1. Consider the function $f(x_1, x_2) = -x_1 + x_2 + \dfrac{-2x_1 - 7x_2 - 6}{x_1 + x_2 + 1}$ on the domain $S = \{(x_1, x_2) \in \Re^2 : x_1 \geq 0, 0 \leq x_2 \leq 4, x_1 - x_2 \leq 4\}$. It is easy to verify that $(0,0)$ is a local minimum point, while the global minimum is attained at $(8, 4)$.

In order to guarantee the local-global property, the results given in Theorem 7.2.3 (see also Exercise 7.14) may be utilized for characterizing the pseudoconvexity of f on D. In this section we shall present a general approach which allows us to characterize the pseudoconvexity of f on an arbitrary open convex set $S \subseteq D$. The obtained results will also allow us to study in Sect. 7.5 the pseudoconvexity of the sum of two linear fractional functions.

The gradient and the Hessian matrix of f are given, respectively, by

$$\nabla f(x) = a + \frac{1}{d^T x + d_0}\left(c - \frac{c^T x + c_0}{d^T x + d_0}d\right) \qquad (7.2)$$

$$\nabla^2 f(x) = \frac{1}{(d^T x + d_0)^2} \left[-(cd^T + dc^T) + 2\frac{c^T x + c_0}{d^T x + d_0} dd^T \right] \qquad (7.3)$$

The following theorem shows that the linear independence of a, c, d implies that f is not pseudoconvex on S, whatever the open convex set S be.

Theorem 7.3.1. *Let f be pseudoconvex on an open convex set $S \subseteq D$. Then, the vectors a, c, d are linearly dependent.*

Proof. The thesis clearly holds if $n = 2$. Let $n \geq 3$ and assume that $rank[a, c, d] = 3$. We have $\nabla f(x) \neq 0$, $\forall x \in S$, and $rank[\nabla f(x), a, d] = 3$, $\forall x \in S$. Let $A = \begin{bmatrix} \nabla^T f(x) \\ a^T \\ d^T \end{bmatrix}$; for every fixed $x \in S$ the linear map $A : \Re^n \to \Re^3$ is surjective so that there exist w_1, w_2 such that $\nabla f(x)^T w_1 = 0$, $a^T w_1 = 0$, $d^T w_1 < 0$ and $\nabla f(x)^T w_2 = 0$, $a^T w_2 > 0$, $d^T w_2 = 0$.
By setting $w = w_1 + w_2$ we have $\nabla f(x)^T w = 0$, $a^T w > 0$, $d^T w < 0$.
We shall prove that $\nabla f(x)^T w = 0$ implies $w^T \nabla^2 f(x) w < 0$, $\forall x \in S$.
The equality $\nabla f(x)^T w = 0$ implies

$$a^T w + \frac{c^T w}{d^T x + d_0} - \frac{c^T x + c_0}{(d^T x + d_0)^2} d^T w = 0. \qquad (7.4)$$

If $x \in S$ is such that $c^T x + c_0 = 0$, from (7.4) $c^T w = -a^T w(d^T x + d_0) < 0$, so that $w^T \nabla^2 f(x) w < 0$.
Consider now the case $c^T x + c_0 \neq 0$. If $c^T w = 0$, from (7.4) we have $\frac{c^T x + c_0}{d^T x + d_0} d^T w = a^T w(d^T x + d_0)$ so that

$$w^T \nabla^2 f(x) w = \frac{2}{(d^T x + d_0)^2} a^T w \, d^T w(d^T x + d_0) < 0.$$

If $c^T w \neq 0$, from (7.4) we have $c^T w = \frac{c^T x + c_0}{d^T x + d_0} d^T w - a^T w(d^T x + d_0)$ so that

$$w^T \nabla^2 f(x) w = \frac{2}{d^T x + d_0} a^T w \, d^T w < 0.$$

To sum up, condition $\nabla f(x)^T w = 0$ implies that $w^T \nabla^2 f(x) w < 0$ for all $x \in S$ and, consequently, f is not pseudoconvex on S (see Theorem 3.4.7) and this contradicts the assumption. The linear dependence of a, c, d follows. $\square$

The following theorem characterizes the maximal open domains of pseudoconvexity of function f.

Theorem 7.3.2. *Consider the function f. The following conditions hold:*
(i) if $a = \alpha d$, $\alpha \geq 0$, then f is pseudoconvex on D;
(ii) if $c = \gamma d$, $c_0 - \gamma d_0 \geq 0$, then f is pseudoconvex on D;

(iii) if $a = \alpha d$, $\alpha < 0$, and $c = \gamma d$, $c_0 - \gamma d_0 < 0$, then f is pseudoconvex on every open convex set S such that:

$$S \subseteq \{x \in \Re^n : d^T x + d_0 > d_0^*\}$$

or

$$S \subseteq \{x \in \Re^n : 0 < d^T x + d_0 < d_0^*\}$$

where $d_0^* = \sqrt{\dfrac{c_0 - \gamma d_0}{\alpha}}$;

(iv) if $c = \beta a + \gamma d$, $\beta > 0$ and $rank[a, d] = 2$, then f is pseudoconvex on every open convex set S such that:

$$S \subseteq \{x \in \Re^n : \beta a^T x + c_0 - \gamma d_0 > 0, \; d^T x + d_0 > 0\};$$

(v) if $c = \beta a + \gamma d$, $\beta < 0$ and $rank[a, d] = 2$, then f is pseudoconvex on every open convex set S such that:

$$S \subseteq \{x \in \Re^n : \beta a^T x + c_0 - \gamma d_0 > 0, \; d^T x + d_0 + \beta > 0\}$$

or

$$S \subseteq \{x \in \Re^n : \beta a^T x + c_0 - \gamma d_0 < 0, \; 0 < d^T x + d_0 < -\beta\}.$$

In any other case f is not pseudoconvex on $S \subseteq D$ whatever the open convex set S be.

Proof. Taking into account Theorem 7.3.1, we must analyze the exhaustive cases $rank[a, c, d] = 1$, $rank[a, c, d] = 2$.

• $rank[a, c, d] = 1$.

Let $a = \alpha d$, $c = \gamma d$. We have

$$\nabla f(x) = \frac{1}{(d^T x + d_0)^2} [\alpha(d^T x + d_0)^2 - (c_0 - \gamma d_0)]d$$

$$\nabla^2 f(x) = \frac{2}{(d^T x + d_0)^3} (c_0 - \gamma d_0) dd^T.$$

If $c_0 - \gamma d_0 \geq 0$, then $\nabla^2 f(x)$ is positive semidefinite on D so that f is convex (in particular, pseudoconvex) on D.

Consider now the case $c_0 - \gamma d_0 < 0$.

If $\alpha \geq 0$, then $\nabla f(x) \neq 0$ for all $x \in D$ so that $\nabla f(x)^T v = 0$ implies that $v^T \nabla^2 f(x) v = 0$ and, consequently, f is pseudoconvex on D.

If $\alpha < 0$, then $\nabla f(x) = 0$ for all $x \in D^* = \{x \in \Re^n : d^T x + d_0 = d_0^*\}$, where $d_0^* = \sqrt{\dfrac{c_0 - \gamma d_0}{\alpha}}$; choosing v such that $d^T v \neq 0$, we have $v^T \nabla^2 f(x) v < 0$ for every $x \in D^*$, so that f is not pseudoconvex on every open convex set S such that $S \cap D^* \neq \emptyset$, while it is pseudoconvex on every open convex set $S \subset D$ such that $S \cap D^* = \emptyset$. Consequently, (iii) holds.

• $rank[a, c, d] = 2$.

The following two exhaustive cases occur:

(a) $a = \alpha d$ and $rank[c, d] = 2$;

(b) $c = \beta a + \gamma d$ and $rank[a, d] = 2$.

(a) We have

$$\nabla f(x) = \frac{1}{d^T x + d_0} \left[c + \left(\alpha(d^T x + d_0) - \frac{c^T x + c_0}{d^T x + d_0} \right) d \right]$$

The linear independence of c and d implies that $\nabla f(x) \neq 0$, $\forall x \in D$. For every $v \in \Re^n$ such that $\nabla f(x)^T v = 0$, we have $v^T \nabla^2 f(x) v = \frac{2}{d^T x + d_0} \alpha(d^T v)^2$.

Consequently, if $\alpha \geq 0$, then f is pseudoconvex on D, while, in the case $\alpha < 0$, f is not pseudoconvex on every open convex set $S \subseteq D$ since we can choose v such that $d^T v \neq 0$.

Note that if $a = \alpha d$, $\alpha \geq 0$, then f is pseudoconvex on D either when $rank[a, c, d] = 1$ or when $rank[a, c, d] = 2$, and thus (i) holds.

(b) We have

$$\nabla f(x) = \frac{1}{d^T x + d_0} \left[(d^T x + d_0 + \beta)\, a - \frac{\beta a^T x + c_0 - \gamma d_0}{d^T x + d_0}\, d \right]$$

$$\nabla^2 f(x) = \frac{1}{(d^T x + d_0)^2} \left[-\beta(ad^T + da^T) + 2\frac{\beta a^T x + c_0 - \gamma d_0}{d^T x + d_0}\, dd^T \right].$$

It follows that $\nabla f(x) = 0$ if and only if $\beta < 0$ and $x \in \Gamma$ where

$$\Gamma = \{x \in \Re^n : d^T x + d_0 + \beta = 0,\ \beta a^T x + c_0 - \gamma d_0 = 0\}.$$

Since $rank[a, d] = 2$, for every $x \in \Gamma$ the Hessian matrix $\nabla^2 f(x)$ is indefinite so that f is not pseudoconvex on every open convex set S such that $S \cap \Gamma \neq \emptyset$. If $\nabla f(x) \neq 0$ and $d^T x + d_0 + \beta = 0$, then $\nabla f(x)^T v = 0$ implies that $d^T v = 0$, so that $v^T \nabla^2 f(x) v = 0$. If $\nabla f(x) \neq 0$ and $d^T x + d_0 + \beta \neq 0$, then $\nabla f(x)^T v = 0$ implies that $a^T v = \dfrac{\beta a^T x + c_0 - \gamma d_0}{(d^T x + d_0 + \beta)(d^T x + d_0)} d^T v$, so that

$$v^T \nabla^2 f(x) v = \frac{2}{(d^T x + d_0)^2} \frac{\beta a^T x + c_0 - \gamma d_0}{d^T x + d_0 + \beta} (d^T v)^2.$$

Consequently, if $\beta > 0$, then $\nabla f(x) \neq 0$ for all $x \in D$ and f is pseudoconvex on every open convex set S such that $S \subseteq \{x \in \Re^n : \beta a^T x + c_0 - \gamma d_0 > 0\} \cap D$ and (iv) holds.

If $\beta < 0$, then f is pseudoconvex on S if

$$S \subseteq \{x \in \Re^n : \beta a^T x + c_0 - \gamma d_0 > 0,\ d^T x + d_0 + \beta > 0\}$$

or

$$S \subseteq \{x \in \Re^n : \beta a^T x + c_0 - \gamma d_0 < 0,\ d^T x + d_0 + \beta < 0\} \cap D$$

and (v) holds.

If $\beta = 0$ and $c_0 - \gamma d_0 \geq 0$, then f is convex on D. Consequently, if $c = \gamma d$ and $c_0 - \gamma d_0 \geq 0$, f is convex (in particular, pseudoconvex) on D either when $rank[a, c, d] = 1$ or when $rank[a, c, d] = 2$ so that (ii) holds.
The proof is complete.
$\square$

Corollary 7.3.1. *The function f is pseudoconvex on D if and only if it assumes one of the following forms:*

(i) $f(x) = \alpha d^T x + \dfrac{c^T x + c_0}{d^T x + d_0}, \ \alpha \geq 0;$

(ii) $f(x) = a^T x + \dfrac{c_0 - \gamma d_0}{d^T x + d_0} + \gamma, \ c_0 - \gamma d_0 \geq 0.$

Remark 7.3.1. Note that in case (ii) of the previous Corollary the function f is convex on D.

The following theorem states the results given in Theorem 7.3.2 in terms of pseudoconcavity.

Theorem 7.3.3. *Consider the function f. The following conditions hold.*
(i) If $a = \alpha d$, $\alpha \leq 0$, then f is pseudoconcave on D.
(ii) If $c = \gamma d$, $c_0 - \gamma d_0 \leq 0$, then f is pseudoconcave on D.
(iii) If $a = \alpha d$, $\alpha > 0$, and $c = \gamma d$, $c_0 - \gamma d_0 > 0$, then f is pseudoconcave on every open convex set S such that:

$$S \subseteq \{x \in \Re^n : d^T x + d_0 > d_0^*\}$$

or

$$S \subseteq \{x \in \Re^n : 0 < d^T x + d_0 < d_0^*\}$$

where $d_0^* = \sqrt{\dfrac{c_0 - \gamma d_0}{\alpha}}.$
(iv) If $c = \beta a + \gamma d$, $\beta > 0$ and $rank[a, d] = 2$, then f is pseudoconcave on every open convex set S such that:

$$S \subseteq \{x \in \Re^n : \beta a^T x + c_0 - \gamma d_0 < 0, \ d^T x + d_0 > 0\}.$$

(v) If $c = \beta a + \gamma d$, $\beta < 0$ and $rank[a, d] = 2$, then f is pseudoconcave on every open convex set S such that:

$$S \subseteq \{x \in \Re^n : \beta a^T x + c_0 - \gamma d_0 < 0, \ d^T x + d_0 + \beta > 0\}$$

or

$$S \subseteq \{x \in \Re^n : \beta a^T x + c_0 - \gamma d_0 > 0, \ 0 < d^T x + d_0 < -\beta\}.$$

In any other case f is not pseudoconcave on $S \subseteq D$, whatever the open convex set S be.

By combining Theorems 7.3.2 and 7.3.3, the characterization of the pseudo-linearity of f is achieved.

Theorem 7.3.4. *Consider the function* f. *The following conditions hold:*
(i) if $a = 0$, *then* f *is pseudolinear on* D;
(ii) if $c = \gamma d$, *with* $c_0 - \gamma d_0 = 0$, *then* f *is pseudolinear on* D;
(iii) if $a = \alpha d$, $c = \gamma d$, *with* $\alpha(c_0 - \gamma d_0) < 0$, *then* f *is pseudolinear on* D;
(iv) if $a = \alpha d$, $c = \gamma d$, *with* $\alpha(c_0 - \gamma d_0) > 0$, *then* f *is pseudolinear on every open convex set* $S \subseteq D$ *such that*

$$S \subseteq \{x \in \Re^n : d^T x + d_0 > d_0^*\}$$

or

$$S \subseteq \{x \in \Re^n : 0 < d^T x + d_0 < d_0^*\}$$

where $d_0^* = \sqrt{\frac{c_0 - \gamma d_0}{\alpha}}$.
In any other case f *is not pseudolinear on* $S \subseteq D$ *whatever the open convex set* S *be.*

We have studied the pseudoconvexity of f by assuming $d^T x + d_0 > 0$; the obtained results can be easily adapted to the case $d^T x + d_0 < 0$, re-writing f as follows: $f(x) = a^T x + \dfrac{-c^T x - c_0}{-d^T x - d_0}$.
In the following example we find the maximal open convex domains of the pseudoconvexity of function f when the denominator is either positive or negative.

Example 7.3.2. Consider the function

$$f(x_1, x_2) = 2x_1 + 32x_2 + \frac{-2x_1 + 3x_2 + 2}{3x_1 + 13x_2 + 1}$$

With respect to the domain $D = \{(x_1, x_2) \in \Re^2 : 3x_1 + 13x_2 + 1 > 0\}$ it is easy to verify that (iv) of Theorem 7.3.2 holds with $\beta = \frac{1}{2}$, $\gamma = -1$, so that f is pseudoconvex on every open convex set $S \subseteq D_1^+ = \{(x_1, x_2) \in \Re^2 : x_1 + 16x_2 + 3 > 0, \ 3x_1 + 13x_2 + 1 > 0\}$.
In order to study the pseudoconvexity of the function f on the half-plane $\{(x_1, x_2) \in \Re^2 : 3x_1 + 13x_2 + 1 < 0\}$, we can re-write f as follows: $f(x_1, x_2) = 2x_1 + 32x_2 + \dfrac{2x_1 - 3x_2 - 2}{-3x_1 - 13x_2 - 1}$. By reapplying again Theorem 7.3.2, case (v) occurs so that f is pseudoconvex on every open convex set contained in $D_1^- = \{(x_1, x_2) \in \Re^2 : x_1 + 16x_2 + 3 < 0, \ 3x_1 + 13x_2 + \frac{3}{2} < 0\}$ or in $D_2^- = \{(x_1, x_2) \in \Re^2 : x_1 + 16x_2 + 3 > 0, \ -\frac{1}{2} < 3x_1 + 13x_2 + 1 < 0\}$.
The maximal domains of the pseudoconvexity of the given function are depicted in Figure 7.1.

Remark 7.3.2. In [55] it has been shown that the ratio (7.1) between a quadratic function and an affine function is pseudolinear on D if and only if it reduces to the sum of a linear and a linear fractional function satisfying (i–iii) of Theorem 7.3.4.

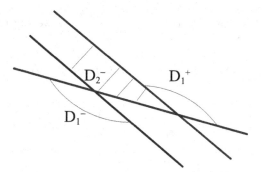

Fig. 7.1. Maximal domains

7.4 Pseudoconvexity and the Charnes–Cooper Variable Transformation

Charnes–Cooper's transformation of variables was originally suggested by Charnes A. and Cooper W. W. in [64] with the aim of transforming a linear fractional optimization problem into an equivalent linear programmming problem (see also Sect. 8.2). Subsequently, this transformation and its generalizations have been utilized in different contexts [13, 248, 256]. According to a recent approach [48, 49], Charnes–Cooper's transformation also turns out to be a useful tool in studying the pseudoconvexity of some classes of functions since it preserves pseudoconvexity.

More precisely, consider Charnes–Cooper's transformation

$$y = \frac{x}{b^T x + b_0} \tag{7.5}$$

where $b \in \Re^n \backslash \{0\}$, $b_0 \in \Re \backslash \{0\}$, $x \in S = \{x \in \Re^n : b^T x + b_0 > 0\}$. This map is a diffeomorphism and its inverse is

$$x = \frac{b_0 y}{1 - b^T y} \tag{7.6}$$

where $y \in S^* = \{y \in \Re^n : \frac{b_0}{1 - b^T y} > 0\}$.

We shall see how Charnes–Cooper's transformation may be utilized in order to study the pseudoconvexity of the following fractional function

$$f(x) = \frac{\frac{1}{2} x^T A x + a^T x + a_0}{(b^T x + b_0)^2}$$

where A is an $n \times n$ symmetric matrix, $a, b \in \Re^n$, $b \neq 0$, $a_0, b_0 \in \Re$, $b_0 \neq 0$, $x \in S = \{x \in \Re^n : b^T x + b_0 > 0\}$.

By performing the transformation $x = \dfrac{b_0 y}{1 - b^T y}$, we obtain the following quadratic function

$$f(x(y)) = \psi(y) = y^T Q y + q^T y + q_0$$

where

$$Q = \frac{1}{2} A - \frac{ab^T + ba^T}{2b_0} + \frac{a_0}{b_0^2} bb^T, \quad q = \frac{1}{b_0}(a - 2\frac{a_0}{b_0} b), \quad q_0 = \frac{a_0}{b_0^2}, \qquad (7.7)$$

$$y \in S^* = \{ y \in \Re^n : \frac{b_0}{1 - b^T y} > 0 \}.$$

From Corollary 5.6.1, the function $f(x)$ is pseudoconvex on S if and only if the quadratic function $\psi(y)$ is pseudoconvex on the half-space S^* or, equivalently, when $\psi(y)$ is convex on $\Re^n$ or is merely pseudoconvex on S^* (see Corollary 6.3.3). In any case, the problem reduces to express the results related to the characterization of the pseudoconvexity of $\psi(y)$ in terms of the initial data of $f(x)$. This has been done in [58], where the following results are given, according to cases $ker A = b^\perp$ and $ker A \neq b^\perp$.

Theorem 7.4.1. *Consider the function f with $A = \delta bb^T$, $\delta \in \Re$.*
Then, f is pseudoconvex on S if and only if one of the following conditions holds:
(i) $a = \gamma b$, $\gamma \in \Re$ and $\delta b_0^2 - 2\gamma b_0 + 2a_0 \geq 0$;
(ii) $a = \gamma b$, $\gamma \in \Re$, $\delta b_0^2 - 2\gamma b_0 + 2a_0 < 0$ and $\gamma \leq \delta b_0$.

Theorem 7.4.2. *If $ker A \neq b^\perp$, the function f is pseudoconvex on the half-space S if and only if A is positive semidefinite on $b^\perp$ and one of the following conditions holds:*
(i) there exists $\alpha \in \Re$ such that $Ab - \frac{\|b\|^2}{b_0} a = \alpha b$ with

$$\alpha \geq \frac{b_0 b^T a - 2 \|b\|^2 a_0}{b_0^2}; \qquad (7.8)$$

(ii) $Ab - \frac{\|b\|^2}{b_0} a \neq \alpha b$ for every $\alpha \in \Re$, there exist $a^*, b^* \in \Re^n$ such that $Ab^* = b$, $Aa^* = a$, $b^* \in b^\perp$, $b^T a^* = b_0$ and

$$a^{*T} a \leq 2a_0; \qquad (7.9)$$

(iii) $Ab - \frac{\|b\|^2}{b_0} a \neq \alpha b$ for every $\alpha \in \Re$, there exist $a^*, b^* \in \Re^n$ such that $Ab^* = b$, $Aa^* = a$, $b^{*T} b \neq 0$ and

$$a_0 - \frac{a^{*T} a}{2} + \frac{1}{2b^T b^*} \left(b_0 - b^T a^* \right)^2 \geq 0; \qquad (7.10)$$

(iv) $Ab - \frac{\|b\|^2}{b_0} a \neq \alpha b$ for every $\alpha \in \Re$ and there exist $\mu^* \in \Re$, $a^* \in \Re^n$ such that $a = Aa^* + \mu^* b$, $b \notin Im A$ and

$$a_0 - \mu^* b_0 - \frac{1}{2} a^{*T} A a^* \geq 0. \qquad (7.11)$$

Remark 7.4.1. The pseudoconvexity of f on S implies that A has at most one negative eigenvalue (see Exercise 7.4).

Remark 7.4.2. When $ker A \neq b^{\perp}$, then f is pseudoconvex on S if and only if the quadratic function $\psi(y)$ is convex on $\Re^n$. The following example shows that the convexity of $\psi(y)$ does not guarantee the convexity of f and, at the same time, points out that Charnes–Cooper's transformation does not preserve convexity.

Example 7.4.1. Consider the function

$$f(x_1, x_2) = \frac{x_1^2 - x_2^2 + x_1 - x_2 + 1}{(x_2 + 1)^2}$$

$(x_1, x_2) \in S = \{(x_1, x_2) \in \Re^2 : x_2 + 1 > 0\}$.

According to (7.7) we have $Q = \begin{bmatrix} 1 & -\frac{1}{2} \\ -\frac{1}{2} & 1 \end{bmatrix}$.

Since Q is positive definite, $\psi(y)$ is convex on $\Re^2$ but $f(x_1, x_2)$ is not convex. In fact, the restriction of f on the line $x_2 = \frac{1}{2}x_1$ is $\varphi(x_1) = f(x_1, \frac{1}{2}x_1) = \frac{3x_1^2 + 2x_1 + 4}{(x_1 + 2)^2}$, so that $\varphi''(x_1) = \frac{4(8 - 5x_1)}{(x_1 + 2)^4}$. Consequently, φ is concave for $x_1 > \frac{8}{5}$ and convex for $-2 < x_1 < \frac{8}{5}$.

Remark 7.4.3. As we have pointed out, Charnes–Cooper's transformation may be a useful tool in studying the pseudoconvexity of new classes of generalized fractional functions which turn out to be equivalent (under this transformation) to classes of functions whose pseudoconvexity has already been characterized.

Charnes–Cooper's transformation may also be applied for reducing a class of functions into one for which the study of pseudoconvexity is easier to carry on. This kind of approach has been used, for instance, in [59], where the ratio between a quadratic form and the cube of an affine function has been transformed into the product between a quadratic form and the cube of an affine function.

7.5 Sum of Two Linear Fractional Functions

In the class of generalized convex functions, particular attention is devoted to the sum of the ratios of linear fractional functions because of its various applications in economic as well as in non-economic fields [114, 246, 250, 256]. Unfortunately, for the sum of two or more ratios, none of the properties of single ratio linear fractional function are still true. In particular, the sum may have several local not global minimum points. For this reason, the study of the pseudoconvexity of this particular class of functions becomes relevant. Results regarding this topic may be found in [28, 44, 48, 49].

In this section we shall focus our attention on the sum of two linear fractional functions whose pseudoconvexity can be studied by means of Charnes–Cooper's transformation.

Consider the function

$$h(x) = \frac{m^T x + m_0}{p^T x + p_0} + \frac{q^T x + q_0}{b^T x + b_0}$$

defined on $S = \{x \in \Re^n : p^T x + p_0 > 0, \ b^T x + b_0 > 0\}$, where $m, p, q, b \in \Re^n$, $p, b \neq 0$ and $m_0, p_0, q_0, b_0 \in \Re$, $p_0, b_0 \neq 0$.

The following theorems characterize the pseudoconvexity of h on S.

Theorem 7.5.1. *Consider the function h where $rank[p, b] = 2$. Then, h is pseudoconvex on S if and only if one of the following conditions holds.*
(i) There exists $\alpha \geq 0$ such that

$$p_0 m - m_0 p = \alpha(p_0 b - b_0 p). \tag{7.12}$$

(ii) There exists $\gamma \in \Re$ such that

$$p_0 q - q_0 p = \gamma(p_0 b - b_0 p), \quad \frac{q_0 - \gamma b_0}{p_0} \geq 0. \tag{7.13}$$

(iii) There exist $\beta > 0$, $\delta \in \Re$, $\lambda_1 \geq 0$, $\lambda_2 \geq 0$ such that

$$p_0 q - q_0 p = \beta(p_0 m - m_0 p) + \delta(p_0 b - b_0 p) \tag{7.14}$$

$$\beta(p_0 m - m_0 p) = \lambda_1(-p) + \lambda_2(p_0 b - b_0 p) \tag{7.15}$$

$$\frac{q_0 - \delta b_0 - (\lambda_1 + \lambda_2 b_0)}{p_0} \geq 0. \tag{7.16}$$

Proof. By applying Charnes–Cooper's transformation $y = \dfrac{x}{p^T x + p_0}$ whose inverse is $x = \dfrac{p_0 y}{1 - p^T y}$, function h is transformed into

$$\psi(y) = \frac{(p_0 m - m_0 p)^T y}{p_0} + \frac{(p_0 q - q_0 p)^T y + q_0}{(p_0 b - b_0 p)^T y + b_0} + \frac{m_0}{p_0}$$

while domain S is transformed into

$$S^* = \left\{ y \in \Re^n : \frac{1 - p^T y}{p_0} > 0, \ \frac{(p_0 b - b_0 p)^T y + b_0}{p_0} > 0 \right\}.$$

From Corollary 5.6.1, h is pseudoconvex on S if and only if ψ is pseudoconvex on S^*, so that the study of pseudoconvexity of h can be performed by applying Theorem 7.3.2 to the function ψ which can be expressed as follows:

$$\psi(y) = \frac{(p_0 m - m_0 p)^T y}{p_0} + \frac{\left(\frac{p_0 q - q_0 p}{p_0}\right)^T y + \frac{q_0}{p_0}}{\left(\frac{p_0 b - b_0 p}{p_0}\right)^T y + \frac{b_0}{p_0}} + \frac{m_0}{p_0}.$$

Since $rank[p, b] = 2$, the hyperplanes $1 - p^T y = 0$, $(p_0 b - b_0 p)^T y + b_0 = 0$ are non-parallel, so that case (iii) and case (v) of Theorem 7.3.2 cannot occur. Consequently, ψ is pseudoconvex on S^* if and only if (i) or (ii) or (iv) of Theorem 7.3.2 holds. As a result:

(i–ii) follow from (i) and (ii) of Theorem 7.3.2, respectively;

(iii) From (iv) of Theorem 7.3.2 we have (7.14).

Now we shall prove that S^* is contained in

$$D = \left\{ y \in \Re^n : \frac{\beta(p_0 m - m_0 p)^T y + q_0 - \gamma b_0}{p_0} > 0, \; \frac{(p_0 b - b_0 p)^T y + b_0}{p_0} > 0 \right\}$$

if and only if (7.15) and (7.16) hold.

Consider the cone $C = \left\{ y \in \Re^n : -\frac{p^T y}{p_0} \geq 0, \; \frac{(p_0 b - b_0 p)^T y}{p_0} \geq 0 \right\}$ and let $\bar{y}$ be such that $1 - p^T \bar{y} = 0$, $(p_0 b - b_0 p)^T \bar{y} + b_0 = 0$; $S^* = \bar{y} + C \subseteq D$ if and only if the half-space $\frac{\beta}{p_0}(p_0 m - m_0 p)^T y \geq 0$ supports C at the origin (this is equivalent to (7.15)) and $\frac{1}{p_0}[\beta(p_0 m - m_0 p)^T \bar{y} + q_0 - \delta b_0] \geq 0$. This last inequality is equivalent to (7.16) since $\frac{1}{p_0}[\beta(p_0 m - m_0 p)^T \bar{y} + q_0 - \delta b_0] = \frac{1}{p_0}[\lambda_1(1 - p^T \bar{y}) + \lambda_2[(p_0 b - b_0 p)^T \bar{y} + b_0] - (\lambda_1 + \lambda_2 b_0) + q_0 - \delta b_0] = \frac{1}{p_0}[q_0 - \delta b_0 - (\lambda_1 + \lambda_2 b_0)] \geq 0$. The proof is complete.   □

Consider now the case $rank[p, b] = 1$. The function h assumes the form

$$h_1(x) = \frac{m^T x + m_0}{p^T x + p_0} + \frac{q^T x + q_0}{k p^T x + b_0}.$$

When $k = 0$, h_1 reduces to the sum of a linear and a linear fractional function whose pseudoconvexity has been characterized in Sect. 7.3. When $k = \frac{b_0}{p_0}$, h_1 reduces to a linear fractional function which is pseudolinear. For these reasons and in order to avoid the trivial case $S = \emptyset$, in what follows we assume $k > 0$ or $k < 0$ and $p_0 k - b_0 \neq 0$.

The following theorem holds.

Theorem 7.5.2. *Consider the function h where $b = kp$, $k \neq 0$, and assume $S \neq \emptyset$. Then, h is pseudoconvex on S if and only if one of the following conditions holds.*

(i) There exists $\alpha \geq 0$ such that

$$p_0 m - m_0 p = \alpha(k p_0 - b_0)p. \tag{7.17}$$

(ii) There exists $\gamma \in \Re$ such that

$$p_0 q - q_0 p = \gamma(k p_0 - b_0)p, \quad \frac{q_0 - \gamma b_0}{p_0} \geq 0. \tag{7.18}$$

(iii) There exist $\alpha < 0$, $\gamma \in \Re$, such that $p_0 m - m_0 p = \alpha(k p_0 - b_0)p$, $p_0 q - q_0 p = \gamma(k p_0 - b_0)p$ with $\frac{q_0 - \gamma b_0}{p_0} < 0$. Furthermore, $k \geq \sqrt{\dfrac{q_0 - \gamma b_0}{p_0 \alpha}}$ if $p_0 k - b_0 < 0$, or $k \leq \sqrt{\dfrac{q_0 - \gamma b_0}{p_0 \alpha}}$ if $p_0 k - b_0 > 0$.

Proof. By applying Charnes–Cooper's transformation $y = \dfrac{x}{p^T x + p_0}$ whose inverse is $x = \dfrac{p_0 y}{1 - p^T y}$, the function h is transformed into

$$\psi(y) = \frac{(p_0 m - m_0 p)^T y}{p_0} + \frac{(p_0 q - q_0 p)^T y + q_0}{(p_0 k - b_0)p^T y + b_0} + \frac{m_0}{p_0}$$

while domain S is transformed into

$$S^* = \left\{ y \in \Re^n : \frac{1 - p^T y}{p_0} > 0, \ \frac{(p_0 k - b_0)p^T y + b_0}{p_0} > 0 \right\}.$$

Since $1 - p^T y = 0$, $(p_0 k - b_0)p^T y + b_0 = 0$ are parallel hyperplanes, case (iv) and case (v) of Theorem 7.3.2 cannot occur, so that ψ is pseudoconvex on S^* if and only if (i) or (ii) or (iii) of Theorem 7.3.2 holds.

Consequently (i) and (ii) follow from (i) and (ii) of Theorem 7.3.2, respectively, while (iii) follows from (iii) of Theorem 7.3.2, taking into account that

$$S^* \subseteq \left\{ y \in \Re^n : \frac{(p_0 k - b_0)p^T y + b_0}{p_0} > \sqrt{\frac{q_0 - \gamma b_0}{p_0 \alpha}} \right\}$$

if and only if $p_0 k - b_0 < 0$ and $k \geq \sqrt{\dfrac{q_0 - \gamma b_0}{p_0 \alpha}}$, while

$$S^* \subseteq \left\{ y \in \Re^n : 0 < \frac{(p_0 k - b_0)p^T y + b_0}{p_0} < \sqrt{\frac{q_0 - \gamma b_0}{p_0 \alpha}} \right\}$$

if and only if $p_0 k - b_0 > 0$ and $k \leq \sqrt{\dfrac{q_0 - \gamma b_0}{p_0 \alpha}}$. □

Remark 7.5.1. It is important to note that regarding characterization of pseudoconvexity given in Theorem 7.5.1 and in Theorem 7.5.2, there is no difference between using $y = \dfrac{x}{p^T x + p_0}$ and $y = \dfrac{x}{b^T x + b_0}$.

In terms of pseudoconcavity we have the following results.

Theorem 7.5.3. *Consider the function* h *where* $rank[p, b] = 2$. *Then,* h *is pseudoconcave on* S *if and only if one of the following conditions holds.*
(i) There exists $\alpha \leq 0$ *such that*

$$p_0 m - m_0 p = \alpha(p_0 b - b_0 p). \tag{7.19}$$

(ii) There exists $\gamma \in \Re$ *such that*

$$p_0 q - q_0 p = \gamma(p_0 b - b_0 p), \quad \frac{q_0 - \gamma b_0}{p_0} \leq 0. \tag{7.20}$$

(iii) There exist $\beta > 0$, $\delta \in \Re$, $\lambda_1 \leq 0$, $\lambda_2 \leq 0$ *such that*

$$p_0 q - q_0 p = \beta(p_0 m - m_0 p) + \delta(p_0 b - b_0 p) \tag{7.21}$$

$$\beta(p_0 m - m_0 p) = \lambda_1(-p) + \lambda_2(p_0 b - b_0 p) \tag{7.22}$$

$$\frac{q_0 - \delta b_0 - (\lambda_1 + \lambda_2 b_0)}{p_0} \leq 0. \tag{7.23}$$

Theorem 7.5.4. *Consider the function* h *where* $b = kp$, $k \neq 0$, *and assumes* $S \neq \emptyset$. *Then,* h *is pseudoconcave on* S *if and only if one of the following conditions holds.*
(i) There exists $\alpha \leq 0$ *such that*

$$p_0 m - m_0 p = \alpha(kp_0 - b_0)p. \tag{7.24}$$

(ii) There exists $\gamma \in \Re$ *such that*

$$p_0 q - q_0 p = \gamma(kp_0 - b_0)p, \quad \frac{q_0 - \gamma b_0}{p_0} \leq 0. \tag{7.25}$$

(iii) There exist $\alpha > 0$, $\gamma \in \Re$ *such that* $p_0 m - m_0 p = \alpha(kp_0 - b_0)p$, $p_0 q - q_0 p = \gamma(kp_0 - b_0)p$ *with* $\frac{q_0 - \gamma b_0}{p_0} > 0$. *Furthermore,* $k \geq \sqrt{\dfrac{q_0 - \gamma b_0}{p_0 \alpha}}$ *if* $p_0 k - b_0 < 0$,

or $k \leq \sqrt{\dfrac{q_0 - \gamma b_0}{p_0 \alpha}}$ *if* $p_0 k - b_0 > 0$.

By combining the characterization of pseudoconvexity and pseudoconcavity of the function of h, the following result on pseudolinearity is obtained.

Theorem 7.5.5. *The function* h *is pseudolinear on* S *if and only if one of the following conditions holds.*
(i) $p_0 m - m_0 p = 0$.
(ii) There exists $\gamma \in \Re$ *such that* $p_0 q - q_0 p = \gamma(p_0 b - b_0 p)$ *with* $q_0 - \gamma b_0 = 0$.

(iii) There exist $\alpha,\ \gamma \in \Re$, *such that:*
$$p_0 m - m_0 p = \alpha(p_0 b - b_0 p), \quad p_0 q - q_0 p = \gamma(p_0 b - b_0 p) \ \text{ with } \ \alpha \frac{q_0 - \gamma b_0}{p_0} < 0.$$
(iv) there exist $k,\ \alpha,\ \gamma \in \Re,\ k \neq 0$, *such that:*
$$b = kp, \quad p_0 m - m_0 p = \alpha(p_0 b - b_0 p), \quad p_0 q - q_0 p = \gamma(p_0 b - b_0 p), \quad \alpha \frac{q_0 - \gamma b_0}{p_0} > 0,$$
and $k \geq \sqrt{\dfrac{q_0 - \gamma b_0}{p_0 \alpha}}$ *if* $p_0 k - b_0 < 0$, *or* $k \leq \sqrt{\dfrac{q_0 - \gamma b_0}{p_0 \alpha}}$ *if* $p_0 k - b_0 > 0$.

Proof. The proof follows taking into account of the results related to pseudoconvexity given in Theorem 7.5.1 and in Theorem 7.5.2 and of the ones related to pseudoconcavity given in Theorem 7.5.3 and in Theorem 7.5.4. □

Example 7.5.1. Consider the function

$$h(x_1, x_2) = \frac{2x_1 + 4x_2 + 4}{2x_1 - 3x_2 + 5} + \frac{\frac{2}{5}x_1 - \frac{3}{5}x_2 + 2}{x_1 + 2x_2 + 1}$$

on $S = \{(x_1, x_2) \in \Re^2 : 2x_1 - 3x_2 + 5 > 0, \ x_1 + 2x_2 + 1 > 0\}$.
We have:

$$p_0 q - q_0 p = (-2, 3)^T, \quad p_0 m - m_0 p = (2, 32)^T, \quad p_0 b - b_0 p = (3, 13)^T$$

so that (iii) of Theorem 7.5.1, holds with $\beta = \frac{1}{2}$, $\gamma = -1$, $\lambda_1 = \lambda_2 = 1$, and thus the given function is pseudoconvex on S.
The obtained result can also be achieved by applying Charnes–Cooper's transformation $z_1 = \dfrac{x_1}{2x_1 - 3x_2 + 5}$, $z_2 = \dfrac{x_2}{2x_1 - 3x_2 + 5}$ and its inverse $x_1 = \dfrac{5z_1}{1 - 2z_1 + 3z_2}$, $x_2 = \dfrac{5z_2}{1 - 2z_1 + 3z_2}$.
The transformed function $f(z_1, z_2) = 2z_1 + 3z_2 + \dfrac{-2z_1 + 3z_2 + 2}{3z_1 + 13z_2 + 1} + 4$ is pseudoconvex on $H_1^+ = \{(z_1, z_2) : z_1 + 16z_2 + 3 > 0, \ 3z_1 + 13z_2 + 1 > 0\}$ (see Example 7.3.2). It is easy to verify that the feasible set S is transformed into $S^* = \{(z_1, z_2) : 1 - 2z_1 + 3z_2 > 0, \ 3z_1 + 13z_2 + 1 > 0\}$, which is contained in H_1^+.

7.6 Exercises

7.1. Consider the quadratic fractional function $f(x) = \dfrac{\frac{1}{2}x^T Q x + q^T x + q_0}{d^T x + d_0}$

where $Q = \begin{bmatrix} -1 & 2 & -1 \\ 2 & -1 & -1 \\ -1 & -1 & 2 \end{bmatrix}$, $q = \begin{pmatrix} -3 \\ 3 \\ 0 \end{pmatrix}$, $d = \begin{pmatrix} -\sqrt{3} \\ 3 + 2\sqrt{3} \\ -3 - \sqrt{3} \end{pmatrix}$, $d_0 = -3 - 3\sqrt{3}$,

$q_0 \in \Re$.
Find q_0 such that f is pseudoconvex on $D = \{x : d^T x + d_0 > 0\}$.

7.2. Consider the quadratic fractional function $f(x) = \dfrac{\frac{1}{2}x^T Q x + q^T x + q_0}{d^T x + d_0}$

where $Q = \begin{bmatrix} -1 & 2 & -1 \\ 2 & -1 & -1 \\ -1 & -1 & 2 \end{bmatrix}$, $q = \begin{pmatrix} -4 \\ -1 \\ 5 \end{pmatrix}$, $d = \begin{pmatrix} -2 \\ 1 \\ 1 \end{pmatrix}$, $d_0 = 4$, $q_0 \in \Re$.

Find q_0 such that f is pseudoconvex on $D = \{x : d^T x + d_0 > 0\}$.

7.3. Consider the function $g(x) = \dfrac{(a^T x + a_0)(ka^T x + b_0) + c^T x + c_0}{d^T x + d_0}$, $d \neq 0$.

Prove that g is pseudoconvex on $S = \{x \in \Re^n : d^T x + d_0 > 0\}$ if and only if one of the following conditions holds:
(i) $k \geq 0$;
(ii) $k < 0$, $\exists \gamma, \delta$ such that $c = \gamma a$, $d = \delta a$ and $k(a_0\delta - d_0)^2 \geq \delta[d_0\gamma - c_0\delta + (a_0 k - b_0)(a_0\delta - d_0)]$.

7.4. Consider the function $f(x) = \dfrac{\frac{1}{2}x^T A x + a^T x + a_0}{(b^T x + b_0)^2}$, where A is an $n \times n$ symmetric matrix, $a, b \in \Re^n$, $b \neq 0$, $a_0, b_0 \in \Re$, $b_0 \neq 0$. By assuming that f is pseudoconvex on $S = \{x \in \Re^n : b^T x + b_0 > 0\}$, show that A has at most one negative eigenvalue.

7.5. Give an example which shows that the ratio between a quadratic convex function and the square of an affine function is not convex.

7.6. Consider the function $f(x) = \dfrac{\frac{1}{2}x^T A x + a^T x + a_0}{(b^T x + b_0)^2}$, where A is positive definite, $b \neq 0$, $b_0 \neq 0$.
Show that f is pseudoconvex on $S = \{x \in \Re^n : b^T x + b_0 > 0\}$ if and only if
$$2a_0 - a^T A^{-1}a + \frac{(b_0 - b^T A^{-1}a)^2}{b^T A^{-1}b} \geq 0.$$

7.7. Give an example which shows that the pseudoconvexity of the function $f(x) = \dfrac{\frac{1}{2}x^T A x}{(b^T x + b_0)^2}$ does not imply the convexity of f even if A is positive definite.

7.8. Consider $f(x) = \dfrac{\frac{1}{2}x^T A x + a^T x + a_0}{(b^T x + b_0)^2}$, where A is a nonsingular symmetric matrix, $b \neq 0$, $b_0 \neq 0$.
Show that f is pseudoconvex on $S = \{x \in \Re^n : b^T x + b_0 > 0\}$ if and only if one of the following conditions holds:
(i) $b^T A^{-1}b = 0$ and $2a_0 \geq a^T A^{-1}a$;
(ii) $b^T A^{-1}b \neq 0$ and $2a_0 - a^T A^{-1}a + \dfrac{(b_0 - b^T A^{-1}a)^2}{b^T A^{-1}b} \geq 0$.

7.9. Consider $f(x) = \dfrac{a^T x + a_0}{(b^T x + b_0)^2}$, $b \neq 0$, $b_0 \neq 0$. Show that f is pseudo-convex on $S = \{x \in \Re^n : b^T x + b_0 > 0\}$ if and only if one of the following conditions holds:

(i) $a = kb$ and $a_0 - kb_0 \geq 0$;

(ii) $a = kb$, $k \leq 0$ and $a_0 - kb_0 < 0$.

In particular, show that in case (ii) f is a concave function while in case (i) f is convex when $k > 0$.

7.10. Find a_0 such that the function

$$f(x_1, x_2, x_3) = \frac{\frac{1}{2}x_1^2 + x_2^2 + \frac{3}{2}x_3^2 + 2x_1x_2 + x_1 + 2x_2 + a_0}{(x_1 + 1)^2}$$

is pseudoconvex on $S = \{x \in \Re^3 : x_1 + 1 > 0\}$.

7.11. Find a_0 such that the function

$$f(x_1, x_2, x_3, x_4) = \frac{\frac{1}{2}x_1^2 + 2x_2^2 - x_3^2 + \frac{1}{2}x_4^2 + 2x_1x_2 + x_1 + 2x_2 + x_4 + a_0}{(2x_1 + 4x_2 - 2\sqrt{2}x_3 + 2)^2}$$

is pseudoconvex on $S = \{x \in \Re^3 : 2x_1 + 4x_2 - 2\sqrt{2}x_3 + 2 > 0\}$.

7.12. Find b_0 such that the function

$$f(x_1, x_2, x_3) = \frac{x_1^2 + x_2^2 - x_3^2 + 2x_1x_2 + x_1 + x_2 - x_3 + 1}{(x_3 + b_0)^2}$$

is pseudoconvex on $S = \{x \in \Re^3 : x_3 + b_0 > 0\}$.

7.13. Find a_2 such that the function

$$f(x_1, x_2, x_3) = \frac{x_1^2 + x_3^2 + x_1 + a_2x_2 + x_3 + 1}{(x_2 + 1)^2}$$

is pseudoconvex on $S = \{x \in \Re^3 : x_2 + 1 > 0\}$.

7.14. By specifying Theorem 7.2.3, show that $h(x) = a^T x + \dfrac{c^T x + c_0}{d^T x + d_0}$, $d \neq 0$, is pseudoconvex on $H^+ = \{x \in \Re^n : d^T x + d_0 > 0\}$ if and only if one of the following conditions holds:

(i) $a = kd$, $k \geq 0$;

(ii) There exists $t \in \Re$ such that $c = td$ and $c_0 \geq td_0$.

7.15. Find the maximal domains of pseudoconvexity of the following fractional functions, by assuming the positivity of the denominator:

(a) $f(x_1, x_2) = 3x_1 + 2x_2 + \dfrac{9x_1 - 4x_2 + 15}{x_1 + 4x_2 + 3}$;

(b) $f(x_1, x_2) = -5x_1 + x_2 + \dfrac{20x_1 - 7x_2 + 10}{2x_1 - x_2 + 8}$;

(c) $f(x_1, x_2) = -6x_1 - 10x_2 + \dfrac{9x_1 + 15x_2 + 4}{3x_1 + 5x_2 + 2}$;

(d) $f(x_1, x_2) = -x_1 + 3x_2 + \dfrac{x_1 + x_2 - 1}{-2x_1 + 6x_2 + 7}$.

7.16. Find the maximals domain of pseudoconvexity of the following parametric fractional functions, by assuming the positivity of the denominator.

(a) $f(x_1, x_2) = 4x_1 + (6 + \theta)x_2 + \dfrac{2x_1 + 4x_2 - 3}{2x_1 + 3x_2 + 5}$;

(b) $f(x_1, x_2) = -x_1 - x_2 + \dfrac{x_1 + x_2 + c_0}{2x_1 + 2x_2 + 3}$.

7.17. Verify the pseudoconvexity of the functions

(a) $f(x_1, x_2) = \dfrac{2x_1 - x_2 + 2}{x_1 + 2x_2 + 4} + \dfrac{x_1 + 2x_2 + 5}{3x_1 + x_2 + 5}$,

(b) $g(x_1, x_2) = \dfrac{7x_1 - x_2 + 8}{x_1 + 2x_2 + 4} + \dfrac{-2x_1 + x_2 + 1}{3x_1 + x_2 + 5}$,

on the domain $S = \{(x_1, x_2) \in \Re^2 : x_1 + 2x_2 + 4 > 0, \ 3x_1 + x_2 + 5 > 0\}$.

7.18. Find m_0 such that $f(x_1, x_2) = \dfrac{2x_1 + m_0}{x_1 - x_2 + 3} + \dfrac{x_1 - 3x_2 + 7}{x_1 + x_2 + 1}$ is pseudoconvex on $S = \{(x_1, x_2) \in \Re^2 : x_1 - x_2 + 3 > 0, \ x_1 + x_2 + 1 > 0\}$.

7.7 References

Bajalinov E. B. [14], Barros A. I. [15], Bykadorov I. A. [28], Cambini A., Crouzeix J. P., and Martein L. [44], Cambini A., Martein L., and Schaible S. [48, 49], Cambini R., and Carosi L. [52, 55], Carosi L., and Martein L. [58, 59], Charnes A., and Cooper W. W. [64], Charnes A., Cooper W. W., and Rhodes E. [66], Charnes A., Cooper W. W., and Lewin A. Y. [67], Colantoni C. S., Manes R. P., and Whinston A. [69], Craven B. D. [78], Frenk J., and Schaible S. [114], Hirche J. [140], Khurana A., and Arora S. R. [166], Lo A., and Mackinlay C. [190], Mjelde K. M. [214], Schaible S. [241, 244, 246, 255, 256], Stancu-Minasian I. M. [263, 264, 265, 266, 268, 269].

8

Sequential Methods for Generalized Convex Fractional Programs

8.1 Introduction

In spite of the relevance of the role played by generalized convexity in mathematical programming, research in finding efficient numerical methods for solving generalized convex optimization problems has not yet been sufficiently developed. The only text-book in which solution methods for pseudolinear functions and generalized convex quadratic functions are proposed is Martos' [211].

In this chapter we shall deal with generalized fractional problems which can be solved by means of simplex-like procedures and which have the advantage of finding the optimal solution of the problem through a finite number of iterations.

Due to the very important role that linear fractional problems have had in the development of fractional programming, and in the various methods that have appeared in the literature, in Sect. 8.2 we shall present the theoretical properties together with the main solution methods.

In Sect. 8.3 we shall present the theoretical properties and a sequential method for solving a problem whose objective function is the sum between a linear and a linear fractional function which also allows us, together with Charnes–Cooper's transformation, to solve a problem that has as an objective the sum of two linear fractional functions.

In Sect. 8.4 we shall consider two generalized multiplicative problems.

The algorithms described are based on the common idea of associating with the problem a suitable parametric program which is easier to solve. This kind of approach and some of its applications will be described in a general form in Sect. 8.5.

8.2 The Linear Fractional Problem

Linear fractional functions are perhaps the most popular for modelling objectives in Optimization after, of course, linear and quadratic functions.
A linear fractional problem consists of maximizing (minimizing) the ratio of two affine functions subject to linear constraints. From an analytical point of view, a linear fractional problem may be described in the following standard form:

$$P_{LF} : \max_{x \in S} \left(f(x) = \frac{c^T x + c_0}{d^T x + d_0} \right), \ S = \{ x \in \Re^n : Ax = b, \ x \geq 0 \}$$

where $c, d \in \Re^n$, $c_0, d_0 \in \Re$, A is an $m \times n$ real matrix, $b \in \Re^m$. We shall assume $d \neq 0$, $d_0 \neq 0$, $d^T x + d_0 > 0$ for all $x \in S$, and $rank A = m < n$.
Since the objective function f is pseudolinear, a local maximum point for problem P_{LF} is also global and at least one optimal solution (if one exists) is reached at a vertex of S (see Corollary 4.6.2).
When the feasible set S is unbounded, the existence of an optimal solution for P_{LF} is not ensured; in this regard, the following theorem holds.

Theorem 8.2.1. *Consider the linear fractional problem P_{LF} and let L be the supremum of f on S.*
(i) $L = \max_{x \in S} f(x)$ if and only if there exists a vertex $x^0 \in S$ such that $L = f(x^0)$;
(ii) if the supremum L is not attained, then there exists an extreme direction u such that $L = \lim_{t \to +\infty} f(x^0 + tu)$, where $x^0 \in S$.
(iii) $L = +\infty$ if and only if there exists an extreme direction u such that $d^T u = 0$, $c^T u > 0$.

Proof. (i) If the supremum L is attained as a maximum, then there exists a feasible point $\bar{x}$ such that $f(\bar{x}) = L$. Consider the linear problem

$$\bar{P} : \ max(c^T x + c_0), x \in \bar{S} = S \cap \{ x \in \Re^n : d^T x + d_0 = d^T \bar{x} + d_0 \}.$$

Obviously $\bar{x}$ is an optimal solution for $\bar{P}$ and, because of the linearity of the problem, the maximum is reached at a vertex $\hat{x}$ of $\bar{S}$ which belongs to an edge s of S starting from a vertex $x^0 \in S$. Let $x = x^0 + tu, t \geq 0$, be the equation of the ray containing s and consider the restriction $\varphi(t) = f(x^0 + tu), t \geq 0$.
The derivative $\varphi'(t) = \dfrac{c^T u (d^T x^0 + d_0) - d^T u (c^T x^0 + c_0)}{(t d^T u + d^T x^0 + d_0)^2}$ is constant in sign so that, taking into account that $\hat{x}$ belongs to the edge s, we necessarily have $\varphi'(t) \leq 0$, $\forall t \geq 0$. If $\varphi'(t) < 0$, $\forall t \geq 0$, we have $\hat{x} = x^0$; if $\varphi'(t) = 0$, $\forall t \geq 0$, f is constant on s, so that the vertex x^0 is optimal for P_{LF}.
The converse statement is obvious.
(ii) Let $\{x_n\} \subset S$ be a sequence such that $\{f(x_n)\}$ converges to L and consider the following sequence of linear problems:

$$P_n : \max f(x), \ x \in S_n = S \cap \{ x \in \Re^n : d^T x + d_0 = d^T x_n + d_0 \}.$$

It is well-known that the supremum of the linear problem P_n is not finite if and only if there exists a feasible point x^0 and an extreme direction u, such that $\lim_{t \to +\infty} f(x^0 + tu) = +\infty$. In this case the supremum of P_{LF} is $+\infty$, too, and (ii) holds; furthermore, $\lim_{t \to +\infty} f(x^0 + tu) = +\infty$ implies that $d^T u = 0$, so that the extreme direction u is feasible for any S_n and thus the supremum of P_n is $+\infty$ for every n.

Consider now the case of a finite supremum for every n; this supremum is attained at a vertex y_n of S_n which belongs to an edge of S. Taking into account that $f(y_n) \geq f(x_n)$, we have $\lim_{n \to +\infty} f(y_n) = L$. Since L is not attained, $\{y_n\}$ is necessarily divergent in norm and $f(y_n) \neq L$ for all n. Since S has a finite number of edges (in particular, half-lines), there exists a subsequence of $\{y_n\}$, which, without loss of generality, we can assume to be the same sequence, contained in a half-line whose equation is of the kind $x = x^0 + tu$, $t \geq 0$, where x^0 is a vertex of S. Let t_n be such that $y_n = x^0 + t_n u$. We have $\lim_{n \to +\infty} f(y_n) = \lim_{t_n \to +\infty} f(x^0 + t_n u) = L$ and (ii) holds.

(iii) If L is not attained, from (ii) there exist an extreme direction u and a feasible point x^0 such that $L = \lim_{t \to +\infty} f(x^0 + tu) = \lim_{t \to +\infty} \dfrac{tc^T u + c^T x^0 + c_0}{td^T u + d^T x^0 + d_0}$. Obviously, we have $L = +\infty$ if and only if $d^T u = 0$ and $c^T u > 0$.   □

Corollary 8.2.1. *Let D be the set of optimal solutions for the linear problem $\min_{x \in S}(d^T x + d_0)$. Then, the supremum L of P_{LF} is $+\infty$ if and only if D is unbounded and $\sup_{x \in D}(c^T x + c_0) = +\infty$.*

Proof. Note that $D \neq \emptyset$ since the linear function $d^T x + d_0$ is lower bounded on S. Let $x^0 \in D$ and assume $L = +\infty$. From (iii) of Theorem 8.2.1 there exists an extreme direction u such that $d^T u = 0$, $c^T u > 0$, so that the half-line $x = x^0 + tu, t \geq 0$, is contained in D and we have $\sup_{x \in D}(c^T x + c_0) = +\infty$.

Conversely, there exists a ray $s \subseteq D$ of equation $x = x^0 + tu, t \geq 0$, such that $d^T u = 0$ and $\sup_{x \in s}(c^T x + c_0) = +\infty$. This last equality implies that $c^T u > 0$ so that, from (iii) of Theorem 8.2.1, $L = +\infty$.   □

Remark 8.2.1. When a linear fractional problem has no optimal solutions, it may happen that the supremum is finite.

Consider, for instance, problem P_{LF} with $f(x_1, x_2) = \dfrac{x_1 + 2x_2}{x_2 + 2}$ and $S = \{(x_1, x_2) \in \Re^2 : x_1 - x_2 \leq 2, x_1 \geq 0, x_2 \geq 0\}$. It can be verified that $u^T = (1, 1)$ is an extreme direction and the supremum (not attained) of the problem is $L = \lim_{t \to +\infty} f(\bar{x} + tu) = 3$ for every feasible point $\bar{x} \in S$.

Because of the potentially broad applications of linear fractional programming, several solution methods have been suggested. The interested reader is referred to the book [269] and excellent bibliographies collected by Stancu–Minasian in [264, 265, 266, 268] and by Schaible in [114, 249].

Below, we shall present a selection of such methods.

The pseudolinearity of the objective function of the linear fractional problem P_{LF} implies that the maximum value (if it is attained) is reached at a vertex. Due to this property, it is possible to solve P_{LF} by using a simplex-like procedure, i.e., a technique similar to the Simplex Method in linear programming. We shall use the following standard notations.

Given a basic feasible solution x^i (which is a vertex of S) and the corresponding basis B, we shall partition the matrix A as $A = [B : N]$ and the vectors x, c and d as $x^T = (x_B^T, x_N^T)$, $c^T = (c_B^T, c_N^T)$, $d^T = (d_B^T, d_N^T)$.

Set:
$$\bar{c}_0 = c_B^T B^{-1} b + c_0, \quad \bar{d}_0 = d_B^T B^{-1} b + d_0,$$
$$\bar{c}_N^T = c_N^T - c_B^T B^{-1} N, \quad \bar{d}_N^T = d_N^T - d_B^T B^{-1} N,$$
$\bar{c}_j$ and $\bar{d}_j$ the j-th component of $\bar{c}_N$ and $\bar{d}_N$, respectively.

The function f, expressed in terms of the basic and non-basic variables, is given by $f(x_B, x_N) = f(B^{-1}b - B^{-1}Nx_N, x_N) = \dfrac{\bar{c}_N^T x_N + \bar{c}_0}{\bar{d}_N^T x_N + \bar{d}_0}$.

8.2.1 Isbell–Marlow's Algorithm

The first sequential method for solving the linear fractional problem P_{LF} was suggested by Isbell and Marlow [144]. The basic idea of the method, which works on a compact feasible region, is to generate a finite sequence of vertices $x^1, ..., x^h$, starting from an initial basic feasible solution x^0, such that
$$f(x^0) < f(x^1) < ... < f(x^h) = \max_{x \in S} f(x).$$
Each vertex x^{i+1}, $i = 0, ..., h-1$, is an optimal solution for the linear problem $\max_{x \in S} \psi^i(x)$ where $\psi^i(x) = c^T x + c_0 - f(x^i)(d^T x + d_0)$.

The method is based on the theoretical property stated in the following theorem.

Theorem 8.2.2. x^* is optimal for problem P_{LF} if and only if it is optimal for the linear problem P^* : $\max_{x \in S}[\psi^*(x) = c^T x + c_0 - f(x^*)(d^T x + d_0)]$.

Proof. Let x^* be optimal for P_{LF} and assume, by contradiction, the existence of $\bar{x} \in S$ such that $\psi^*(\bar{x}) > \psi^*(x^*) = 0$. Then, we have $f(\bar{x}) > f(x^*)$ which contradicts the optimality of x^*.

Conversely, if x^* is optimal for P^*, then $\psi^*(x) \le \psi^*(x^*) = 0$, $\forall x \in S$, so that $f(x) \le f(x^*)$, $\forall x \in S$, and this implies the optimality of x^* for P_{LF}. □

Theorem 8.2.2 implies that if the vertex x^{i+1} is such that $\psi^i(x^{i+1}) > 0$, then $f(x^{i+1}) > f(x^i)$, while if $\psi^i(x^{i+1}) = 0$ then x^{i+1} is an optimal solution for P_{LF}.

Isbell–Marlow's algorithm can be summarized as follows:

Step 0. Find a basic feasible solution x^0. Set $i = 0, x^i = x^0$ and go to Step 1.

Step 1. Find an optimal solution x^{i+1} for the linear problem $\max_{x \in S} \psi^i(x)$. If $\psi^i(x^{i+1}) = 0$, Stop: x^{i+1} is an optimal solution for P_{LF}, otherwise go to Step 2.

Step 2. Set $i = i + 1$ and go to Step 1.

In order to point out how Isbell–Marlow's method works, as well as the others which will be described in the next subsections, we shall refer to the following linear fractional problem in the standard form.

$$
\begin{cases}
\max f(x_1, ..., x_6) = \dfrac{3x_1 - x_2 - 22}{x_1 + 2x_2 + 2} \\
x_1 - 2x_2 + x_3 = 3 \\
5x_1 + 3x_2 + x_4 = 54 \\
x_2 + x_5 = 8 \\
-2x_1 + x_2 + x_6 = 4 \\
x_i \geq 0, \ i = 1, .., 6
\end{cases}
\tag{8.1}
$$

Starting from the initial basic solution $x^0 = (0, 0, 3, 54, 8, 4)^T$, we must solve the problem $\max_{x \in S} [\psi^0(x) = 3x_1 - x_2 - 22 - f(x^0)(x_1 + 2x_2 + 2) = 14x_1 + 21x_2]$.

By applying the Simplex Algorithm we obtain the following optimal table.

	-252	0	0	0	$-\frac{63}{5}$	$-\frac{14}{5}$	0
x_3	13	0	0	1	$-\frac{1}{5}$	$\frac{13}{5}$	0
x_6	8	0	0	0	$\frac{2}{5}$	$-\frac{11}{5}$	1
x_1	6	1	0	0	$\frac{1}{5}$	$-\frac{3}{5}$	0
x_2	8	0	1	0	0	1	0

The new basic solution is $x^1 = (6, 8, 13, 0, 0, 8)^T$ and, since $\psi^0(x^1) = 252 > 0$, we must solve the linear problem $\max_{x \in S} \psi^1(x)$, where $\psi^1(x) = 3x_1 - x_2 - 22 - f(x^1)(x_1 + 2x_2 + 2) = \frac{7}{2}x_1 - 21$.

Starting from the basic solution x^1, we get the following optimal table corresponding to the basic solution $x^2 = (9, 3, 0, 0, 5, 19)^T$.

	$-\frac{21}{2}$	0	0	$-\frac{21}{26}$	$-\frac{7}{13}$	0	0
x_5	5	0	0	$\frac{5}{13}$	$-\frac{1}{13}$	1	0
x_6	19	0	0	$\frac{11}{13}$	$\frac{3}{13}$	0	1
x_1	9	1	0	$\frac{3}{13}$	$\frac{2}{13}$	0	0
x_2	3	0	1	$-\frac{5}{13}$	$\frac{1}{13}$	0	0

Since $\psi^1(x^2) = \frac{21}{2} > 0$, we must solve the linear problem $\max_{x \in S} \psi^2(x)$, where $\psi^2(x) = 3x_1 - x_2 - 22 - f(x^2)(x_1 + 2x_2 + 2) = \frac{49}{17}x_1 - \frac{21}{17}x_2 - \frac{378}{17}$. Starting from the basic solution x^2, we get the following optimal table corresponding to the basic solution $x^3 = x^2 = (9, 3, 0, 0, 5, 19)^T$.

	0	0 0	$-\frac{252}{221}$	$-\frac{77}{221}$	0 0		
x_5	5	0 0	$\frac{5}{13}$	$-\frac{1}{13}$	1 0		
x_6	19	0 0	$\frac{11}{13}$	$\frac{3}{13}$	0 1		
x_1	9	1 0	$\frac{3}{13}$	$\frac{2}{13}$	0 0		
x_2	3	0 1	$-\frac{5}{13}$	$\frac{1}{13}$	0 0		

Since $\psi^2(x^3) = 0$, x^3 is the optimal solution for the given linear fractional problem.

8.2.2 Charnes–Cooper's Algorithm

The idea of the method suggested by Charnes and Cooper in [64] is to use a variable transformation in order to reduce the linear fractional problem to a linear one.
Let

$$y = tx, \quad t = \frac{1}{d^T x + d_0} \tag{8.2}$$

When S is bounded, t is positive for all $x \in S$ so that, by applying (8.2) to problem P_{LF}, we obtain the following linear problem:

$$P_L : \max_{(y,t) \in S_L} (c^T y + c_0 t)$$

where $S_L = \{(y, t) \in \Re^{n+1} : Ay - bt = 0, \ d^T y + d_0 t = 1, \ y \geq 0, \ t \geq 0\}$.
Problem P_{LF} and problem P_L are equivalent in the sense stated in the following theorem.

Theorem 8.2.3. *Consider problem P_{LF} and assume that the feasible set S is bounded.*
(i) If $\hat{x}$ is an optimal solution for P_{LF}, then $(\hat{y}, \hat{t})$ is an optimal solution for P_L where $\hat{t} = \dfrac{1}{d^T \hat{x} + d_0}$, $\hat{y} = \hat{t}\hat{x}$.

(ii) If $(\hat{y}, \hat{t})$ is an optimal solution for P_L, then $\hat{t} > 0$ and $\hat{x} = \dfrac{\hat{y}}{\hat{t}}$ is an optimal solution for P_{LF}.

Proof. Firstly, note that if $(y, t) \in S_L$, then $t > 0$. If not, let $(\bar{y}, 0) \in S_L$ and consider a sequence $\{(y_n, t_n)\} \subset S_L$ such that $y_n \to \bar{y}$, $t_n \to 0^+$. Then, the sequence $\{x_n = \frac{y_n}{t_n}\} \subset S$ is such that $\lim_{n \to +\infty} \| x_n \| = \lim_{n \to +\infty} \frac{1}{t_n} \| y_n \| = +\infty$ and this contradicts the boundedness of S. Consequently, to any element $(y, t) \in S_L$ there corresponds an element $x = \dfrac{y}{t} \in S$ and vice versa.

Since $\dfrac{c^T x + c_0}{d^T x + d_0} = \dfrac{c^T y + c_0 t}{d^T y + d_0 t} = c^T y + c_0 t$, $\forall\ x \in S$, $\forall\ (y,t) \in S_L$, $y = tx$,

$t = \dfrac{1}{d^T x + d_0}$, (i) and (ii) follow.   □

Now, we shall apply Charnes–Cooper's algorithm for solving Problem (8.1) whose feasible region is bounded.

Set $y = tx$, $t = \dfrac{1}{x_1 + 2x_2 + 2}$. We have the following linear problem

$$\begin{cases} max\ (3y_1 - y_2 - 22t) \\ y_1 - 2y_2 + y_3 - 3t & = 0 \\ 5y_1 + 3y_2 + y_4 - 54t & = 0 \\ y_2 + y_5 - 8t & = 0 \\ -2y_1 + y_2 + y_6 - 4t & = 0 \\ y_1 + 2y_2 + 2t & = 1 \\ y_i \geq 0,\ t \geq 0,\ i = 1, .., 6 \end{cases} \qquad (8.3)$$

Starting from the initial basic feasible solution $(y^0, t_0) = (0, 0, \frac{3}{2}, 27, 4, 2, \frac{1}{2})$, we obtain the following table.

	11	14	21	0	0	0	0	0
y_3	$\frac{3}{2}$	$\frac{5}{2}$	1	1	0	0	0	0
y_4	27	32	57	0	1	0	0	0
y_5	4	4	9	0	0	1	0	0
y_6	2	0	5	0	0	0	1	0
t	$\frac{1}{2}$	$\frac{1}{2}$	1	0	0	0	0	1

By applying the usual simplex algorithm we obtain the following tables.

	$\frac{13}{5}$	14	0	0	0	0	$-\frac{21}{5}$	0
y_3	$\frac{11}{10}$	$\frac{5}{2}$	0	1	0	0	$-\frac{1}{5}$	0
y_4	$\frac{21}{5}$	32	0	0	1	0	$-\frac{57}{5}$	0
y_5	$\frac{2}{5}$	4	0	0	0	1	$-\frac{9}{5}$	0
y_2	$\frac{2}{5}$	0	1	0	0	0	$\frac{1}{5}$	0
t	$\frac{1}{10}$	$\frac{1}{2}$	0	0	0	0	$-\frac{1}{5}$	1

	$\frac{6}{5}$	0	0	0	0	$-\frac{7}{2}$	$\frac{21}{10}$	0
y_3	$\frac{17}{20}$	0	0	1	0	$-\frac{5}{8}$	$\frac{37}{40}$	0
y_4	1	0	0	0	1	-8	3	0
y_1	$\frac{1}{10}$	1	0	0	0	$\frac{1}{4}$	$-\frac{9}{20}$	0
y_2	$\frac{2}{5}$	0	1	0	0	0	$\frac{1}{5}$	0
t	$\frac{1}{20}$	0	0	0	0	$-\frac{1}{8}$	$\frac{1}{40}$	1

	value							
	$\frac{1}{2}$	0	0	0	$-\frac{7}{10}$	$\frac{21}{10}$	0	0
y_3	$\frac{13}{24}$	0	0	1	$-\frac{37}{120}$	$\frac{221}{120}$	0	0
y_6	$\frac{1}{3}$	0	0	0	$\frac{1}{3}$	$-\frac{8}{3}$	1	0
y_1	$\frac{1}{4}$	1	0	0	$\frac{3}{20}$	$-\frac{19}{20}$	0	0
y_2	$\frac{1}{3}$	0	1	0	$-\frac{1}{15}$	$\frac{8}{15}$	0	0
t	$\frac{1}{24}$	0	0	0	$-\frac{1}{120}$	$-\frac{7}{120}$	0	1

	value							
	$-\frac{2}{17}$	0	0	$-\frac{252}{221}$	$-\frac{77}{221}$	0	0	0
y_5	$\frac{5}{17}$	0	0	$\frac{120}{221}$	$-\frac{37}{221}$	1	0	0
y_6	$\frac{19}{17}$	0	0	$\frac{320}{221}$	$-\frac{25}{221}$	0	1	0
y_1	$\frac{9}{17}$	1	0	$\frac{114}{221}$	$-\frac{2}{221}$	0	0	0
y_2	$\frac{3}{17}$	0	1	$-\frac{64}{221}$	$\frac{5}{221}$	0	0	0
t	$\frac{1}{17}$	0	0	$\frac{7}{221}$	$-\frac{4}{221}$	0	0	1

The basic feasible solutions generated by Charnes–Cooper's algorithm are:
$(y^0, t_0) = (0, 0, \frac{3}{2}, 27, 4, 2, \frac{1}{2})$, $(y^1, t_1) = (0, \frac{2}{5}, \frac{11}{10}, \frac{21}{5}, \frac{2}{5}, 0, \frac{1}{10})$,

$(y^2, t_2) = (\frac{1}{10}, \frac{2}{5}, \frac{17}{20}, 1, 0, 0, \frac{1}{20})$, $(y^3, t_3) = (\frac{1}{4}, \frac{1}{3}, \frac{13}{24}, 0, 0, \frac{1}{3}, \frac{1}{24})$,

$(y^4, t_4) = (\frac{9}{17}, \frac{3}{17}, 0, 0, \frac{5}{17}, \frac{19}{17}, \frac{1}{17})$.

The last basic solution is optimal for the linear problem P_L.
The basic feasible solutions of the feasible region S associated with the previous ones are:
$x^0 = \frac{y^0}{t_0} = (0, 0, 3, 54, 8, 4)^T$, $x^1 = \frac{y^1}{t_1} = (0, 4, 11, 42, 4, 0)^T$,

$x^2 = \frac{y^2}{t_2} = (2, 8, 17, 20, 0, 0)^T$, $x^3 = \frac{y^3}{t_3} = (6, 8, 13, 0, 0, 8)^T$,

$x^4 = \frac{y^4}{t_4} = (9, 3, 0, 0, 5, 19)^T$.

The last basic solution is optimal for the linear fractional problem (8.1).

8.2.3 Martos' Algorithm

Martos' algorithm, suggested in [207], together with Charnes–Cooper's are the best known sequential methods for solving a linear fractional problem. The algorithm works on a compact feasible set and generates a finite sequence of vertices corresponding to increasing levels of the objective function, the last of which is optimal for the problem. The optimality of a vertex x^i is tested by means of the necessary and sufficient condition (8.4) which is equivalent, taking into account the introduced notations, to the one stated in Theorem 4.7.2

$$\bar{\gamma}_N^T = \bar{d}_0 \bar{c}_N^T - \bar{c}_0 \bar{d}_N^T \leq 0. \tag{8.4}$$

The algorithm may be described as follows.
Step 0. Compute a basic feasible solution x^0; go to Step 1.
Step 1. Compute $\bar{\gamma}_N$ and set $J = \{j : \gamma_j > 0\}$. If $J = \emptyset$, Stop, x^0 is an optimal solution. Otherwise, select k such that $\bar{\gamma}_k = \max_{i \in J}\{\bar{\gamma}_i\}$; go to Step 2.
Step 2. The non-basic variable x_k enters the basis by means of a primal simplex iteration. Let x^0 be the new basic solution and go to Step 1.

Now, we shall apply Martos' algorithm to solve Problem (8.1).
The simplex table associated with the vertex $x^0 = (0, 0, 3, 54, 8, 4)^T$ is the

following

	22	3	-1	0	0	0	0
	-2	1	2	0	0	0	0
x_3	3	1	-2	1	0	0	0
x_4	54	5	3	0	1	0	0
x_5	8	0	1	0	0	1	0
x_6	4	-2	1	0	0	0	1

where in the first row and in the second row we can read the reduced costs $\bar{c}_N$ and $\bar{d}_N$ of the numerator and of the denominator of the objective function, respectively.

We have $\bar{\gamma}_N^T = 2(3, -1) + 22(1, 2) = (28, 42)$, $J = \{1, 2\}$. Since $\bar{\gamma}_2 > \bar{\gamma}_1$, the non-basic variable x_2 enters the basis. We obtain:

	26	1	0	0	0	0	1
	-10	5	0	0	0	0	-2
x_3	11	-3	0	1	0	0	2
x_4	42	11	0	0	1	0	-3
x_5	4	2	0	0	0	1	-1
x_2	4	-2	1	0	0	0	1

We have $\bar{\gamma}_N^T = 10(1, 1) + 26(5, -2) = (140, -42)$, $J = \{1\}$, so that the non-basic variable x_1 enters the basis. We obtain:

	24	0	0	0	0	$-\frac{1}{2}$	$\frac{3}{2}$
	-20	0	0	0	0	$-\frac{5}{2}$	$\frac{1}{2}$
x_3	17	0	0	1	0	$\frac{3}{2}$	$\frac{1}{2}$
x_4	20	0	0	0	1	$-\frac{11}{2}$	$\frac{5}{2}$
x_1	2	1	0	0	0	$\frac{1}{2}$	$-\frac{1}{2}$
x_2	8	0	1	0	0	1	0

We have $\bar{\gamma}_N^T = 20(-\frac{1}{2}, \frac{3}{2}) + 24(-\frac{5}{2}, \frac{1}{2}) = (-70, 42)$, $J = \{6\}$, so that the non-basic variable x_6 enters the basis. We obtain:

	12	0	0	0	$-\frac{3}{5}$	$\frac{14}{5}$	0
	-24	0	0	0	$-\frac{1}{5}$	$-\frac{7}{5}$	0
x_3	13	0	0	1	$-\frac{1}{5}$	$\frac{13}{5}$	0
x_6	8	0	0	0	$\frac{2}{5}$	$-\frac{11}{5}$	1
x_1	6	1	0	0	$\frac{1}{5}$	$-\frac{3}{5}$	0
x_2	8	0	1	0	0	1	0

We have $\bar{\gamma}_N^T = 24(-\frac{3}{5}, \frac{14}{5}) + 12(-\frac{1}{5}, -\frac{7}{5}) = (-\frac{84}{5}, \frac{252}{5})$, $J = \{5\}$, so that the non-basic variable x_5 enters the basis. We obtain:

	-2	0	0	$-\frac{14}{13}$	$-\frac{5}{13}$	0	0
	-17	0	0	$\frac{7}{13}$	$-\frac{4}{13}$	0	0
x_5	5	0	0	$\frac{5}{13}$	$-\frac{1}{13}$	1	0
x_6	19	0	0	$\frac{11}{13}$	$\frac{3}{13}$	0	1
x_1	9	1	0	$\frac{3}{13}$	$\frac{2}{13}$	0	0
x_2	3	0	1	$-\frac{5}{13}$	$\frac{1}{13}$	0	0

We have $\bar{\gamma}_N^T = 17(-\frac{14}{13}, -\frac{5}{13}) - 2(\frac{7}{13}, -\frac{4}{13}) = (-\frac{252}{13}, -\frac{77}{13})$. Since $J = \emptyset$, the basic solution $x^4 = (9, 3, 0, 0, 5, 19)^T$ is optimal.

Note that the basic feasible solutions generated by Martos' algorithm are $x^0 = (0, 0, 3, 54, 8, 4)^T$, $x^1 = (0, 4, 11, 42, 4, 0)^T$, $x^2 = (2, 8, 17, 20, 0, 0)^T$, $x^3 = (6, 8, 13, 0, 0, 8)^T$, $x^4 = (9, 3, 0, 0, 5, 19)^T$.

Remark 8.2.2. It is worth noting that Wagner and Yuan in [277] show that Martos' algorithm is equivalent to Charnes–Cooper's algorithm in the sense that both methods lead to an identical sequence of pivot operations starting from the same basic feasible solution. This property can be verified by means of problem (8.1) solved in Sect. 8.2.2 with Charnes–Cooper's algorithm and in this Subsection with Martos' algorithm.

Bitran in [26] shows that Martos's algorithm is better than Isbell–Marlow's algorithm in terms of the number of pivot operations.

8.2.4 Cambini–Martein's Algorithm

This method differs from Martos' algorithm as regards the choice of the entering variable in the pivot operation. An important advantage to the new method with respect to the previous ones, is that the suggested choice of the entering variable allows us to solve the linear fractional problem for every feasible region (bounded or unbounded).

In order to describe the algorithm, we shall introduce the concept of *optimal level solution* which will be developed in a general setting in Sect. 8.5.

A point $\bar{x} \in S$ is said to be an optimal level solution if it solves the linear problem $max(c^T x + c_0)$, $x \in S \cap \{x \in \Re^n : d^T x + d_0 = d^T \bar{x} + d_0\}$.

Obviously, any optimal solution for the linear fractional problem (if one exists) is also an optimal level solution.

Cambini–Martein's algorithm, suggested in [33], generates a finite sequence of vertices $x^0, ..., x^h$ which are optimal level solutions which correspond to increasing levels of the objective function f, such that $f(x^h) = \underset{x \in S}{max} f(x)$ or the supremum of the linear fractional problem is the limit $\underset{t \to +\infty}{lim} f(x^h + tu)$, where u is a suitable extreme direction starting from x^h.

In correspondence to a basic feasible solution x^i, using the usual notations, consider the vector $\bar{\gamma}_N$ given by (8.4) and the set $J = \{j : \bar{\gamma}_j > 0\}$, where $\bar{\gamma}_j$ is the j-th component of $\bar{\gamma}_N$; the following theorem holds.

Theorem 8.2.4. *Let x^i be a vertex of S which is also an optimal level solution. Assume $J \neq \emptyset$ and $\bar{d}_j > 0$ for all $j \in J$. Then:*
(i) Every point of the edge s^k corresponding to the non-basic variable x_k is an optimal level solution if and only if the index k is such that

$$\frac{\bar{c}_k}{\bar{d}_k} = \underset{j \in J}{max} \frac{\bar{c}_j}{\bar{d}_j} \tag{8.5}$$

(ii) The problem $max \dfrac{c^T x + c_0}{d^T x + d_0}$, $x \in S \cap \{x \in \Re^n : d^T x = d^T \bar{x}\}$ is equivalent to the problem $max \dfrac{c^T x + c_0}{d^T x + d_0}$, $x \in S \cap \{x \in \Re^n : d^T x \leq d^T \bar{x}\}$ for every $\bar{x} \in s^k$.

Proof. (i) Let $H = \{x \in \Re^n : d^T x + d_0 = \bar{d}_0 + \theta\}$, $\theta > 0$ and consider the linear problem $max(c^T x + c_0)$, $x \in S \cap H$.

Let w^j be the intersection between the hyperplane H and the edge s^j starting from the vertex x^i. The value of the function $c^T x + c_0$ at w^j is $\dfrac{\bar{c}_j}{\bar{d}_j}\theta + \bar{c}_0$.

Since $h \notin J$ implies that the function f is not increasing along the edge s^h, we have $f(w^h) \leq f(x^i) < f(w^j)$ for all $h \notin J$ and for all $j \in J$, so that

$$\underset{x \in S \cap H}{max} f(x) = \underset{j \in J}{max} f(w^j) = \frac{1}{\bar{d}_0 + \theta} \underset{j \in J}{max} (\frac{\bar{c}_j}{\bar{d}_j}\theta + \bar{c}_0).$$

Taking into account that $\theta > 0$, in the last equality the maximum is reached at an index k, thus verifying (8.5).

(ii) The thesis follows by noting that the function f is increasing along the edge s^k. $\qquad\square$

The algorithm starts solving the linear problem $P_0 : \min\limits_{x \in S}(d^T x + d_0)$. Taking into account Corollary 8.2.1, the supremum of P_{LF} is $+\infty$ if and only if $\sup\limits_{x \in D}(c^T x + c_0) = +\infty$, otherwise an optimal solution x^0 of $\max\limits_{x \in D}(c^T x + c_0)$ is an optimal level solution corresponding to the minimum value m of the denominator. Because of this, if $J \neq \emptyset$, every increasing direction corresponds to an increasing level of the denominator so that $\bar{d}_j > 0$ for all $j \in J$ and, consequently, (8.5) can be applied in order to find the index k. The non-basic variable x_k enters the basis by performing a pivot operation and a new optimal level solution is found. If it is not possible to perform a pivot operation, then the algorithm stops.

In general, by applying Theorem 8.2.4, a finite sequence of optimal level solutions $x^0, .., x^i, .., x^h$ is generated (h may be equal to zero).

With respect to x^h, if $J = \emptyset$, then it is an optimal solution for P_{LF}, otherwise the edge s^k associated with the index k determined by (8.5) is a ray.

The generated sequence of optimal level solutions and the ray s^k verify the following properties, where $X = \bigcup\limits_{i=0}^{h-1} [x^i, x^{i+1}] \cup s^k$:

- For every $\bar{x} \in X$, $f(\bar{x}) = maxf(x), x \in S \cap \{x \in \Re^n : d^T x \leq d^T \bar{x}\}$;
- If $d^T \hat{x} < d^T \bar{x}$, $\hat{x}, \bar{x} \in X$, then $f(\hat{x}) < f(\bar{x})$;
- For every feasible level ξ of the denominator there exists $\bar{x} \in X$ such that $f(\bar{x}) = maxf(x), x \in S \cap \{x \in \Re^n : d^T x + d_0 = \xi\}$.

If x^h is not an optimal solution of P_{LF}, then the supremum is not attained and we have

$$sup\limits_{x \in S} f(x) = sup\limits_{x \in s^k} f(x) = \lim\limits_{x_k \to +\infty} \frac{\bar{c}_k x_k + \bar{c}_0}{\bar{d}_k x_k + \bar{d}_0} = \frac{\bar{c}_k}{\bar{d}_k}.$$

In fact, assume the existence of $x^* \in S$ such that $f(x^*) > sup\limits_{x \in s^k} f(x)$ and consider the level $\xi^* = d^T x^* + d_0$. Let $\hat{x} \in X$ be an optimal solution for the problem $maxf(x), x \in S \cap \{x \in \Re^n : d^T x + d_0 = \xi^*\}$. Since f is increasing on X, we have $sup\limits_{x \in s^k} f(x) > f(\hat{x}) \geq f(x^*)$ and this is a contradiction.

The main steps of the algorithm may be described as follows.

Step 0. Solve the problem $P_0 : \min\limits_{x \in S}(d^T x + d_0)$. If the optimal vertex x^0 of P_0 is unique, then set $i = 0$ and go to Step 1. Otherwise, solve the problem $P_1 : max(c^T x + c_0), x \in S \cap \{x \in \Re^n : d^T x = d^T x^0\}$. If P_1 does not have optimal solutions, then Stop: $sup\limits_{x \in S} f(x) = +\infty$. Otherwise let x^0 be an optimal vertex of P_1. Set $i = 0$ and go to Step 1.

Step 1. Compute $\bar{\gamma}_N = \bar{d}_0 \bar{c}_N - \bar{c}_0 \bar{d}_N$ and set $J = \{j : \bar{\gamma}_j > 0\}$. If $J = \emptyset$, Stop: x^i is an optimal solution for P_{LF}; otherwise let k be such that $\dfrac{\bar{c}_k}{\bar{d}_k} = max\limits_{j \in J} \dfrac{\bar{c}_j}{\bar{d}_j}$ and go to Step 2.

Step 2. Compute $u^k = B^{-1}N^k$. If $u^k \leq 0$, then Stop: $\sup_{x \in S} f(x) = \dfrac{\bar{c}_k}{\bar{d}_k}$, otherwise perform a primal simplex iteration with x_k as an entering variable. Let x^{i+1} be the new basic solution. Set $i = i + 1$ and go to Step 1.

Consider Problem (8.1) again. The denominator $x_1 + x_2 + 2$ of the objective function reaches its minimum value at $x^0 = (0, 0, 3, 54, 8, 4)^T$ which is the only solution so that it is the initial optimal level solution. The simplex table associated with x^0 is

	22	3	−1	0	0	0	0
	−2	1	2	0	0	0	0
x_3	3	1	−2	1	0	0	0
x_4	54	5	3	0	1	0	0
x_5	8	0	1	0	0	1	0
x_6	4	−2	1	0	0	0	1

where in the first row and in the second row we can read the reduced costs $\bar{c}_N$ and $\bar{d}_N$ of the numerator and of the denominator of the objective function, respectively.

We have $\bar{\gamma}_N^T = (28, 42)$, $J = \{1, 2\}$. Since $\max_{j \in J} \dfrac{\bar{c}_j}{\bar{d}_j} = \dfrac{\bar{c}_1}{\bar{d}_1}$, the non-basic variable x_1 enters the basis. We obtain

	13	0	5	−3	0	0	0
	−5	0	4	−1	0	0	0
x_1	3	1	−2	1	0	0	0
x_4	39	0	13	−5	1	0	0
x_5	8	0	1	0	0	1	0
x_6	10	0	−3	2	0	0	1

We have $\bar{\gamma}_N^T = 5(5, -3) + 13(4, -1) = (77, -28)$, $J = \{2\}$, so that the non-basic variable x_2 enters the basis. We obtain

	−2	0	0	$-\frac{14}{13}$	$-\frac{5}{13}$	0	0
	−17	0	0	$\frac{7}{13}$	$-\frac{4}{13}$	0	0
x_1	9	1	0	$\frac{3}{13}$	$\frac{2}{13}$	0	0
x_2	3	0	1	$-\frac{5}{13}$	$\frac{1}{13}$	0	0
x_5	5	0	0	$\frac{5}{13}$	$-\frac{1}{13}$	1	0
x_6	19	0	0	$\frac{11}{13}$	$\frac{3}{13}$	0	1

We have $\bar{\gamma}_N^T = 17(-\frac{14}{13}, -\frac{5}{13}) - 2(\frac{7}{13}, -\frac{4}{13}) = (-\frac{252}{13}, -\frac{77}{13})$.
Since $J = \emptyset$, $x^2 = (9, 3, 0, 0, 5, 19)^T$ is an optimal solution for the problem.

Remark 8.2.3. As we have already pointed out, Cambini–Martein's algorithm differs from Martos' algorithm in the choice of the entering variable in the pivot operation. By means of this different choice the average number of vertices examined by Cambini–Martein's algorithm is, almost always, lower than the number of vertices generated by Martos' algorithm (see [102]).

8.2.5 The Case of an Unbounded Feasible Region

As we have already remarked, Cambini–Martein's algorithm works on every feasible set. In this regard, consider the following two examples related to linear fractional problems having an unbounded feasible region. In the former an optimal solution exists, while in the latter the supremum is finite but not attained.

Example 8.2.1. Consider the linear fractional problem

$$\begin{cases} max \ \dfrac{2x_1 + 3x_2 - 1}{x_1 + 2x_2 + 2} \\ -6x_1 + x_2 + x_3 = 2 \\ x_1 - 2x_2 + x_4 = 4 \\ 5x_1 - x_2 + x_5 = 47 \\ x_i \geq 0, \ i = 1,..,5 \end{cases}$$

The denominator of the objective function reaches its minimum value at the vertex $x^0 = (0, 0, 2, 4, 47)^T$ which is the only solution so that x^0 is the initial optimal level solution. The simplex table associated with x^0 is

	1	2	3	0	0	0	
	-2	1	2	0	0	0	
x_3	2	-6	1	1	0	0	
x_4	4	1	-2	0	1	0	
x_5	47	5	-1	0	0	1	

We have $\bar{\gamma}_N^T = (5, 8)$, $J = \{1, 2\}$ and $\max\limits_{j \in J} \dfrac{\bar{c}_j}{d_j} = \dfrac{\bar{c}_1}{d_1}$. The non-basic variable x_1 enters the basis. We obtain:

	-7	0		7	0	-2	0
	-6	0		4	0	-1	0
x_3	26	0	-11	1	6	0	
x_1	4	1	-2	0	1	0	
x_5	27	0	9	0	-5	1	

We have $\bar{\gamma}_N^T = (14, -5)$, $J = \{2\}$, so that the non-basic variable x_2 enters the basis. We obtain:

	-28	0 0 0	$\frac{17}{9}$	$-\frac{7}{9}$
	-18	0 0 0	$\frac{11}{9}$	$-\frac{4}{9}$
x_3	59	0 0 1	$-\frac{1}{9}$	$\frac{11}{9}$
x_1	10	1 0 0	$-\frac{1}{9}$	$\frac{2}{9}$
x_2	3	0 1 0	$-\frac{5}{9}$	$\frac{1}{9}$

We have $\bar{\gamma}_N^T = (-\frac{2}{9}, -\frac{14}{9})$. Since $J = \emptyset$, the vertex $x^2 = (10, 3, 59, 0, 0)^T$ is the optimal solution for the given problem.

Example 8.2.2. Consider the linear fractional problem

$$\begin{cases} max \ \dfrac{2x_1 + 3x_2 - 1}{x_1 + 2x_2 + 2} \\ -6x_1 + x_2 + x_3 = 2 \\ x_1 - 2x_2 + x_4 = 4 \\ x_i \geq 0, \ i = 1, .., 4 \end{cases}$$

Starting from the initial optimal level solution $x^0 = (0, 0, 2, 4)^T$, the algorithm generates the optimal level solution $x^1 = (4, 0, 26, 0)^T$ which corresponds to the following simplex table.

	-7	0	7	0	-2
	-6	0	4	0	-1
x_3	26	0	-11	1	6
x_1	4	1	-2	0	1

We have $\bar{\gamma}_N^T = (14, -5)$, $J = \{2\}$. Since the column associated with x_2 is $(-11, -2)^T$, it is not possible to perform a primal simplex iteration and the given problem does not have any optimal solutions. The supremum is $\dfrac{\bar{c}_2}{\bar{d}_2} = \dfrac{7}{4}$ which is reached along the ray of equation $x = x^1 + ku, \ k \geq 0$, where $u^T = (2, 1, 11, 0)$.

Remark 8.2.4. Regarding Charnes–Cooper's algorithm, when the feasible region is unbounded it may happen that a point of the kind $(\bar{y}, 0)$ becomes feasible for the linear problem P_L. In this case Charnes–Cooper's transformation is meaningless. Nevertheless, when $(\bar{y}, 0)$ is an optimal solution for P_L (this happens if and only if the supremum of P_{LF} is finite and not attained), it is possible to establish a connection between $(\bar{y}, 0)$ and an extreme direction u of S such that $sup f(x) = \lim_{k \to +\infty} f(\bar{x} + ku)$, $\bar{x} \in S$. This connection will be shown by solving the linear fractional problem given in Example 8.2.2.

By means of Charnes–Cooper's transformation, the problem is transformed into the following linear problem:

$$P_L : \begin{cases} max\ (2y_1 + 3y_2 - t) \\ -6y_1 + y_2 + y_3 - 2t = 0 \\ y_1 - 2y_2 + y_4 - 4t = 0 \\ y_1 + 2y_2 + 2t \quad = 1 \\ y_i \geq 0,\ t \geq 0,\ i = 1, .., 4 \end{cases}$$

Starting from the initial basic solution $(0, 0, 1, 2, \frac{1}{2})^T$, we obtain the following simplex table.

	$\frac{1}{2}$	$\frac{5}{2}$	4	0	0	0
y_3	1	-5	3	1	0	0
y_4	2	3	2	0	1	0
t	$\frac{1}{2}$	$\frac{1}{2}$	1	0	0	1

By applying the simplex algorithm, the following tables are obtained.

	$-\frac{5}{6}$	$\frac{55}{6}$	0	$-\frac{4}{3}$	0	0
y_2	$\frac{1}{3}$	$-\frac{5}{3}$	1	$\frac{1}{3}$	0	0
y_4	$\frac{4}{3}$	$\frac{19}{3}$	0	$-\frac{2}{3}$	1	0
t	$\frac{1}{6}$	$\frac{13}{6}$	0	$-\frac{1}{3}$	0	1

	$-\frac{20}{13}$	0	0	$\frac{1}{13}$	0	$-\frac{55}{13}$
y_2	$\frac{6}{13}$	0	1	$\frac{1}{13}$	0	$\frac{10}{13}$
y_4	$\frac{11}{13}$	0	0	$\frac{4}{13}$	1	$-\frac{38}{13}$
y_1	$\frac{1}{13}$	1	0	$-\frac{2}{13}$	0	$\frac{6}{13}$

	$-\frac{7}{4}$	0	0	0	$-\frac{1}{4}$	$-\frac{7}{2}$
y_2	$\frac{1}{4}$	0	1	0	$-\frac{1}{4}$	$\frac{3}{2}$
y_3	$\frac{11}{4}$	0	0	1	$\frac{13}{4}$	$-\frac{19}{2}$
y_1	$\frac{1}{2}$	1	0	0	$\frac{1}{2}$	-1

The optimal solution is $(\bar{y}, 0)$ where $\bar{y}^T = (\frac{1}{2}, \frac{1}{4}, \frac{11}{4}, 0)$. Every vector u proportional to $\bar{y}$ is a direction such that $\lim\limits_{k \to +\infty} f(\bar{x} + ku) = \sup\limits_{x \in S} f(x), \bar{x} \in S$ and the optimal value $\frac{7}{4}$ of P_L is the supremum of P_{LF}.

Remark 8.2.5. Unlike Charnes–Cooper's algorithm, Isbell–Marlow's algorithm and Martos' algorithm cannot be applied in the unbounded case. In fact, by applying Isbell–Marlow's algorithm to the linear fractional problem given in Example 8.2.1, starting from $x^0 = (0, 0, 2, 4, 47)^T$, we have $\sup\limits_{x \in S} \psi^0(x) = +\infty$ even if the linear fractional problem has optimal solutions.

With respect to Martos' algorithm, starting from x^0 and taking into account that $\bar{\gamma}_2 > \bar{\gamma}_1$, the non-basic variable x_2 enters the basis obtaining the following simplex table.

	-5	20	0	-3	0	0
	-6	13	0	-2	0	0
x_2	2	-6	1	1	0	0
x_4	8	-11	0	2	1	0
x_5	49	-1	0	1	0	1

We have $\bar{\gamma}_N^T = (55, -8)$, $J = \{1\}$ but the variable x_1 cannot enter the basis. The value $\frac{5}{6}$ is the limit of the linear fractional function along the feasible ray of equation $x = x^1 + ku$, $k \geq 0$, where $x^1 = (0, 2, 0, 8, 49)^T$ and $u = (0, 6, 0, 11, 1)^T$. This value is not the supremum of the function which is attained as a maximum at $x^2 = (10, 3, 59, 0, 0)^T$.

Consequently, the algorithm suggested by Isbell and Marlow and the one suggested by Martos do not process the linear fractional problem in the unbounded case.

8.3 A Generalized Linear Fractional Problem

In this section we shall present a sequential method for solving an optimization problem whose objective function is the sum between a linear and a linear fractional function. This method, together with the use of Charnes–Cooper's transformation, will also allow us to solve a problem that has as an objective function the sum of two linear fractional functions (see Sect. 8.3.2).
Consider the problem

$$P : \min \left(f(x) = a^T x + \frac{c^T x + c_0}{d^T x + d_0} \right), \ x \in S = \{x \in \Re^n : Ax = b, \ x \geq 0\}$$

where $a, c, d \in \Re^n$, $d \neq 0$, $c_0, d_0 \in \Re$, A is an $m \times n$ real matrix, $rank A = m < n$, $b \in \Re^m$, and $d^T x + d_0 > 0$ for all $x \in S$.
When $a = 0$, problem P reduces to the linear fractional problem considered in Sect. 8.2. For this reason, we shall assume $a \neq 0$.
Sequential methods for solving problem P without any assumption of generalized convexity on f have been suggested by some authors [34, 173, 202]. Recently, the algorithm suggested in [34] has been specified when f is pseudoconvex [60]. Before describing this algorithm we shall point out some theoretical properties of problem P.

Theorem 8.3.1. *Let ℓ be the infimum of problem P.*
(i) ℓ is attained as a minimum if and only if there exists a feasible point x_0 belonging to an edge of S such that $f(x_0) = \ell$.
(ii) If ℓ is not attained as a minimum, then there exists a feasible point x_0 and an extreme direction u such that $\ell = \lim_{t \to +\infty} f(x_0 + tu)$.
(iii) $\ell = -\infty$ if and only if there exists a feasible point x_0 and an extreme direction u such that $\ell = \lim_{t \to +\infty} f(x_0 + tu) = -\infty$.

Proof. (i) If the infimum is attained as a minimum, then there exists a feasible point $\bar{x}$ such that $\ell = f(\bar{x})$. Consider the problem

$$\bar{P} : \min f(x), \ x \in \bar{S} = S \cap \{x \in \Re^n : d^T x + d_0 = d^T \bar{x} + d_0\}.$$

Obviously $\bar{x}$ is an optimal solution for $\bar{P}$ and since $\bar{P}$ is a linear problem the minimum is also reached at a vertex of $\bar{S}$ which belongs to an edge of S. The converse statement is obvious.
(ii) The proof is similar to the one given in (ii) of Theorem 8.2.1.
(iii) This follows immediately from (i) and (ii). □

The pseudoconvexity of f on $D = \{x \in \Re^n : d^T x + d_0 > 0\}$ is equivalent to stating (see Corollary 7.3.1) that f assumes one of the following forms:

(I) $f(x) = \alpha d^T x + \dfrac{c^T x + c_0}{d^T x + d_0}, \quad \alpha > 0$

(II) $f(x) = a^T x + \dfrac{c_0 - \gamma d_0}{d^T x + d_0} + \gamma, \quad c_0 - \gamma d_0 > 0.$

The functions in (I) and (II) behave differently with respect to the infimum, in the sense that when the infimum is not attained, in the first case it is $-\infty$, while in the second case it may also be finite. In this regard, the following theorems hold.

Theorem 8.3.2. *Consider problem P where $a = \alpha d$, $\alpha > 0$. Then the infimum of P is $-\infty$ if and only if there exists an extreme direction u such that $d^T u = 0$, $c^T u < 0$. In any other case the infimum is attained as a minimum.*

Proof. Let x_0 be a feasible point and let u be an extreme direction. Consider the restriction $f(x_0 + tu) = \alpha t d^T u + \alpha d^T x_0 + \dfrac{t c^T u + c^T x_0 + c_0}{t d^T u + d^T x_0 + d_0}$. Since $x = x_0 + tu \in S$, $\forall t \geq 0$, we necessarily have $d^T u \geq 0$.

We have $\lim_{t \to +\infty} f(x_0 + tu) = +\infty$ if and only if $d^T u > 0$ or $d^T u = 0, c^T u > 0$; the limit is $-\infty$ if and only if $d^T u = 0, c^T u < 0$ and it is finite if and only if $d^T u = c^T u = 0$. In this last case, $f(x_0 + tu) = f(x_0)$ for all $t \geq 0$. The thesis follows from Theorem 8.3.1. □

Theorem 8.3.3. *Consider problem P where $c = \gamma d$, $c_0 - \gamma d_0 > 0$.*
(i) The infimum of P is $-\infty$ if and only if there exists an extreme direction u such that $a^T u < 0$.
(ii) The infimum is finite and not attained as a minimum if and only if there exists an extreme direction u such that $a^T u = 0$, $d^T u > 0$, and there does not exist an extreme direction v such that $a^T v < 0$.
In any other case the infimum is attained as a minimum.

Proof. Let u be an extreme direction and consider the restriction

$$f(x_0 + tu) = t a^T u + a^T x_0 + \frac{c_0 - \gamma d_0}{t d^T u + d^T x_0 + d_0} + \gamma$$

where x_0 is a feasible point.
We have $\lim_{t \to +\infty} f(x_0 + tu) = +\infty$ if and only if $a^T u > 0$; the limit is $-\infty$ if and only if $a^T u < 0$ and it is finite and less than $f(x_0)$ if and only if $a^T u = 0$, $d^T u > 0$. The thesis follows from Theorem 8.3.1. □

8.3.1 Sequential Methods

The obtained theoretical properties allow us to establish an algorithm for solving problem P. The idea of the algorithm is to associate with problem P the following linear parametric problem:

$$P(\theta): \min_{x \in S(\theta)} \left(a^T x + \frac{c^T x + c_0}{d^T x + d_0} \right), \quad S(\theta) = S \cap \{x \in \Re^n : d^T x + d_0 = \delta_0 + \theta\}$$

where $\delta_0 = \min_{x \in S} (d^T x + d_0)$.

By setting $\Theta = \{\theta : S(\theta) \neq \emptyset\}$, we have

$$\inf_{x \in S} f(x) = \inf_{\theta \in \Theta} \inf_{x \in S(\theta)} z(\theta) \tag{8.6}$$

where in case (I)

$$z(\theta) = \alpha(\delta_0 - d_0 + \theta) + \frac{\psi(\theta)}{\delta_0 + \theta}, \quad \psi(\theta) = \min_{x \in S(\theta)} (c^T x + c_0)$$

while in case (II)

$$z(\theta) = \phi(\theta) + \frac{c_0 - \gamma d_0}{\delta_0 + \theta} + \gamma, \quad \phi(\theta) = \min_{x \in S(\theta)} a^T x.$$

In both the cases problem P can be solved by means of a simplex-like procedure based on a suitable post-optimality analysis. Note that (8.6) implies a decreasing (increasing) value of $f(x)$ corresponding to a decreasing (increasing) value of $z(\theta)$, so that to a local minimum for $z(\theta)$ there corresponds a local minimum for $f(x)$ which is also global for the pseudoconvexity of the function.

Below we shall describe a sequential method for solving problem P in case (I) and a sequential method for solving problem P in case (II).

• Case I $a = \alpha d$, $\alpha > 0$.
By taking into account Theorem 8.3.2, the infimum of problem P is $-\infty$ if and only if the infimum of the affine function $c^T x + c_0$ on $S(0)$ is $-\infty$. If this is not the case, consider the parametric problem

$$P_I(\theta): \min_{x \in S(\theta)} (c^T x + c_0)$$

Let $x_0 \in S$ be an optimal vertex for $P_I(0)$, and set $x_0 = (x_{B_0}, 0)$ where B_0 is the set of indices associated with its basic variables. By applying sensitivity analysis we find $(x_{B_0}(\theta), 0) = (x_{B_0} + \theta u_{B_0}, 0)$ which is optimal for $P(\theta)$ for every θ belonging to the stability interval $[\theta_0, \theta_1] = \{\theta : x_{B_0}(\theta) \geq 0\}$. If $z'(0) \geq 0$, then $(x_{B_0}, 0)$ is an optimal solution for P. If there exists $\bar{\theta} \in [\theta_0, \theta_1]$ such that $z'(\bar{\theta}) = 0$, then $(x_{B_0}(\bar{\theta}), 0)$ is an optimal solution for P, otherwise for $\theta > \theta_1$ the feasibility is lost and it is restored by applying a dual simplex iteration. We then find a new stability interval and we repeat the analysis. By proceeding in this way, a finite sequence of basis $B_k, k = 0, 1, ...,$ and a finite sequence of stability intervals $[\theta_k, \theta_{k+1}], k = 0, 1, ...,$ are generated.

With the usual notations, corresponding to the basis B_k, we have

$$(x_{B_k}(\theta), 0) = (x_{B_k} + \theta u_{B_k}, 0), \quad \psi(\theta) = c_{B_k}^T x_{B_k} + \theta c_{B_k}^T u_{B_k} + c_0, \quad \theta \in [\theta_k, \theta_{k+1}]$$

so that

$$z(\theta) = \alpha(\delta_0 - d_0 + \theta) + \frac{c_{B_k}^T x_{B_k} + \theta c_{B_k}^T u_{B_k} + c_0}{\delta_0 + \theta}, \ \theta \in [\theta_k, \theta_{k+1}]$$

$$z'(\theta) = \alpha + \frac{\xi_{B_k}}{(\delta_0 + \theta)^2}, \ \xi_{B_k} = \delta_0 c_{B_k}^T u_{B_k} - c_{B_k}^T x_{B_k} - c_0, \ \theta \in [\theta_k, \theta_{k+1}]$$

The algorithm can be described as follows:

Step 0. Solve the linear problem $\min_{x \in S}(d^T x + d_0)$ and let δ_0 be its optimal value. Solve problem $P_I(0) : \ \min(c^T x + c_0), \ x \in S \cap \{x : d^T x + d_0 = \delta_0\}$. If $P_I(0)$ has no solutions, then Stop: $\inf_{x \in S} f(x) = -\infty$.

Otherwise let x_0 be an optimal solution for $P_I(0)$ which is also an optimal solution for Problem $P(\theta_0)$ with $\theta_0 = 0$. Set $k = 0$ and go to Step 1.

Step 1. Determine $[\theta_k, \theta_{k+1}]$ the stability interval associated with the optimal solution $(x_{B_k}(\theta_k), 0) = (x_{B_k} + \theta_k u_{B_k}, 0)$ for $P(\theta_k)$. Compute $\xi_{B_k} = c_{B_k}^T u_{B_k}(\delta_0) - c_{B_k}^T x_{B_k} - c_0$. If $\xi_{B_k} \geq 0$, Stop: $(x_{B_k} + \theta_k u_{B_k}, 0)$ is an optimal solution for P, otherwise go to Step 2.

Step 2. Compute $\widetilde{\theta} = -\delta_0 + \sqrt{-\dfrac{\xi_{B_k}}{\alpha}}$.

If $\widetilde{\theta} \in [\theta_k, \theta_{k+1}]$, Stop: $(x_{B_k} + \widetilde{\theta} u_{B_k}, 0)$ is an optimal solution for P, otherwise let i be such that $x_{B_{k_i}} + \theta_{k+1} u_{B_{k_i}} = 0$. Perform a dual simplex iteration, set $k = k + 1$ and go to Step 1.

Example 8.3.1. Consider the following problem

$$\begin{cases} \min \left(x_1 + x_2 + \dfrac{-80x_1 - 60x_2 + 1}{x_1 + x_2 + 1} \right) \\ x_1 - 4x_2 + x_3 = 2 \\ x_2 + x_4 = 2 \\ x_i \geq 0, \ i = 1, .., 4 \end{cases}$$

Step 0. $x_0 = (0, 0, 2, 2)^T$ is the only solution for the problem $\min_{x \in S}(x_1 + x_2 + 1)$ and we have $\delta_0 = 1$; go to Step 1.

Step 1. With respect to the problem $P_I(\theta)$ the simplex-table associated with x_0 is given by

	−1	−80	−60	0	0	0
x_3	2	1	−4	1	0	0
x_4	2	0	1	0	1	0
x_5	θ	1	1	0	0	1

In order to restore the optimality of x_0, we perform a pivot operation on the underlined element. We obtain

	$-1 + 80\theta$	0	20	0	0	80
x_3	$2 - \theta$	0	−5	1	0	−1
x_4	2	0	1	0	1	0
x_1	θ	1	1	0	0	1

The stability interval is $[0, 2]$, $\xi_{B_0} = -81$. Since $\xi_{B_0} < 0$, go to Step 2.

Step 2. We have $\tilde{\theta} = -\delta_0 + \sqrt{-\frac{\xi_{B_0}}{\alpha}} = -1 + 9 = 8$. Since $\tilde{\theta} > 2$, we perform a pivot operation on the underline element, according to the dual simplex algorithm. We obtain

	$7 + 76\,\theta$	0	0	4	0	76
x_2	$-\frac{2}{5} + \frac{1}{5}\theta$	0	1	$-\frac{1}{5}$	0	$\frac{1}{5}$
x_4	$\frac{12}{5} - \frac{1}{5}\theta$	0	0	$\frac{1}{5}$	1	$-\frac{1}{5}$
x_1	$\frac{2}{5} + \frac{4}{5}\theta$	1	0	$\frac{1}{5}$	0	$\frac{4}{5}$

and go to Step 1.

Step 1. The stability interval is $[2, 12]$ and $\xi_{B_1} = -69$. Since $\xi_{B_1} < 0$, go to Step 2.

Step 2. We have $\tilde{\theta} = -\delta_0 + \sqrt{-\frac{\xi_{B_1}}{\alpha}} = -1 + \sqrt{69} \in [2, 12]$, so that $(-\frac{2}{5} + \frac{4}{5}\sqrt{69}, -\frac{3}{5} + \frac{1}{5}\sqrt{69}, 0, \frac{13}{5} - \frac{1}{5}\sqrt{69})$ is the optimal solution for problem P.

- Case II $c = \gamma d$, $c_0 - \gamma d_0 > 0$.

If the infimum of the linear function $a^T x$ is $-\infty$, then the infimum of problem P is $-\infty$, too, otherwise consider the parametric problem

$$P_{II}(\theta) : \min_{x \in S(\theta)} a^T x$$

Starting from a vertex $x_0 \in S$ which is an optimal solution for $P_{II}(0)$, and referring to the notations introduced before and with respect to the stability interval $[\theta_k, \theta_{k+1}]$, we have

$$z(\theta) = \frac{c_0 - \gamma d_0}{\delta_0 + \theta} + a_{B_k}^T x_{B_k} + \theta a_{B_k}^T u_{B_k} + \gamma, \quad \theta \in [\theta_k, \theta_{k+1}]$$

$$z'(\theta) = a_{B_k}^T u_{B_k} - \frac{c_0 - \gamma d_0}{(\delta_0 + \theta)^2}, \quad \theta \in [\theta_k, \theta_{k+1}]$$

If $a_{B_k}^T u_{B_k} < 0$ and $\theta_{k+1} = +\infty$, then the infimum of the problem is $-\infty$; if $a_{B_k}^T u_{B_k} = 0$ and $\theta_{k+1} = +\infty$, then the infimum is $a_{B_k}^T x_{B_k}$ and it is not attained as a minimum;

if $a_{B_k}^T u_{B_k} > 0$, we have $z'(\tilde{\theta}) = 0$ with $\tilde{\theta} = -\delta_0 + \sqrt{\frac{c_0 - \gamma d_0}{a_{B_k}^T u_{B_k}}}$. If $\tilde{\theta} < \theta_k$, then

$(x_{B_k}(\theta_k), 0)$ is an optimal solution for P; if $\tilde{\theta} \in [\theta_k, \theta_{k+1}]$, then $(x_{B_k}(\tilde{\theta}), 0)$ is an optimal solution for P.

In any other case, we consider the vertex $(x_{B_k}(\theta_{k+1}), 0)$ and we apply a dual simplex iteration in order to find a new stability interval; we repeat the analysis.

The algorithm can be described as follows:

Step 0. Solve the problem $\min\limits_{x \in S}(d^T x + d_0)$ and let δ_0 be the optimal value.
Solve problem $P_{II}(0)$: $\min a^T x$, $x \in S \cap \{x : d^T x + d_0 = \delta_0\}$. If $P_{II}(0)$ has no solutions, Stop: $\inf\limits_{x \in S} f(x) = -\infty$.
Otherwise let x_0 be an optimal solution for Problem $P_{II}(0)$ which is also an optimal solution for Problem $P(\theta_0)$ with $\theta_0 = 0$. Set $k = 0$, and go to Step 1.
Step 1. Determine $[\theta_k, \theta_{k+1}]$ the stability interval associated with the optimal solution $(x_{B_k}(\theta_k), 0) = (x_{B_k} + \theta_k u_{B_k}, 0)$ of $P(\theta_k)$. Compute $a_{B_k}^T u_{B_k}$. If $a_{B_k}^T u_{B_k} < 0$ and $\theta_{k+1} = +\infty$, Stop: the infimum of problem P is $-\infty$; if $a_{B_k}^T u_{B_k} < 0$ and θ_{k+1} is finite, go to Step 2; if $a_{B_k}^T u_{B_k} = 0$ and $\theta_{k+1} = +\infty$, Stop: the infimum of P is $a_{B_k}^T x_{B_k}$ and it is not attained as a minimum; if $a_{B_k}^T u_{B_k} = 0$ and θ_{k+1} is finite, go to Step 2; if $a_{B_k}^T u_{B_k} > 0$, go to Step 3.
Step 2. Let i be such that $x_{B_{k_i}} + \theta_{k+1} u_{B_{k_i}} = 0$. Perform a dual simplex iteration, set $k = k + 1$ and go to Step 1.

Step 3. Compute $\widetilde{\theta} = -\delta_0 + \sqrt{\dfrac{c_0 - \gamma d_0}{a_{B_k}^T u_{B_k}}}$.

If $\widetilde{\theta} \in [\theta_k, \theta_{k+1}]$, Stop: $(x_{B_k} + \widetilde{\theta} u_{B_k}, 0)$ is an optimal solution for P; if $\widetilde{\theta} < \theta_k$, Stop: $(x_{B_k} + \theta_k u_{B_k}, 0)$ is an optimal solution for P; if $\widetilde{\theta} > \theta_{k+1}$, let i be such that $x_{B_{k_i}} + \theta_{k+1} u_{B_{k_i}} = 0$. Perform a dual simplex iteration, set $k = k + 1$ and go to Step 1.

Example 8.3.2. Consider the following problem

$$
\begin{cases}
\min \left(-2x_1 + 6x_2 + \dfrac{12}{x_1 + 2x_2 + 1} \right) \\
x_1 - 2x_2 + x_3 = 3 \\
-x_1 + x_2 + x_4 = 1 \\
x_i \geq 0, \ i = 1, ..., 4
\end{cases}
$$

Step 0. The linear problem $\min\limits_{x \in S}(x_1 + 2x_2 + 1)$ has the only solution $x_0 = (0, 0, 3, 1)^T$, and we have $\delta_0 = 1$; go to Step 1.
Step 1. With respect to problem $P_{II}(0)$ the simplex-table associated with x_0 is given by

		0	-2	6	0	0	0
x_3	3	1	-2	1	0	0	
x_4	1	-1	1	0	1	0	
x_5	θ	1	2	0	0	1	

In order to restore the optimality of x_0, we perform a pivot operation on the underlined element. We obtain

		2θ	0	10	0	0	2
x_3	$3 - \theta$	0	-4	1	0	-1	
x_4	$1 + \theta$	0	3	0	1	1	
x_1	θ	1	2	0	0	1	

The stability interval is $[0,3]$. Since $a_{B_0}^T u_{B_0} = -2 < 0$, go to Step 2.

Step 2. We perform a pivot operation on the underline element according to the dual simplex algorithm. We obtain

	$\frac{15}{2} - \frac{1}{2}\theta$	0 0	$\frac{5}{2}$ 0 $-\frac{1}{2}$
x_2	$-\frac{3}{4} + \frac{1}{4}\theta$	0 1	$-\frac{1}{4}$ 0 $\frac{1}{4}$
x_4	$\frac{13}{4} + \frac{1}{4}\theta$	0 0	$\frac{3}{4}$ 1 $\frac{1}{4}$
x_1	$\frac{3}{2} + \frac{1}{2}\theta$	1 0	$\frac{1}{2}$ 0 $\frac{1}{2}$

Go to Step 1.

Step 1. The stability interval is $[3, +\infty]$. Since $a_{B_1}^T u_{B_1} = \frac{1}{2} > 0$, go to Step 3.

Step 3. It results that $\widetilde{\theta} = -\delta_0 + \sqrt{\frac{c_0 - \gamma d_0}{a_{B_1}^T u_{B_1}}} = -1 + 2\sqrt{6} \in [3, +\infty]$.

Then, $(1+\sqrt{6}, -1+\frac{1}{2}\sqrt{6}, 0, 3+\frac{1}{2}\sqrt{6})^T$ is the optimal solution for the problem.

Example 8.3.3. (The infimum is finite but not attained).
Consider the following problem

$$\begin{cases} min \left(-3x_1 + 4x_2 + \dfrac{5}{x_1 + 2x_2 + 1} \right) \\ 3x_1 - 4x_2 + x_3 = 6 \\ -x_1 + x_2 + x_4 = 4 \\ x_i \geq 0, \ i = 1, ..., 4 \end{cases}$$

Step 0. $x_0 = (0, 0, 6, 4)^T$ is the only solution for problem $\min_{x \in S}(x_1 + 2x_2 + 1)$ and we have $\delta_0 = 1$; go to Step 1.

Step 1. With respect to $P_{II}(0)$ the simplex-table associated with x_0 is given by

	0	-3	4 0 0 0
x_3	6	3	-4 1 0 0
x_4	4	-1	1 0 1 0
x_5	θ	$\underline{1}$	2 0 0 1

In order to restore the optimality of x_0, we perform a pivot operation on the underlined element. We obtain

	3θ	0	10 0 0 3
x_3	$6 - 3\theta$	0	$\underline{-10}$ 1 0 -3
x_4	$4 + \theta$	0	3 0 1 1
x_1	θ	1	2 0 0 1

The stability interval is $[0, 2]$. Since $a_{B_0}^T u_{B_0} = -3 < 0$, go to Step 2.

Step 2. We perform a pivot operation on the underline element, according to the dual simplex algorithm. We obtain

	6	0 0	1 0 0
x_2	$-\frac{3}{5}+\frac{3}{10}\theta$	0 1	$-\frac{1}{10}$ 0 $\frac{3}{10}$
x_4	$\frac{29}{5}+\frac{1}{10}\theta$	0 0	$\frac{3}{10}$ 1 $\frac{1}{10}$
x_1	$\frac{6}{5}+\frac{2}{5}\theta$	1 0	$\frac{1}{5}$ 0 $\frac{2}{5}$

Go to Step 1.

Step 1. The stability interval is $[2, +\infty]$. Since $a_{B_1}^T u_{B_1} = 0$, the infimum of the problem is 6 and it is not attained as a minimum.

8.3.2 The Sum of Two Linear Fractional Functions

The optimization of a sum of linear ratios arises in various areas, such as multi-stage stochasting shipping [4], layered manifacturing [195, 196], cluster analysis [228], multiobjective bond portfolio [177], and combinatorial optimization [227] (for several others economic applications see [114, 256]). This kind of problems is difficult to be solved since it does not have any generalized convexity properties; it has attracted the interest of researchers for a number of years and different approaches for solve them have been proposed. For instance, in [106] an approach in the so-called image space is suggested, in [68, 96, 181, 187] branch-and-bound procedures are given, and in [23] suitable transformations are used which reduce the problem in another one simpler to be handle.

For the particular case of the sum of two linear ratios, some others algorithms have been proposed, also in the framework of bicriteria problems, where a compromise solution is sought (see for all [41, 34]).

In this section we shall present a simplex-like procedure under pseudoconvexity assumptions on the sum of two linear fractional functions.

When the sum of two linear ratio is pseudoconvex on the feasible set, the problem

$$P : \min_{x \in S} \left(h(x) = \frac{m^T x + m_0}{p^T x + p_0} + \frac{q^T x + q_0}{b^T x + b_0} \right), \quad S = \{x \in \Re^n : Ax = b, \ x \geq 0\}$$

may be transformed by means of Charnes–Cooper's transformation into an equivalent one with the sum of a linear and a linear fractional function as objective. Consequently, we may use the simple simplex-like procedures described before in this section for determing the optimal solution (if one exsits) for P.

At this regard consider the following problem

$$P : \begin{cases} \min \left(\frac{x_1 + x_2}{1 - x_1} + \frac{1 - 81x_1 - 60x_2}{x_2 + 1} \right) \\ 3x_1 - 4x_2 + x_3 = 2 \\ 2x_1 + x_2 + x_4 = 2 \\ x_i \geq 0, \ i = 1, ..4 \end{cases}$$

It is easy to verify that $p^T x + p_0 = 1 - x_1 > 0$, $b^T x + b_0 = x_2 + 1 > 0$, for all $x = (x_1, x_2, x_3, x_4)^T \in S$. The objective function is pseudoconvex on S for (i) of Theorem 7.5.1; by applying Charnes–Cooper's transformation $y_i = \dfrac{x_i}{1 - x_1}$, $i = 1, .., 4$, whose inverse is $x_i = \dfrac{y_i}{1 + y_1}$, $i = 1, .., 4$, we obtain the problem

$$P^* : \begin{cases} min \left(y_1 + y_2 + \dfrac{1 - 80y_1 - 60y_2}{y_1 + y_2 + 1} \right) \\ y_1 - 4y_2 + y_3 = 2 \\ y_2 + y_4 = 2 \\ y_i \geq 0, \ i = 1, ..4 \end{cases}$$

By referring to Example 8.3.1, the optimal solution for P^* is given by $(y_1, y_2, y_3, y_4) = (-\frac{2}{5} + \frac{4}{5} \sqrt{69}, -\frac{3}{5} + \frac{1}{5} \sqrt{69}, 0, \frac{13}{5} - \frac{1}{5} \sqrt{69})$, so that the optimal solution for P is given by $(x_1, x_2, x_3, x_4) = (\frac{-2+4\sqrt{69}}{3+4\sqrt{69}}, \frac{-3+\sqrt{69}}{3+4\sqrt{69}}, 0, \frac{13-\sqrt{69}}{3+4\sqrt{69}})$.

8.4 Generalized Linear Multiplicative Programs

In this section we shall consider the following class of generalized linear multiplicative problems:

$$P : \min_{x \in S} \ \left[f(x) = c^T x + (a^T x + a_0)(d^T x + d_0)^p \right]$$

where $S = \{x \in \Re^n : Ax = b, \ x \geq 0\}$, $c, a, d \in \Re^n$, $a, d \neq 0$, $a_0, d_0 \in \Re$, $p \in \Re \backslash \{0\}$, A is an $m \times n$ real matrix, $rank A = m < n$. We shall study problem P in the case $p = 1$ and in the case $c = 0$.

8.4.1 The Sum of a Linear Function and the Product of Two Affine Functions

Consider the problem

$$P_1 : \min_{x \in S} \ \left[f_1(x) = c^T x + (a^T x + a_0)(d^T x + d_0) \right]$$

As we have seen in Sect. 6.6, function f_1 is not generalized convex so that problem P_1 may have several local minimum points which are not global. Methods for solving problem P_1 have been proposed in [173, 174, 257]. In particular in [174, 257], the problem is solved by means of the following parametric linear programming problem.

$$P_1(\theta) : z(\theta) = \min_{x \in S(\theta)} \ \left[c^T x + (\xi + \theta)(a^T x + a_0) \right]$$

where $S(\theta) = S \cap \{x \in \Re^n : d^T x + d_0 = \xi + \theta\}$, and ξ is a feasible level, i.e., $S(0) \neq \emptyset$.

The algorithm proposed in [174] finds the optimal solution for P_1 by solving $P_1(\theta)$ for all the feasible levels, from the minimum level $\xi_{min} = \min_{x \in S}(d^T x + d_0)$ to the maximum level $\xi_{max} = \max_{x \in S}(d^T x + d_0)$, assuming the compactness of S.

The algorithm proposed in [257] works for every feasible region (bounded or unbounded) and finds the optimal solution for P_1 (if one exists) by generating a finite sequence of local minimum points, the last of which is the global one. This algorithm may be easily adapted to the case where f_1 is pseudoconvex on S.

Taking into account Corollary 6.6.1 and Theorem 6.13, we shall consider the case which corresponds to the linear independence of a, d since, when a, d are linearly dependent, P_1 reduces to a problem which is very easy to solve.

According to Corollary 6.6.1, we can suppose that the affine function $d^T x + d_0$ is lower bounded on S.

Set $\xi_0 = \min_{x \in S}(d^T x + d_0)$ and consider the parametric problem $P_1(\theta)$ with $\xi = \xi_0$.

If $P_1(0)$ does not have any solutions, then $\inf_{x \in S} f_1(x) = -\infty$, otherwise let x^0 be a basic optimal solution for $P_1(0)$ and let B_0 and N_0 be the set of indices associated with its basic variables and with its non-basic variables, respectively. By applying sensitivity analysis we find $(x_{B_0}(\theta), 0) = (x_{B_0} + \theta u_{B_0}, 0)$ which is optimal for $P_1(\theta)$ for every θ belonging to the interval $[0, \theta_1] = F_0 \cap O_0$, where $F_0 = \{\theta \in \Re : x_{B_0} + \theta u_{B_0} \geq 0\}$, and $O_0 = \{\theta \in \Re : \bar{c}_{N_0} + (\xi_0 + \theta)\bar{a}_{N_0} \geq 0\}$.

If $z'(0) \geq 0$, then $(x_{B_0}, 0)$ is optimal for P_1; if there exists $\theta' \in (0, \theta_1]$ such that $z'(\theta') = 0$, then $(x_{B_0}(\theta'), 0)$ is optimal for P_1, otherwise, if $\theta_1 = +\infty$, $\inf_{x \in S} f_1(x) = -\infty$, while if θ_1 is finite, the feasibility or the optimality is lost and it is restored by applying a dual simplex iteration or a primal simplex iteration. By means of this procedure, a finite sequence of basis B_k and a finite sequence of intervals $[\theta_k, \theta_{k+1}]$ are generated.

In correspondence to the basis B_k, with the usual notations, set:

$x^k(\theta) = (x_{B_k}(\theta), 0) = (x_{B_k} + \theta u_{B_k}, 0)$;

$\bar{f}_{N_k} = \bar{c}_{N_k} + (\xi_0 + \theta)\bar{a}_{N_k}$;

$F_k = \{\theta \in \Re : x_{B_k} + \theta u_{B_k} \geq 0\}$;

$O_k = \{\theta \in \Re : \bar{f}_{N_k} \geq 0\}$.

The optimal value function $z(\theta)$ is of the form $z(\theta) = \lambda\theta^2 + \mu\theta + \gamma$, where $\lambda = a_{B_k}^T u_{B_k}$, $\mu = c_{B_k}^T u_{B_k} + a_{B_k}^T x_{B_k} + \xi_0 a_{B_k}^T u_{B_k}$; if $\lambda \neq 0$, let $\theta' = -\frac{\mu}{2\lambda}$ be the critical point of $z(\theta)$.

The algorithm may be described as follows:

Step 0. Solve the linear problem $\min_{x \in S}(d^T x + d_0)$ and let ξ_0 be its optimal value. Solve problem $P_1(0)$ with $\xi = \xi_0$. If $P_1(0)$ does not have solutions, Stop: $\inf_{x \in S} f_1(x) = -\infty$, otherwise let x^0 be a basic optimal solution for $P_1(0)$, set $\theta_0 = 0$, $k = 0$, and go to Step 1.

Step 1. If $z'(\theta_k) \geq 0$, then Stop: $(x_{B_k}(0), 0)$ is an optimal solution for P_1, otherwise, calculate F_k, O_k and θ'. If $\theta' \in F_k \cap O_k$, then Stop: $(x_{B_k}(\theta'), 0)$ is an optimal solution for P_1, otherwise go to Step 2.

Step 2. If $\theta_{k+1} = \sup(F_k \cap O_k) = +\infty$, then Stop: $\inf\limits_{x \in S} f_1(x) = -\infty$, otherwise go to Step 3.

Step 3. If $F_k \cap O_k = F_k$, apply a dual simplex iteration, set $k = k+1$ and go to Step 1. If $F_k \cap O_k = O_k$, apply a primal simplex iteration, set $k = k+1$ and go to Step 1.

Example 8.4.1. Consider the problem

$$\begin{cases} min \quad [-6x_1 - 12x_2 + (-x_1 - 4x_2 + 42)(x_1 + x_2 - 11)] \\ -x_1 + x_2 + x_3 = 16 \\ x_1 + 3x_2 + x_4 = 56 \\ -4x_1 - 25x_2 + x_5 = -250 \\ x_i \geq 0, \ i = 1, .., 4 \end{cases}$$

By setting $d = (1, 1)^T$, $a = (-1, -4)^T$ we have $c = \alpha a + \beta d$, $\alpha = 2$, $\beta = -4$; since ii) of Corollary 6.6.2 holds, the objective function is pseudoconvex on the feasible set S.

The problem $\min\limits_{x \in S}(x_1 + x_2 - 11)$ has as its only solution $x^0 = (0, 10, 6, 26, 0)^T$ and the corresponding minimum value is $\xi_0 = -1$.

The parametric linear programming problem becomes:

$$P_1(\theta): \quad \min\limits_{x \in S(\theta)} \quad [(-\theta - 5)x_1 + (-4\theta - 8)x_2 + 42\theta - 42],$$

$$S(\theta) = S \cap \{x \in \Re^5 : x_1 + x_2 - 11 = -1 + \theta\}$$

We have $x^0(\theta) = (0, 10 + \theta, 6 - \theta, 26 - 3\theta, 25\theta)^T$ and $z(\theta) = -4\theta^2 - 6\theta - 122$. It results that $z'(0) < 0$, $O_0 = [0, +\infty)$, $F_0 = [0, 6]$, $\theta' = -\frac{3}{4} \notin F_0 \cap O_0 = F_0$; by means of a dual simplex iteration the variable x_1 enters the basis. We have $x^1(\theta) = (-3 + \frac{1}{2}\theta, 13 + \frac{1}{2}\theta, 0, 20 - 2\theta, 63 + \frac{29}{2}\theta)^T$ and $z(\theta) = -\frac{5}{2}\theta^2 - \frac{27}{2}\theta - 131$. It results that $z'(6) < 0$, $O_1 = [6, +\infty)$, $F_1 = [6, 10]$, $\theta' = -\frac{27}{10} \notin F_1 \cap O_1 = F_1$; by means of a dual simplex iteration the variable x_3 enters the basis. We have $x^2(\theta) = (-13 + \frac{3}{2}\theta, 23 - \frac{1}{2}\theta, -20 + 2\theta, 0, 273 - \frac{13}{2}\theta)^T$ and $z(\theta) = \frac{1}{2}\theta^2 - \frac{13}{2}\theta - 161$. It results that $z'(10) < 0$, $O_2 = [10, +\infty)$, $F_2 = [10, 46]$, $\theta' = \frac{81}{2}$. Since $\theta' \in F_2 \cap O_2 = [10, 46]$, $x^2(\theta') = (\frac{191}{4}, \frac{11}{4}, 61, 0, \frac{39}{4})^T$ is the optimal solution for the problem.

8.4.2 The Product Between an Affine Function and the Power of an Affine Function

Consider the problem

$$P_2: \inf\limits_{x \in S} \left[f_2(x) = (a^T x + a_0)(d^T x + d_0)^p \right], \quad S = \{x \in \Re^n : Ax = b, \ x \geq 0\}$$

In order to consider the objective function for every value of the exponent p, it is necessary to have $d^T x + d_0 > 0$ for all $x \in S$. Furthermore, to avoid trivial cases, we shall assume the linear independence of the vectors a, d.

Before to present a sequential method for solving problem P_2, we shall study, firstly, the pseudoconvexity of the function f_2.

Theorem 8.4.1. *Consider the function $f_2(x) = (a^T x + a_0)(d^T x + d_0)^p$ where a, d are linearly independent, $p \in \Re$, $p \neq 0$. Then, f_2 is pseudoconvex on the half-space $D^+ = \{x \in \Re^n : d^T x + d_0 > 0\}$ if and only if one of the following conditions holds:*
(i) $p \in [-1, 0)$, and $a^T x + a_0 > 0$ for all $x \in D^+$;
(ii) $p \in (-\infty, -1] \cup (0, +\infty)$, and $a^T x + a_0 < 0$ for all $x \in D^+$.

Proof. We have $\nabla f_2(x) = (d^T x + d_0)^{p-1} \left[(d^T x + d_0) a + p(a^T x + a_0) d \right]$, so that the linear independence of a, d implies that $\nabla f_2(x) \neq 0$, $x \in D^+$.

The function f_2 is pseudoconvex on D^+ if and only if $v^T \nabla f_2(x) = 0$ implies that $v^T \nabla^2 f_2(x) v \geq 0$ (see Corollary 3.4.1).

We have $v^T \nabla f_2(x) = 0$ if and only if $a^T v = -\dfrac{p(a^T x + a_0) d^T v}{d^T x + d_0}$, so that $v^T \nabla^2 f_2(x) v = p(-p-1)(d^T x + d_0)^{p-1}(a^T x + a_0)(d^T v)^2$.

Since the linear independence of a, d implies the existence of v such that $d^T v \neq 0$ and $v^T \nabla f(x) = 0$, it follows that $v^T \nabla^2 f_2(x) v \geq 0$ if and only if (i) or (ii) holds.   □

An algorithm for solving problem P_2 has been proposed in [30, 31, 198, 199] when p is an integer and in [51] for all $p \neq 0$. Now, under the pseudoconvexity assumption, we shall present two different methods for solving problem P_2. The former is based on a parametric approach, while the latter extends the approach given in Sect. 8.2.4, i.e., primal simplex iterations are performed according to the rule (8.5) where now "max" is replaced by "min".

A Parametric Approach

Consider the following parametric linear problem associated with problem P_2:

$$P_2(\theta) : z(\theta) = \min_{x \in S(\theta)} \ (\xi_0 + \theta)^p (a^T x + a_0), \ S(\theta) \tag{8.7}$$

$$= S \cap \{x \in \Re^n : d^T x + d_0 = \xi_0 + \theta\}$$

where $\xi_0 = \min_{x \in S}(d^T x + d_0)$.

Note that ξ_0 exists since the affine function $d^T x + d_0$ is lower bounded on S. If $P_2(0)$ does not have any solutions (this may happen in case (ii) of Theorem 8.4.1), then $\inf_{x \in S} f_2(x) = -\infty$, otherwise let x^0 be a basic optimal solution for $P_2(0)$ and let B_0 be the set of indices associated with its basic variables. By applying sensitivity analysis we find $(x_{B_0}(\theta), 0) = (x_{B_0} + \theta u_{B_0}, 0)$ which

is optimal for $P_2(\theta)$ for every θ belonging to the interval $[0, \theta_1] = F_0$, where
$F_0 = \{\theta \in \Re : x_{B_0} + \theta u_{B_0} \geq 0\}$.
If $z'(0) \geq 0$, then $(x_{B_0}, 0)$ is optimal for P_2; if there exists $\theta' \in (0, \theta_1]$ such
that $z'(\theta') = 0$, then $(x_{B_0}(\theta'), 0)$ is optimal for P_2, otherwise, if $\theta_1 = +\infty$,
$\inf_{x \in S} f_2(x) = \lim_{\theta \to +\infty} z(\theta)$, while, if θ_1 is finite, the feasibility is lost and it is
restored by applying a dual simplex iteration. By means of this procedure,
a finite sequence of basis B_k and a finite sequence of intervals $[\theta_k, \theta_{k+1}]$ are
generated.
The optimal value function $z(\theta)$ is of the form $z(\theta) = (\xi_0 + \theta)^p(\alpha\theta + \beta)$ and its
derivative is $z'(\theta) = (\xi_0 + \theta)^{p-1}((p+1)\alpha\theta + p\beta + \alpha\xi_0)$, where $\alpha = a_{B_k}^T u_{B_k}$,
$\beta = a_{B_k}^T x_{B_k} + a_0$.
The algorithm may be described as follows (note that the value of the limit
$\lim_{\theta \to +\infty} z(\theta)$ changes according to the case $p > 0$ or $p < -1$).
Step 0. Solve the linear problem $\min_{x \in S}(d^T x + d_0)$ and let ξ_0 be its optimal
value. Solve problem $P_2(0)$ with $\xi = \xi_0$. If $P_2(0)$ does not have any solutions,
Stop: $\inf_{x \in S} f_2(x) = -\infty$, otherwise let x^0 be a basic optimal solution for $P_2(0)$,
set $\theta_0 = 0$, $k = 0$, and go to Step 1.
Step 1. If $z'(\theta_k) \geq 0$, then Stop: $(x_{B_k}(0), 0)$ is an optimal solution for P_2,
otherwise calculate F_k and θ'. If $\theta' \in F_k$, then Stop: $(x_{B_k}(\theta'), 0)$ is an optimal
solution for P_2, otherwise go to Step 2.
Step 2. If θ_{k+1} is finite, apply a dual simplex iteration, set $k = k+1$ and go
to Step 1, otherwise Stop: $\inf_{x \in S} f_2(x) = -\infty$ if $p > 0$, $\inf_{x \in S} f_2(x) = 0$ if $p < -1$.

Example 8.4.2. Consider the problem

$$\begin{cases} min(-x_1 - x_2 + 6)(x_1 + 2x_2 - 9)^2 \\ x_1 - x_2 + x_3 = 16 \\ 2x_1 + 3x_2 + x_4 = 42 \\ -7x_1 - 10x_2 + x_5 = -70 \\ x_i \geq 0, \ i = 1, .., 5 \end{cases}$$

The objective function is pseudoconvex on the feasible region S since we have
$S \subset \{(x_1, x_2) \in \Re^2 : -x_1 - x_2 + 6 < 0, \ x_1 + 2x_2 - 9 > 0\}$.
The problem $\min_{x \in S}(x_1 + 2x_2 - 9)$ has as its only solution $x^0 = (10, 0, 6, 22, 0)^T$
and the corresponding minimum value is $\xi_0 = 1$.
The parametric linear programming problem becomes:

$$P_2(\theta) : \min_{x \in S(\theta)} (1 + \theta)^2(-x_1 - x_2 + 6),$$

$$S(\theta) = S \cap \{x \in \Re^5 : x_1 + 2x_2 - 9 = 1 + \theta\}$$

We have $x^0(\theta) = (10 + \theta, 0, 6 - \theta, 22 - 2\theta, 7\theta)^T$ and $z(\theta) = (1 + \theta)^2(-4 - \theta)$. It
results that $z'(0) < 0$, $F_0 = [0, 6]$, $\theta' = -3 \notin F_0$; by means of a dual simplex
iteration, the variable x_2 enters the basis.

We have $x^1(\theta) = (14 + \frac{1}{3}\theta, -2 + \frac{1}{3}\theta, 0, 20 - \frac{5}{3}\theta, 8 + \frac{17}{3}\theta)^T$ and $z(\theta) = (1 + \theta)^2(-6 - \frac{2}{3}\theta)$. It results that $z'(6) < 0$, $F_1 = [6, 12]$, $\theta' = -\frac{19}{3} \notin F_1$; by means of a dual simplex iteration, the variable x_3 enters the basis. We have $x^2(\theta) = (54 - 3\theta, -22 + 2\theta, -60 + 5\theta, 0, 88 - \theta)^T$, $z(\theta) = (1 + \theta)^2(-26 + \theta)$. It results that $z'(12) < 0$, $F_2 = [12, 18]$, $\theta' = 17$.

Since $\theta' \in F_2$, $x^2(\theta') = (3, 12, 25, 0, 71)^T$ is the optimal solution for the problem.

A Non-parametric Approach

Using the notations introduced in Sect. 8.2, the matrix A is partitioned as $A = [B : N]$ and the vectors x, a and d as $x^T = (x_B^T, x_N^T)$, $a^T = (a_B^T, a_N^T)$, $d^T = (d_B^T, d_N^T)$.

Set:
$\bar{a}_0 = a_B^T B^{-1} b + a_0$, $\bar{d}_0 = d_B^T B^{-1} b + d_0$,
$\bar{a}_N^T = a_N^T - a_B^T B^{-1} N$, $\bar{d}_N^T = d_N^T - d_B^T B^{-1} N$,
$\bar{a}_j$ and $\bar{d}_j$ denote the j-th component of $\bar{a}_N$ and $\bar{d}_N$, respectively.

In correspondence to the non-basic variable x_{N_k}, let
$\varphi(x_{N_k}) = (\bar{a}_{N_k} x_{N_k} + \bar{a}_0)(\bar{d}_{N_k} x_{N_k} + \bar{d}_0)^p$.

We have $\varphi'(x_{N_k}) = (\bar{d}_{N_k} x_{N_k} + \bar{d}_0)^{p-1}[(p+1)\bar{a}_{N_k}\bar{d}_{N_k} x_{N_k} + \bar{a}_{N_k}\bar{d}_0 + p\bar{d}_{N_k}\bar{a}_0]$.

Denote with $x_{N_k}^*$ the critical point of $\varphi(x_{N_k})$ and let $\bar{x}_{N_k} = \min\limits_{u_j > 0} \frac{x_{B_j}}{\bar{u}_j}$, where u_j is the j-th component of the column $u = B^{-1}N_k$, where N_k is the column of N associated with x_{N_k}. By convention, set $\bar{x}_{N_k} = +\infty$ if $B^{-1}N_k \leq 0$. The main steps of the algorithm may be described as follows.

Step 0. Solve the linear problem $\min\limits_{x \in S}(d^T x + d_0)$ and let ξ_0 be its optimal value. Solve the problem $P^* : \min(a^T x + a_0)$, $x \in S \cap \{x \in \Re^n : d^T x + d_0 = \xi_0\}$. If P^* does not have any optimal solutions, then Stop: $\inf f_2(x) = -\infty$. Otherwise, let x^0 be an optimal vertex for P^*. Set $i = 0$ and go to Step 1.

Step 1. Set $J = \{j : \bar{d}_j > 0\}$. If $J = \emptyset$, Stop: x^i is an optimal solution for P_2; otherwise, let k be such that $\frac{\bar{a}_k}{\bar{d}_k} = \min\limits_{j \in J} \frac{\bar{a}_j}{\bar{d}_j}$ and go to Step 2.

Step 2. If $u \leq 0$, then Stop: the infimum is $-\infty$ or 0 according to the case $p > 0$, $p < -1$; otherwise, calculate $\bar{x}_{N_k}$ and $x_{N_k}^*$. If $x_{N_k}^* \in [0, \bar{x}_{N_k}]$, then Stop: $(x_B - x_{N_k}^* B^{-1}N_k, 0)$ is the optimal solution for P_2, otherwise, perform a primal simplex iteration with x_{N_k} as the entering variable. Let x^{i+1} be the new basic solution. Set $i = i + 1$ and go to Step 1.

Example 8.4.3. Consider the problem given in Example 8.4.2 again. The simplex-table associated with $x^0 = (10, 0, 6, 22, 0)^T$ is given by

4	0	$\frac{3}{7}$	0	0	$-\frac{1}{7}$
-1	0	$\frac{4}{7}$	0	0	$\frac{1}{7}$
x_3 6	0	$-\frac{17}{7}$	1	0	$\frac{1}{7}$
x_4 22	0	$\frac{1}{7}$	0	1	$\frac{2}{7}$
x_1 10	1	$\frac{10}{7}$	0	0	$-\frac{1}{7}$

where in the first row and in the second row we can read the reduced costs $\bar{a}_N$ and $\bar{d}_N$, respectively. We have: $J = \{2,5\}$, $\min\limits_{j\in J}\dfrac{\bar{a}_j}{\bar{d}_j} = \dfrac{\bar{a}_5}{\bar{d}_5}$, $\bar{x}_5 = 42$, $z(x_5) = (-\frac{1}{7}x_5 - 4)(\frac{1}{7}x_5 + 1)^2$, and $x_5^* = -21 \notin [0,42]$.
The non-basic variable x_5 enters the basis by means of a primal simplex iteration.

10	0	-2	1	0 0
-7	0	3	-1	0 0
x_5 42	0	-17	7	0 1
x_4 10	0	5	-2	1 0
x_1 16	1	-1	1	0 0

We have: $J = \{2\}$, $\bar{x}_2 = 2$, $z(x_2) = (-2x_2 - 10)(3x_2 + 7)^2$, and $x_2^* = -\frac{37}{9} \notin [0,2]$. The non-basic variable x_2 enters the basis by means of a primal simplex iteration.

14	0 0	$\frac{1}{5}$	$\frac{2}{5}$	0
-13	0 0	$\frac{1}{5}$	$-\frac{3}{5}$	0
x_5 76	0 0	$\frac{1}{5}$	$\frac{17}{5}$	1
x_2 2	0 1	$-\frac{2}{5}$	$\frac{1}{5}$	0
x_1 18	1 0	$\frac{3}{5}$	$\frac{1}{5}$	0

We have: $J = \{3\}$, $\bar{x}_3 = 30$, $z(x_3) = (\frac{1}{5}x_3 - 14)(\frac{1}{5}x_3 + 13)^2$, and $x_3^* = 25$. Since $x_3^* \in [0,30]$, the optimal solution for the problem is $x^* = (18,2,0,0,76)^T + 25(-\frac{3}{5}, \frac{2}{5}, 0, 0, -\frac{1}{5})^T = (3,12,25,0,0,71)^T$.

8.5 The Optimal Level Solutions Method

The algorithms described in Sect. 8.3 and in Sect. 8.4 are based on the common idea of associating with an optimization problem a suitable parametric program which easier to solve. This kind of approach has been suggested in many papers (see for instance [29, 30, 31, 32, 33, 34, 35, 145, 280, 281]) and leads to efficient solution methods. For this reason, in this section we shall present the general approach given in [101].
Consider the following problem

$$P : \inf_{x \in S} [\Phi(x) = F(x, g(x))]$$

where S is a nonempty set of $\Re^n$, g is a continuous function defined on S and F is a continuous function defined on $\{(x, \xi) : x \in \Re^n,\ \xi = g(x)\}$.

By means of the following variable transformation

$$T : S \rightarrow S \times \Re,\quad x \rightarrow (x, \xi),\ \xi = g(x)$$

problem P is transformed into the following one

$$P^* : \begin{cases} \inf\ F(x, \xi) \\ x \in S \\ g(x) = \xi,\quad \xi \in \Xi = g(S). \end{cases}$$

Problems P and P^* are equivalent in the sense that P has an optimal solution $\bar{x}$ if and only if P^* has an optimal solution $(\bar{x}, \bar{\xi})$ with $\bar{\xi} = g(\bar{x})$.

Obviously, transformation T is useful when the transformed problem is easier to handle than the original one.

In order to point out some relationships between P and P^* we shall give the following definitions.

Definition 8.5.1. *A real number ξ is said to be a feasible level for P if $\xi \in \Xi$ or, equivalently, if there exists $\bar{x} \in S$ such that $g(\bar{x}) = \xi$.*
The set of all the feasible levels of P will be called the feasible levels set for P.

Definition 8.5.2. *The point $\bar{x} \in S$ is said to be an optimal level solution for P corresponding to the level ξ if $\bar{x}$ is an optimal solution for the problem*

$$P(\xi) : \begin{cases} \inf\ F(x, \xi),\ x \in S(\xi) \\ S(\xi) = \{x \in S : g(x) = \xi\}. \end{cases}$$

The whole set of optimal level solutions corresponding to the same level ξ is denoted by L_ξ and $L = \bigcup_{\xi \in \Xi} L_\xi$ is called the optimal level solutions set for P.

Let us note that if there exists a feasible level ξ such that the infimum of $P(\xi)$ is $-\infty$, then the infimum of problem P is $-\infty$, too, and vice versa. In general, we have

$$\inf_{x \in S} \Phi(x) = \inf_{\xi \in \Xi}\ \inf_{x \in S(\xi)}\ F(x, \xi).$$

If problem P has optimal solutions, then

$$\min_{x \in S} \Phi(x) = \min_{\xi \in \Xi}\ \inf_{x \in S(\xi)}\ F(x, \xi).$$

Note that problem P may have optimal solutions even if the infimum of $P(\xi)$ is finite and not attained, as is shown in the following example.

Example 8.5.1. Consider problem P where $\Phi(x_1, x_2) = \dfrac{-x_1}{x_2 + 1} + x_2 - x_1$ and
$S = \{(x_1, x_2) : x_1 - x_2 \le 4, \ x_1 \ge 0, \ x_2 \ge 0\}$.
By setting $g(x_1, x_2) = x_2 - x_1$, problem P is transformed into the following
parametric problem

$$P^* : \begin{cases} inf \left(\dfrac{-x_1}{x_2 + 1} + \xi \right) \\ (x_1, x_2) \in S, \ x_2 - x_1 = \xi, \ \xi \in [-4, +\infty). \end{cases}$$

In correspondence to the feasible level $\xi = 0$, the infimum of problem $P(0)$ is
-1 and it is not attained. Nevertheless, $(4, 0)$ is the optimal solution for P,
corresponding to the level $\xi = -4$.

When $L_\xi \ne \emptyset$ for all $\xi \in \Xi$, we have

$$\min_{x \in S} \Phi(x) = \min_{\xi \in \Xi} \min_{x \in S(\xi)} F(x, \xi). \tag{8.8}$$

Set $z(\xi) = \min_{x \in S(\xi)} F(x, \xi)$.
The relation (8.8) points out that to a global minimum point $\bar{x}$ for P there
corresponds the minimum level $\xi = g(\bar{x})$ for the function $z(\xi)$, $\xi \in \Xi$ and vice
versa. When $\bar{x}$ is a non global local minimum point, then the corresponding
level $\bar{\xi}$ is not necessarily a local minimum for $z(\xi)$ (see [101]). Fortunately,
the converse statement is true under suitable assumptions, as is stated in the
following theorem.

Theorem 8.5.1. *Consider the parametric problem P^* and assume that Ξ is
an interval and $L_\xi \ne \emptyset$ for all $\xi \in \Xi$. If $\bar{\xi}$ is a local minimum point for the
optimal value function $z(\xi)$, then each point $\bar{x} \in L_{\bar{\xi}}$ is a local minimum point
for P. Furthermore, if S is a convex set and $\Phi(x)$ is semistrictly quasiconvex
on S, then each point $\bar{x} \in L_{\bar{\xi}}$ is a global minimum point for P.*

Proof. Let $\epsilon > 0$ be such that $z(\xi) \ge z(\bar{\xi})$ for all $\xi \in (\bar{\xi} - \epsilon, \bar{\xi} + \epsilon)$. Since
$g(\bar{x}) = \bar{\xi}$, the continuity of g implies the existence of a neighborhood $I(\bar{x})$ of
$\bar{x}$ such that $g(x) \in (\bar{\xi} - \epsilon, \bar{\xi} + \epsilon)$ for all $x \in I(\bar{x})$. Assume, by contradiction,
the existence of $x^* \in I(\bar{x})$ such that $\Phi(x^*) < \Phi(\bar{x})$ and let $\xi^* = g(x^*)$. Then,
we have $z(\xi^*) \le F(x^*, \xi^*) = \Phi(x^*) < \Phi(\bar{x}) = z(\bar{\xi})$, and this contradicts the
local optimality of $\bar{\xi}$.
The last statement follows from Theorem 4.5.1.   □

The general approach that we have described allows us to include several
parametric algorithms in the framework of the optimal level solutions method.
Recently in [53, 54], this method has been proposed in order to solve a nonconvex optimization problem by means of a parametric convex program.
In [53] the following optimization problem is considered:

$$P : \inf_{x \in S} \left[f(x) = \frac{1}{2} x^T Q x + q^T x - (d^T x)^2 \right], \quad S = \{x \in \Re^n : Ax \geq b\}$$

where Q is a symmetric positive definite $n \times n$ matrix, $q, d \in \Re^n$, A is an $m \times n$ matrix, $b \in \Re^m$.

The objective function f is a d.c. function, i.e., the difference of convex functions, so that P is not in general a generalized convex problem.

Problem P is solved by means of the optimal level solutions approach, by considering the following parametric strictly convex quadratic problem:

$$P(\xi) : \inf_{x \in S(\xi)} \left(\frac{1}{2} x^T Q x + q^T x - (\xi)^2 \right), \quad S(\xi) = S \cap \{x \in \Re^n : d^T x = \xi\}$$

In [54] the following nonlinear multiplicative problem is studied:

$$P : \inf_{x \in S} \left[f(x) = \left(\frac{1}{2} x^T Q x + q^T x + q_0 \right) \left(d^T x + d_0 \right)^p \right],$$

where $S = \{x \in \Re^n : Ax \geq b\}$, Q is a symmetric positive definite $n \times n$ matrix, $q, d \in \Re^n$, $q_0, d_0, p \in \Re$, $p \neq 0$, A is an $m \times n$ matrix, $b \in \Re^m$.

The objective function f is not in general a generalized convex function. Nevertheless, the following parametric strictly convex quadratic problem is associated with P:

$$P(\xi) : \inf_{x \in S(\xi)} \left(\frac{1}{2} x^T Q x + q^T x + q_0 \right) \xi^p, \quad S(\xi) = S \cap \{x \in \Re^n : d^T x + d_0 = \xi\}$$

The study of generalized convexity of the described problems is still an open problem; its characterization would allow us to improve the suggested methods.

Finally, we shall point out that also the problem of minimizing the ratio of a quadratic and an affine function (see Sect. 7.2) can be embedded in the optimal level solutions method.

Consider the problem

$$P_Q : \inf_{x \in S} \left[f(x) = \frac{\frac{1}{2} x^T Q x + q^T x + q_0}{d^T x + d_0} \right]$$

where $S = \{x \in \Re^n : Ax \geq b\}$, Q is a symmetric $n \times n$ matrix, $q, d \in \Re^n$, $d \neq 0$, $q_0, d_0, p \in \Re$, $p \neq 0$, A is an $m \times n$ matrix, $b \in \Re^m$.

By setting $d^T x + d_0 = \xi$, the following parametric quadratic problem is obtained

$$P_Q(\xi) : \inf_{x \in S_Q(\xi)} \left(\frac{1}{2} x^T Q x + q^T x + q_0 \right) \xi^{-1}$$

where $S_Q(\xi) = S \cap \{x \in \Re^n : d^T x + d_0 = \xi\}$.

By means of a suitable optimality conditions, an optimal solution of problem P_Q can be found making use of any of the known algorithms of parametric quadratic programming. In [200, 201], an algorithm is suggested in the case where the matrix Q is positive definite.

8.6 References

Almogy Y., and Levin O. [5], Bajalinov E. B. [14], Barros A. I. [15], Benson H. P. [22, 23], Bitran G. [26], Bitran G., and Novaes A. G. [25], Cambini A. [29, 31], Cambini A., Carosi L., and Martein L. [43], Cambini A., Martein L., and Pellegrini L. [30], Cambini A., Martein L., and Sodini C. [32], Cambini A., and Martein L. [33, 35, 37, 34], Cambini A., Martein L., and Schaible S. [42], Cambini A., Schaible S., and Sodini C. [38], Cambini A., Marchi A., Martein L., and Schaible S. [40], Cambini A., Martein L., and Stancu-Minasian I. M. [41, 34], Cambini R. [51], Cambini R., and Sodini C. [53, 54, 56, 57], Carosi L., and Martein L. [60], Charnes A., and Cooper W. W. [64, 65], Cheng D. Z., Daescu O., Dai Y., Katoh N., Wu X., and Xu J. [68], Drezner Z., Schaible S., and Simchi-Levi D. [95], Dür M., Horst, R., and Thoai, N. V. [96], Ellero A. [101], Ellero A., and Moretti Tomasin E. [103], Falk, J. E., and Palocsay, S. M. [105, 106], Hirche J. [140, 141], Ibaraki T., Ishii H., Iwase J., Hasegawa T., and Mine H. [142], Ibaraki T. [143], Isbell J. R., and Marlow W. H. [144], Jagannathan, R. [145], Joksch H. C. [153], Konno H., and Kuno T. [173], Konno H., Yajima Y., and Matsui T. [174], Konno H., Kuno T., and Yajima Y. [175], Konno H., and Yajima Y. [176], Konno H., and Yamashita H. [179], Konno H. [182], Martein L., and Pellegrini L. [198, 199, 200], Martein L., and Sodini C. [201], Martein L. [202], Martein L., and Schaible S. [203], Martein L., and Bertolucci V. [205], Martos B. [207, 208, 210, 211], Mjelde K. M. [214], Ottaviani M., and Pacelli G. [220], Ritter K. [232], Schaible S. [246, 249, 250, 255, 256], Schaible S., and Sodini C. [257], Schaible S., and Shi J. [258], Schechter M. [259], Singh C., and Dass B. K. [260], Sodini C. [261, 262], Stancu-Minasian I. M., and Tigan St. [267], Stancu-Minasian I. M. [269], Swarup K. [272], Ujvári M. [276], Wagner H. M., and Yuan J. S. C. [277], Wolf H. [280, 281], Zionts S. [288].

9

Solutions

Chapter 1

1.1 (i) We have $\| x - x_0 \| = | t | \| z - x_0 \| < \| z - x_0 \| \leq R.$

(ii) The proof follows from (i) if x_0 belongs to the segment $[z_1, z_2]$; in the opposite case, the vectors $z_1 - x_0$, $z_2 - x_0$ are linearly independent so that, by the Schwartz inequality, $\| \lambda z_1 + (1-\lambda)z_2 - x_0 \| = \| \lambda(z_1 - x_0) + (1-\lambda)(z_2 - x_0) \| < \lambda \| z_1 - x_0 \| + (1-\lambda) \| z_2 - x_0 \| \leq \lambda R + (1-\lambda)R = R$, so that $\lambda z_1 + (1-\lambda)z_2$ is an interior point for $\lambda \in (0, 1)$.

(iii) This is a direct consequence of (ii).

(iv) Let z be a boundary point of B. If z is not an extreme point, then z belongs to a segment $[z_1, z_2]$ contained in B so that z is an interior point of B, and this contradicts the assumption.

1.2 Let $\{S_i, i \in I\}$ a family of convex sets and let $S = \bigcap_i S_i$. If $x, y \in S$, then $x, y \in S_i$, $\forall i$, so that the convexity of S_i implies $\lambda x + (1 - \lambda)y \in S_i$ for every i and, consequently, $\lambda x + (1 - \lambda)y \in S$.

1.3 (i) $z_1, z_2 \in \Gamma = \alpha S + \beta T$ if and only if there exist $s_1, s_2 \in S, t_1, t_2 \in T$, such that $z_1 = \alpha s_1 + \beta t_1$, $z_2 = \alpha s_2 + \beta t_2$. We have $z = \lambda z_1 + (1 - \lambda)z_2 = \alpha(\lambda s_1 + (1 - \lambda)s_2) + \beta(\lambda t_1 + (1 - \lambda)t_2)$. The convexity of S, T implies that $s^* = \lambda s_1 + (1-\lambda)s_2 \in S$, $t^* = \lambda t_1 + (1-\lambda)t_2 \in T$ for every $\lambda \in [0, 1]$, so that $z = \alpha s^* + \beta t^* \in \Gamma, \forall \lambda \in [0, 1]$.

(ii) Let $(s_1, t_1), (s_2, t_2) \in S \times T$. From the convexity of S and T we have $\lambda(s_1, t_1) + (1-\lambda)(s_2, t_2) = (\lambda s_1 + (1-\lambda)s_2, \lambda t_1 + (1-\lambda)t_2) \in S \times T, \forall \lambda \in [0, 1].$

1.4 (a) Let $z_1, z_2 \in \text{conv} S$. Then, there exist m points $x_1, ..., x_m \in S$ such that $z_1 = \sum_{i=1}^{m} \alpha_i x_i$, $\alpha_i \geq 0$, $i = 1, ..., m$, $\sum_{i=1}^{m} \alpha_i = 1$, and there exist k points $x_{m+1}, ..., x_{m+k} \in S$ such that $z_2 = \sum_{j=m+1}^{m+k} \alpha_j x_j$, $\alpha_j \geq 0$, $j = m + 1, ..., m + k,$

$$\sum_{j=m+1}^{m+k} \alpha_j = 1.$$ We have $z = \lambda z_1 + (1 - \lambda)z_2 = \sum_{i=1}^{m+k} \gamma_i x_i$, where $\gamma_i = \lambda \alpha_i$ for $i = 1, ..., m$, and $\gamma_i = (1 - \lambda)\alpha_i$ for $i = m + 1, ..., m + k$.

It follows that z is a linear combination of elements of S. From the non-negativity of $\lambda, 1 - \lambda, \alpha_i$, it results $\gamma_i \geq 0, i = 1, ..., m + k$.

It remains to be proven that $\sum_{i=1}^{m+k} \gamma_i = 1.$ We have $\sum_{i=1}^{m+k} \gamma_i = \sum_{i=1}^{m} \gamma_i + \sum_{i=m+1}^{m+k} \gamma_i =$

$$\lambda \sum_{i=1}^{m} \alpha_i + (1 - \lambda) \sum_{i=m+1}^{m+k} \alpha_i = \lambda + (1 - \lambda) = 1.$$

(b) This follows by (a) and by Theorem 1.2.2.

1.5 Let $z \in intS$. Then, there exists a ball of radius $\epsilon > 0$ and center z contained in S; in particular, for every direction d, $z + \alpha d \in S$, $\forall \alpha \in (0, \epsilon)$. Let $d = z - x$ and consider the half-line $y = x + t(z - x)$, $t \geq 0$. By setting $t = 1 + \alpha = \mu$, the thesis is achieved.

Assume now that for every $x \in S$ there exists $\mu > 1$ such that $y = x + \mu(z - x) \in S$. We have $z = \frac{1}{\mu}y + (1 - \frac{1}{\mu})x$ with $0 < \frac{1}{\mu} < 1$ and $\frac{1}{\mu} + (1 - \frac{1}{\mu}) = 1$. Consequently, z is a convex combination of two points of S and thus, from Theorem 1.2.3, is an interior point.

1.6 Let $x_0 \in intS$ and assume that there exists an half-line $x = x_0 + tu$, $t \geq 0$, having two boundary points: $y = x_0 + t_1 u$, $z = x_0 + t_2 u$, with $0 < t_1 < t_2$. Then, we have $y = \frac{t_1}{t_2}z + (1 - \frac{t_1}{t_2})x_0$, so that, from Theorem 1.2.3, y is an interior point and this contradicts the assumption.

1.7 (i) This follows by noting that

$$W = \{z \in \Re^n : z = x_1 + \sum_{i=2}^{k+1} c_i(x_i - x_1), c_i \in \Re, i = 2, ..., k + 1\}.$$

(ii) This follows by noting that the maximum number of linearly independent vectors is n.

1.8 (i) From Theorem 1.2.2, S contains the convex hull of points of S; the thesis follows from (i) of Exercise 1.7.

(ii) This is a direct consequence of (i).

1.9 Let k be the dimension of the smallest linear manifold W containing S. From Exercise 1.8, W contains a k-dimensional simplex. Since every point of the form $\lambda_1(x_2 - x_1) + ... + \lambda_k(x_{k+1} - x_k)$, $\lambda_i > 0$, $\sum_{i=1}^{k} \lambda_i = 1$, belongs to the relative interior of the simplex, S has a nonempty relative interior.

1.10 This follows from Theorem 1.2.11 taking into account that there are not extreme directions.

1.11 This follows from Theorem 1.2.11 taking into account that there are not extreme directions and, furthermore, that a polytope has a finite number of vertices.

1.12 Let U and V be convex cones. From (i) of Exercise 1.3, the sum $U + V$ is convex. Furthermore, if $z \in U + V$, there exist $u \in U$ and $v \in V$ such that $z = u + v$. It follows that for each $k \geq 0$, $ku \in U$ and $kv \in V$, so that $ku + kv = k(u + v) = kz$ belongs to $U + V$ for every $k \geq 0$.

1.13 (i) It is sufficient to prove that $C - C$ is a subspace and that $affC$ contains necessarily C and $-C$.
(ii) hint: the dimension of $affC$ is equal to the maximum number of linearly independent vectors contained in C.

1.14 Let $c \in intC$. The thesis follows by noting that $\alpha^T c > 0$ for all $\alpha \in riC^+$ and $\alpha^T c < 0$ for all $\alpha \in riC^-$.

1.15 If $\alpha \in C_1^+ \cap C_2^+$, we have $\alpha^T c_1 \geq 0$ for all $c_1 \in C_1$, and $\alpha^T c_2 \geq 0$ for all $c_2 \in C_2$, so that $\alpha^T(c_1 + c_2) \geq 0$ for all $c_1 \in C_1$, $c_2 \in C_2$ and thus $C_1^+ \cap C_2^+ \subseteq (C_1 + C_2)^+$. Conversely, $\alpha \in (C_1 + C_2)^+$ implies $\alpha^T(c_1 + c_2) \geq 0$ for all $c_1 \in C_1$, $c_2 \in C_2$. Setting $c_2 = 0$, we deduce $\alpha \in C_1^+$; analogously setting $c_1 = 0$, we have $\alpha \in C_2^+$, so that $(C_1 + C_2)^+ \subseteq C_1^+ \cap C_2^+$.

1.16 It is sufficient to apply Theorem 1.2.14.

1.17 The thesis is equivalent to prove that $intU^+ \cap V^- \neq \emptyset$. If not, since $intU^+$ and V^- are convex sets, there exists $\gamma \neq 0, \gamma \in \Re^n$ such that $\gamma^T z \geq 0$ for all $z \in U^+$, and $\gamma^T z \leq 0$ for all $z \in V^-$. It follows that $\gamma \in U^{++} \cap V^{--}$, i.e., $\gamma \in U \cap V$, and this contradicts the assumption.

1.18 The proof is similar to the one given in Exercise 1.17.

1.19 Let $x_1, ..., x_k$ be the vertices of the polyhedron S and let $d_1, ..., d_h$ be the extreme directions. From Theorem 1.2.11, a feasible point x can be expressed as $x = \displaystyle\sum_{i=1}^{k} \alpha_i x_i + \sum_{j=1}^{h} \beta_j d_j, \ \alpha_i \geq 0, \ i = 1, ..., k, \ \sum_{i=1}^{k} \alpha_i = 1, \ \beta_j \geq 0,$ $j = 1, ..., h$.

We have $c^T x = \displaystyle\sum_{i=1}^{k} \alpha_i c^T x_i + \sum_{j=1}^{h} \beta_j c^T d_j$. If $c^T d_{j^*} < 0$ for some j^*, then by setting $\alpha_i = 0$, $i = 1, ..., k$, and $\beta_j = 0$, $j = 1, ..., h$, $j \neq j^*$, we have $\displaystyle\lim_{\beta_{j^*} \to +\infty} \beta_{j^*} c^T d_{j^*} = -\infty$ so that the infimum is $-\infty$. If $c^T d_j \geq 0$, $j = 1, ..., h$, or if there are not extreme directions, by setting $c^T x_s = \min\{c^T x_1, ..., c^T x_k\}$, we have $c^T x \geq \displaystyle\sum_{i=1}^{k} \alpha_i c^T x_s = c^T x_s$, so that the infimum is attained at the vertex x_s.

1.20 If $f(x) = a^T x + b$, we have $f(\lambda x_1 + (1 - \lambda)x_2) = a^T(\lambda x_1 + (1 - \lambda)x_2) + \lambda b + (1 - \lambda)b = \lambda(a^T x_1 + b) + (1 - \lambda)(a^T x_2 + b)$. It follows that both the definitions of convex and concave functions are verified.

Assume now that f is both convex and concave, so that $f(\lambda x_1 + (1 - \lambda)x_2) = \lambda f(x_1) + (1 - \lambda)f(x_2)$, $\forall \lambda \in [0, 1]$. We consider the cases $f(0) = 0, f(0) \neq 0$.

case $f(0) = 0$.

We must prove (i) $f(kx) = kf(x)$, $\forall k \in \Re$, (ii) $f(x_1 + x_2) = f(x_1) + f(x_2)$.

(i) If $k \in [0, 1]$, we have $kx = kx + (1 - k) \cdot 0$, so that $f(kx) = kf(x) + (1 - k)f(0) = kf(x)$. If $k > 1$, we have $x = \frac{1}{k}(kx) + (1 - \frac{1}{k}) \cdot 0$ and thus $f(x) = \frac{1}{k}f(kx)$; consequently $f(kx) = kf(x)$, $\forall k > 1$.

If $k < 0$, we have $0 = \frac{1}{2}(kx) + \frac{1}{2}(-kx)$, so that $0 = f(0) = \frac{1}{2}f(kx) + \frac{1}{2}f(-kx)$ and $f(kx) = -f(-kx)$. Since $-k > 0$, then $f(-kx) = -kf(x)$ and consequently $f(kx) = kf(x)$, $\forall k < 0$ and (i) holds.

(ii) $f(x_1 + x_2) = f(2(\frac{1}{2}x_1 + \frac{1}{2}x_2)) = 2f(\frac{1}{2}x_1 + \frac{1}{2}x_2) = f(x_1) + f(x_2)$. By setting $a_i = f(e^i)$, we have $f(x) = f\left(\sum_{i=1}^{n} x_i e^i\right) = \sum_{i=1}^{n} x_i f(e^i) = \sum_{i=1}^{n} x_i a_i = a^T x$.

case $f(0) \neq 0$.

Let $g(x) = f(x) - f(0)$. We have $g(\lambda x_1 + (1 - \lambda)x_2) = f(\lambda x_1 + (1 - \lambda)x_2) - \lambda f(0) - (1 - \lambda)f(0) = \lambda g(x_1) + (1 - \lambda)g(x_2)$. Since $g(0) = 0$, g is of the kind $g(x) = a^T x$ so that, setting $f(0) = b$, we have $f(x) = a^T x + b$.

1.21 Let x^* be the optimal solution of $d(\lambda z_1 + (1 - \lambda)z_2) = \min_{x \in S} \| \lambda z_1 + (1 - \lambda)z_2 - x \|$. We have $\| \lambda z_1 + (1 - \lambda)z_2 - x^* \| = \| \lambda(z_1 - x^*) + (1 - \lambda)(z_2 - x^*) \| \leq \lambda \| z_1 - x^* \| + (1 - \lambda) \| z_2 - x^* \| \leq \lambda \min_{x \in S} \| z_1 - x \| + (1 - \lambda) \min_{x \in S} \| z_2 - x \| = \lambda d(z_1) + (1 - \lambda)d(z_2)$.

1.22 (a) Since $f_i(\lambda x_1 + (1 - \lambda)x_2) \leq \lambda f_i(x_1) + (1 - \lambda)f_i(x_2)$, $\forall \lambda \in [0, 1]$, we have $z(\lambda x_1 + (1 - \lambda)x_2) = \max_{i \in \{1,..,m\}}\{f_i(\lambda x_1 + (1 - \lambda)x_2)\} \leq \lambda \max_{i \in \{1,..,m\}}\{f_i(x_1)\} + (1 - \lambda) \max_{i \in \{1,..,m\}}\{f_i(x_2)\} = \lambda z(x_1) + (1 - \lambda)z(x_2)$, $\forall \lambda \in [0, 1]$.

(b) The proof is similar to the previous one.

1.23 We have $H(x_1, x_2) = x_1^{\alpha-2} x_2^{\beta-2} \begin{bmatrix} \alpha(\alpha - 1)y^2 & \alpha\beta xy \\ \alpha\beta xy & \beta(\beta - 1)x^2 \end{bmatrix}$. By using the Hessian matrix characterization of convex and concave functions, f is convex if and only if $\alpha(\alpha - 1) \geq 0$, $\beta(\beta - 1) \geq 0$, $| H(x_1, x_2) | = \alpha\beta(1 - \alpha - \beta) \geq 0$, while f is concave if and only if $\alpha(\alpha - 1) \leq 0$, $\beta(\beta - 1) \leq 0$, $| H(x_1, x_2) | \geq 0$. By solving the systems, the thesis is achieved.

1.24 (i) We have $f(\lambda x_1 + (1 - \lambda)x_2) = \sum_{i=1}^{n} \alpha_i f_i(\lambda x_1 + (1 - \lambda)x_2) \leq \sum_{i=1}^{n} \alpha_i(\lambda f_i(x_1) + (1 - \lambda)f(x_2)) = \lambda f(x_1) + (1 - \lambda)f(x_2)$.

(ii) This follows from (i) substituting $\leq$ with $<$.

1.25 We have $f(\lambda x_1 + (1 - \lambda)x_2) \leq \lambda f(x_1) + (1 - \lambda)f(x_2)$; since g is non-decreasing, it results $g(f(\lambda x_1 + (1 - \lambda)x_2)) \leq g(\lambda f(x_1) + (1 - \lambda)f(x_2))$. From the convexity of g, the thesis is achieved.

1.26 (i) and (ii) are obtained applying (i) of Theorem 1.3.4 in the cases $m = 1$ and $m = 2$ with $\alpha_1 = \alpha_2 = 1$. Conversely, if f_1, f_2 are convex functions, then $\alpha_1 f_1(x)$ and $\alpha_2 f_2(x)$ are convex for (i), as well as their sum from (ii).

1.27 $f(x) = x$, $g(x) = -x$ are convex functions but the product $h(x) = -x^2$ is not convex; $f(x) = x$, $g(x) = x$ are concave functions but the product $h(x) = x^2$ is not concave.

1.28 We have $z(\lambda x_1 + (1-\lambda)x_2) = g(A(\lambda x_1 + (1-\lambda)x_2) + b) = g(\lambda(Ax_1 + b) + (1 - \lambda)(Ax_2 + b)) \leq \lambda g(Ax_1 + b) + (1-\lambda)g(Ax_2 + b) = \lambda z(x_1) + (1-\lambda)z(x_2)$.

1.29 $f(x) = e^{z(x)}$ is convex from Theorem 1.3.5; $f(x) = x$ is convex but $z(x) = \log f(x) = \log x$ is not convex.

1.30 Hint: $-f$ is positive and concave (see Example 1.3.5).

1.31 $f(x) = x^2 + 1$ is convex but $h(x) = \frac{1}{f(x)}$ is not concave.

1.32 We have $f(x) \geq f(x_0) + f'(x_0)(x - x_0)$, so that $\lim\limits_{x \to +\infty} f'(x_0)(x - x_0) = +\infty$.

1.33 Let x_0, $x \in S$ with $x \neq x_0$. Since f is convex, too, (1.8) holds, i.e., $f(x) \geq f(x_0) + \nabla f(x_0)^T(x - x_0)$, $\forall x \in S$. By contradiction, assume the existence of $\bar{x} \neq x_0$ such that $f(\bar{x}) = f(x_0) + \nabla f(x_0)^T(x - x_0)$. The strict convexity implies $f(\lambda x_0 + (1-\lambda)\bar{x}) < \lambda f(x_0) + (1-\lambda)f(\bar{x})$ for every $\lambda \in (0, 1)$, so that $f(\lambda x_0 + (1 - \lambda)\bar{x}) < f(x_0) + (1 - \lambda)(\bar{x} - x_0)^T \nabla f(x_0)$. By replacing in (1.8) x with $\lambda x_0 + (1 - \lambda)\bar{x}$, we get a contradiction.
The proof of the converse statement is similar to the convex case.

1.34 (i) $z(tx) = f_1(tx) \cdot f_2(tx) \ldots \cdot f_m(tx) = t^{\alpha_1} f_1(x) \cdot t^{\alpha_2} f_2(x) \ldots \cdot t^{\alpha_m} f_m(x) = t^{\sum_{i=1}^{m} \alpha_i} f_1(x) \cdot f_2(x) \ldots f_m(x) = t^{\sum_{i=1}^{m} \alpha_i} z(x)$.
(ii) $z(tx) = (f_1(tx) + f_2(tx) + \ldots + f_m(tx))^\beta = (t^\alpha f_1(x) + t^\alpha f_2(x) + \ldots + t^\alpha f_m(x))^\beta = t^{\alpha\beta}(f_1(x) + f_2(x) + \ldots + f_m(x))^\beta = t^{\alpha\beta} z(x)$.

1.35 Obviously, f is homogeneous of degree α. For the converse statement, let $f(1) = k$; it results $f(t) = f(t \cdot 1) = t^\alpha f(1) = kt^\alpha$.

1.36 Hint: see the proof given in Theorem 1.4.1

1.37 Assume that f is both linearly homogeneous and convex. From the convexity assumption, we have $f(x) \geq f(x_0) + \nabla f(x_0)^T(x - x_0) = f(x_0) - \nabla f(x_0)^T x_0 + \nabla f(x_0)^T x$. By Euler' Theorem, we have $\nabla f(x_0)^T x_0 = f(x_0)$, so that $f(x) \geq \nabla f(x_0)^T x$. Assume now that f is linearly homogeneous and that $f(x) \geq \nabla f(x_0)^T x$, $\forall x$, $x_0 \in int\Re_+^n$. By Euler' Theorem, we have $\nabla f(x_0)^T x_0 = f(x_0)$, so that $\nabla f(x_0)^T x - \nabla f(x_0)^T x_0 \leq f(x) - f(x_0)$ or, equivalently, $f(x) \geq f(x_0) + \nabla f(x_0)^T(x - x_0)$, $\forall x$, $x_0 \in int\Re_+^n$.

1.38 The relation $x^T \nabla f(x) = f(x)$ implies, by differentiation, that $\nabla f(x) + \nabla^2 f(x)x = \nabla f(x)$, i.e., $\nabla^2 f(x)x = 0$, $\forall x \in int\Re^n_+$. It follows the singularity of $H(x)$.

1.39 The gradient and the Hessian of f are $\nabla f(x) = (1 + \log(a^T x + b))a$, $\nabla^2 f(x) = \frac{aa^T}{a^T x + b}$, respectively, so that the Hessian is semidefinite positive for every x of the domain and, consequently, the function is convex. Since the domain of S is an open set, the set of all global minimum reduces to the set of all critical points given by $S^* = \{x : a^T x + b = e^{-1}\}$.

1.40 Since S is open, the existence of a global maximum x_0 implies that $\nabla f(x_0) = 0$. On the other hand, the convexity of f implies that x_0 is a global minimum, so that f is constant, and this contradicts the assumption.
A convex function may have a local maximum point (consider for instance $f(x) = x^2 - x \mid x \mid$), but it cannot have a strict local maximum.

1.41 We have $\nabla f(x) = Ax$ so that $\nabla f(0) = 0$; the convexity of f implies that $x_0 = 0$ is a global minimum. Conversely, there exists a ball B of center $x_0 = 0$ such that $x^T Ax \geq 0$, $\forall x \in B$. Corresponding to $z \in \Re^n$, there exist $x \in B$, $k \in \Re$ such that $z = kx$. It follows that $f(z) = \frac{1}{2}z^T Az = \frac{1}{2}k^2 x^T Ax \geq 0$, $\forall z \in \Re^n$, so the the quadratic form is positive semidefinite and, consequently, f is convex.

1.42 The convexity of f implies that x_0 is a global minimum if and only if is a critical point. On the other hand, x_0 is a critical point if and only if verifies the system $\nabla f(x_0) = Ax_0 + a = 0$, and this is equivalent to say that $rank A = rank[A, a]$.

1.43 By the first order characterization of convexity, $f(x) \geq f(x_0) + (x - x_0)^T \nabla f(x_0)$, $\forall x \in S$ so that $\nabla f(x_0) = 0$ implies $f(x) \geq f(x_0)$, $\forall x \in S$.

1.44 Hint: see the proof of Theorem 1.5.1 and Exercise 1.43.

Chapter 2

2.1 (a) f is strictly quasiconvex; (b) f is quasiconvex; (c) f is semistrictly quasiconvex; (d) f is quasiconvex and semistrictly quasiconvex. Note that f is not lower semicontinuous.

2.2 The function is quasiconvex.

2.3 Since it is possible to choose $x_1 \in (x_0 - \epsilon, x_0)$, $x_2 \in (x_0, x_0 + \epsilon)$ such that $f(x_1) < \alpha$, $f(x_2) < \alpha$, f is not quasiconvex on $[x_1, x_2]$.

2.4 By referring to Example 2.3.3 we have: (a) f is strictly quasiconvex; (b) f is both quasiconvex and quasiconcave; (c) f is strictly quasiconvex.

2.5 (a) Set $L_{\leq \alpha} = \{x \in S : -\frac{1}{f(x)} \leq \alpha\}$. If $\alpha \geq 0$, then $L_{\leq \alpha} = S$, otherwise $L_{\leq \alpha} = \{x \in S : f(x) \leq -\frac{1}{\alpha}\}$. In each case $L_{\leq \alpha}$ is a convex set.

(b) By referring to Example 2.3.4 we have that $-\frac{1}{f}$ is (strictly) quasiconvex and consequently $\frac{1}{f}$ is (strictly) quasiconcave.

2.6 Set $L_{\leq\alpha} = \{x \in S : \frac{f(x)}{g(x)} \leq \alpha\}$; if $\alpha < 0$, then $L_{\leq\alpha} = \emptyset$. If $\alpha \geq 0$, $L_{\leq\alpha} = \{x \in S : f(x) - \alpha g(x) \leq 0\}$, so that $L_{\leq\alpha}$ is the lower level set of the convex function $h(x) = f(x) - \alpha g(x)$. Consequently $L_{\leq\alpha}$ is a convex set and the thesis follows from Theorem 2.2.3.

2.7 The function z can be viewed as the composition product between the non-negative function $h(x) = \frac{f(x)}{g(x)}$ and the increasing function $s(y) = y^\alpha$. The thesis follows from Theorem 2.3.7 and from (i) of Theorem 2.3.8.

2.8 (i) We must prove that $z(x_1) = \frac{f(x_1)}{g(x_1)} \geq \frac{f(x_2)}{g(x_2)} = z(x_2)$ implies $z(\lambda x_1 + (1-\lambda)x_2) < z(x_1)$, $\lambda \in (0,1)$.
Taking into account the strict convexity of f and the concavity of g, together with their sign, we have $f(\lambda x_1 + (1-\lambda)x_2) < \lambda f(x_1) + (1-\lambda)f(x_2) \leq \lambda f(x_1) + (1-\lambda)\frac{f(x_1)}{g(x_1)}g(x_2) = \frac{f(x_1)}{g(x_1)}(\lambda g(x_1) + (1-\lambda)g(x_2)) \leq \frac{f(x_1)}{g(x_1)}g(\lambda x_1 + (1-\lambda)x_2)$.
It follows $\frac{f(\lambda x_1 + (1-\lambda)x_2)}{g(\lambda x_1 + (1-\lambda)x_2)} < \frac{f(x_1)}{g(x_1)}$, i.e., $z(\lambda x_1 + (1-\lambda)x_2) < z(x_1)$, $\lambda \in (0,1)$.
The proofs of (ii) and (iii) are similar.

2.9 The lower and upper level sets are half-spaces.

2.10 Apply Theorem 2.2.3 and Theorem 2.2.6.

2.11 It is sufficient to note that all the functions are quasiconvex and homogeneous of degree $\alpha \geq 1$.

2.12 Apply Theorem 2.4.1 .

2.13 Hint: see (ii) of Theorem 2.3.8.

2.14 Taking into account Example 2.3.4, we have that $\frac{1}{g}$ is a positive convex function, so that $-z(x) = \frac{-f(x)}{\frac{1}{g(x)}}$ is semistrictly quasiconvex (see (ii) of Theorem 2.3.8).

2.15 The thesis follows from (ii) of Theorem 2.3.8 by noting that an affine function is both convex and concave.

2.16 By contradiction, assume the existence of an interior strict local maximum point x_0 and let x_1 be a feasible point. Then, there exists $\epsilon > 0$ such that $x = x_0 + t(x_1 - x_0) \in S$ and $f(x) < f(x_0)$, $\forall t \in [-\epsilon, \epsilon]$. By setting $y = x_0 - \epsilon(x_1 - x_0)$, $z = x_0 + \epsilon(x_1 - x_0)$, we have $x_0 = \frac{1}{2}y + \frac{1}{2}z$ with $f(x_0) > \max\{f(y), f(z)\}$, and this contradicts the quasiconvexity of f.

2.17 See (b) of Exercise 2.1.

2.18 Let f be a function defined on a convex set $S \subseteq \Re^n$ and let $\varphi(t) = f(x_0 + tu)$, $t \in I$, the restriction of f on a line segment through x_0. The function $\varphi(t)$ is quasiconvex if and only if verifies the implication:

$t_1, t_2 \in I$, $\varphi(t_1) \geq \varphi(t_2) \Rightarrow \varphi(t_1 + \lambda(t_2 - t_1)) \leq \varphi(t_1)$, $\forall \lambda \in [0, 1]$. By setting $x_1 = x_0 + t_1 u$, $x_2 = x_0 + t_2 u$, we have $x_1 + \lambda(x_2 - x_1) = x_0 + t_1 u + \lambda(t_2 - t_1)u$. The thesis follows by noting that $\varphi(t_1) \geq \varphi(t_2)$ and the logical implication $\varphi(t_1 + \lambda(t_2 - t_1)) \Rightarrow \varphi(t_1 + \lambda(t_2 - t_1)) \leq \varphi(t_1)$, $\forall \lambda \in [0, 1]$ are equivalent to $f(x_1) \geq f(x_2)$, and $f(x_1 + \lambda(x_2 - x_1)) \leq f(x_1)$, $\forall \lambda \in [0, 1]$, respectively.

2.19 If f is strictly quasiconvex, then obviously (i) and (ii) hold. Assume now the validity of (i) and (ii). We must prove that $f(x_1) \geq f(x_2)$ implies $f(x_1 + t(x_2 - x_1)) < f(x_1)$, $\forall t \in (0, 1)$. This implication follows from (i) if $f(x_1) > f(x_2)$. Consider the case $f(x_1) = f(x_2)$ and assume, by contradiction, that f is not strictly quasiconvex. Then, taking into account (i), there exist $t^* \in (0, 1)$ such that $f(x_1 + t^*(x_2 - x_1)) = f(x_1) = f(x_2)$. The restriction $\varphi(t) = f(x_1 + t(x_2 - x_1))$, $t \in [0, 1]$ cannot be constant in $[0, 1]$ otherwise any of its point is a global minimum, and this contradicts (ii). It follows that $\varphi(t)$ attains its minimum value at a point $\tilde{t} \in (0, 1)$ such that $\varphi(\tilde{t}) < \varphi(0) = \varphi(t^*) = \varphi(1)$. Consequently, $\varphi(t)$ is not semistrictly quasiconvex on $[0, \tilde{t}]$ if $t^* < \tilde{t}$ or it is not semistrictly quasiconvex on $[\tilde{t}, 1]$ if $t^* > \tilde{t}$.

2.20 Assume that $x_0 \in intS$ is a global maximum. The non-constancy of f implies the existence of $x_1 \in S$ such that $f(x_1) < f(x_0)$; since x_0 is an interior point and the function is lower semicontinuous, there exists $\epsilon > 0$ such that the line segment $I = \{x = x_1 + t(x_0 - x_1), \ t \in [0, 1+\epsilon]\}$ is contained in S and, furthermore, $f(x) > f(x_1)$, $\forall x \in I$. By choosing $t^* \in (1, 1+\epsilon)$ and by setting $x^* = x_1 + t^*(x_0 - x_1)$ we have $x_0 \in [x_1, x^*]$ with $f(x_0) \geq \max\{f(x_1), f(x^*)\} = f(x^*)$, and this contradicts the semistrictly quasiconvexity of f which requires $f(x_0) < f(x^*)$.

2.21 It is sufficient to consider the function $f(x) = 0$ if $x \neq 0$, $f(0) = 1$.

2.22 Assume that $x_0 \in intS$ is an interior local maximum point which is not a local minimum. Then, in a suitable neighbourhood of x_0, there exists a point $x_1 \in S$ such that $f(x_1) < f(x_0)$. We achieve the thesis like as in Exercise 2.20.

2.23 Assume $f(x_1) > f(x_2)$. The convexity of f implies $f(\lambda x_1 + (1 - \lambda)x_2) \leq \lambda f(x_1) + (1 - \lambda)f(x_2) < \lambda f(x_1) + (1 - \lambda)f(x_1) = f(x_1)$.

Chapter 3

3.1 Obviously, a quasiconvex function cannot have semistrict local maximum points. Assume now that φ is not quasiconvex; then, there exist $t_1, t_2 \in I$ and $t_0 \in (t_1, t_2)$ such that $\max\{\varphi(t_1), \varphi(t_2)\} < \varphi(t_0)$. From the generalized Weierstrass' Theorem φ attains maximum value at a point $t^* \in (t_1, t_2)$ for which $\max\{\varphi(t_1), \varphi(t_2)\} < \varphi(t_0) \leq \varphi(t^*)$. Consequently t^* is a semistrict local maximum and this is absurd.

3.2 Consider the function $f(x) = \begin{cases} -x^2 & x \leq 0 \\ 0 & 0 < x \leq 2 \\ -(x-2)^2 & x > 2 \end{cases}$.

The point $x_0 = 1$ is a semistrict local maximum but it is not a strict local maximum point.

3.3 Consider the function $\varphi(t) = \begin{cases} t+2 & t < 0 \\ -t & t \geq 0 \end{cases}$. $\varphi(t)$ is not upper semicontinuous, does not have any kind of maximum points and is not quasiconvex.

3.4 Obviously, a quasiconvex function cannot have semistrict local maximum points. Assume now that φ is not quasiconvex; then there exist $t_1, t_2 \in I$ and $t_0 \in (t_1, t_2)$ such that $\max\{\varphi(t_1), \varphi(t_2)\} < \varphi(t_0)$. From the Weierstrass' Theorem, φ attains maximum value at an interior point $t^* \in (t_1, t_2)$ for which $\max\{\varphi(t_1), \varphi(t_2)\} < \varphi(t_0) \leq \varphi(t^*)$. Consequently, t^* is a semistrict local maximum point with $\varphi'(t^*) = 0$, and this is a contradiction.

3.5 The answer is negative. Consider for instance the quasiconvex function $f(x) = \arctan x$. We have $f'(0) = 1 > 0$ and $\lim\limits_{x \to +\infty} f(x) = \frac{\pi}{2}$.

3.6 See Exercise 3.4.

3.7 (b).

3.8 The thesis follows from Theorem 3.2.9 and from Theorem 3.2.12.

3.9 It is sufficient to note that the derivative of the function $\phi(y) = -\frac{1}{y}$ is positive for all $y \neq 0$, so that $-\frac{1}{f(x)}$ is pseudoconcave (strictly pseudoconcave) if f is pseudoconcave (strictly pseudoconcave). It follows that $\frac{1}{f(x)}$ is pseudoconvex (strictly pseudoconvex).

3.10 Apply Theorem 3.2.11.

3.11 We prove the pseudoconvexity of f by showing the pseudoconvexity of all its restrictions.
The restriction of f on a vertical line is an affine function which is pseudoconvex. The restriction of f on the line $y = mx + q$, $x > -1$, is $\varphi(x) = mx + q + \frac{1}{x+1}$, and its derivative is $\varphi'(x) = \frac{m(x+1)^2 - 1}{(x+1)^2}$. When $m \leq 0$, φ is pseudoconvex since it is decreasing without critical points. If $m > 0$, we have a feasible critical point at $x_0 = -1 + \sqrt{m^{-1}}$ which is a local minimum. Since any restriction is pseudoconvex, f is pseudoconvex.

3.12 Apply Theorem 3.2.8.

3.13 From Theorem 2.4.1 the function is quasiconcave and it is also pseudoconcave since $\nabla f(x) \neq 0$, $\forall x \in int\Re_+^n$.

3.14 From Theorem 2.4.2 the function is quasiconcave and it is also pseudoconcave since $\nabla f(x) \neq 0$, $\forall x \in int\Re_+^n$.

3.15 Since $\log z(x)$ is a concave (in particular, pseudoconcave) function and the derivative of the exponential function is positive, $e^{\log f(x)} = z(x)$ is pseudoconcave (see Theorem 3.2.11).

3.16 (a) This is a particular case of Exercise 3.15; (b) $-z(x)$ is pseudoconcave; (c) $-z(x)$ is strictly pseudoconcave.

3.17 From Theorem 3.17, the function $z_1(x) = \frac{g(x)}{-f(x)}$ is strictly pseudoconvex, so that $z(x) = -\frac{1}{z_1(x)} = \frac{f(x)}{g(x)}$ is strictly pseudoconvex (see also the solution of Exercise 3.9).

3.18 Assume that f is not quasiconvex. Then, there exist $x_1, x_2 \in S$, $t^* \in (0,1)$ such that $f(x^*) > \max\{f(x_1), f(x_2)\}$ with $x^* = x_1 + t^*(x_2 - x_1)$. By setting $\varphi(t) = f(x_1 + t(x_2 - x_1))$, $t \in [0,1]$, the continuity of f implies the existence of $t_1, t_2 \in [0,1]$ such that $t_1 < t^* < t_2$ and $\varphi(t_1) = \varphi(t_2)$. Let $z_1 = x_1 + t_1(x_2 - x_1)$, $z_2 = x_1 + t_2(x_2 - x_1)$ and consider the level set $\Gamma = \{x \in S : f(x) = f(z_1) = f(z_2)\}$. The convexity of Γ implies $[z_1, z_2] \subset \Gamma$, and this is a contradiction since $x^* \in [z_1, z_2]$ and $f(x^*) > f(z_1)$. Similarly, it can be proven that f is quasiconcave so that the thesis holds.
The converse statement follows from (ii) of Theorem 3.3.1.

3.19 Obviously, (i) and (ii) imply the pseudolinearity of f. Let f be pseudolinear. We must prove that $(x_2 - x_1)^T \nabla f(x_1) < 0$ implies $f(x_2) < f(x_1)$. Assume, to get a contradiction, that $f(x_2) \geq f(x_1)$. If $f(x_2) > f(x_1)$ the pseudoconcavity of f implies $(x_2 - x_1)^T \nabla f(x_1) > 0$ and this is absurd. If $f(x_2) = f(x_1)$ the quasilinearity of f implies that f is constant on the line segment $[x_1, x_2]$, so that $(x_2 - x_1)^T \nabla f(x_1) = 0$ and this is absurd. Consequently (i) holds. Similarly it can be proven that $(x_2 - x_1)^T \nabla f(x_1) > 0$ implies $f(x_2) > f(x_1)$, so that the thesis follows.

3.20 We have $\{(x,y) \in int\Re_+^2 : f(x,y) = k\} = \{(x,y) \in int\Re_+^2 : (1-k^2)y^2 = (k(x+2)-1)^2\}$. The level sets of the function are the intersection of the positive orthant with the lines $kx - \sqrt{1-k^2}\, y + 2k - 1 = 0$ if $\frac{1}{2} \leq k < 1$, and with the lines $kx + \sqrt{1-k^2}\, y + 2k - 1 = 0$ if $-1 < k < \frac{1}{2}$.
Since $\nabla f(x,y) \neq 0$, $\forall(x,y) \in int\Re_+^2$, f is pseudolinear on $int\Re_+^2$.

3.21 By setting $f(x,y,z) = k$, we have $(2x + y - z)^2 + (4 - k)(2x + y - z) - 1 - 4k = 0$, i.e., $2x + y - z = \frac{k-4\pm\sqrt{k^2+8k+20}}{2}$. It follows that the feasible level sets are parallel hyperplanes. Since $\nabla f(x,y,z) = (2,1,-1)^T + \frac{1}{(2x+y-z+4)^2}(2,1,-1)^T \neq 0$, f verifies (i) and (ii) of Theorem 3.3.9 so that it is pseudolinear.

3.22 (i) $f(x,y,z) = \alpha + \sqrt{ax + by + cz + \alpha^2}$, $ax + by + cz + \alpha^2 > 0$.
(ii) $f(x,y,z) = \frac{x+y+\sqrt{(x+y+z)(x+y-z)}}{z}$, $(x,y,z) \in int\Re_+^2$, $z < x + y$.

3.23 The level sets of the given functions are the following family of lines:
(a) $2kx - 4y + k^2 + 2k = 0$; (b) $2kx + (4 - 2k)y - k^2 = 0$; (c) $k^2x - y - 2k = 0$;
(d) $\sqrt{1 + k^2}x - ky + 1 = 0$.

3.24 Apply Theorem 3.3.12.

3.25 Hint: see the characterization of quasiconvexity given in Exercise 3.6.

3.26 By setting $u = (u_1, u_2)^T$, we have $u^T \nabla f(x_1, x_2) = 0$ if and only if $u_1(-2x_1 - x_2) + u_2(-x_1) = 0$ or, equivalently, for all $(x_1, x_2) \in int\Re_+^2$, $u_2 = u_1(-2 - \frac{x_2}{x_1})$ so that $u^T \nabla^2 f(x)u = -2u_1^2 - 2u_1u_2 = 2u_1^2(1 + \frac{x_2}{x_1}) > 0$. Theorem 3.4.6 implies that f is pseudoconvex on $int\Re_+^2$ and it is also quasiconvex on $\Re_+^2$ as a consequence of Theorem 2.2.12.

3.27 The bordered Hessian is $D(x_1, x_2) = \begin{bmatrix} 0 & \frac{-1}{x_2+1} & 3 + \frac{x_1+1}{(x_2+1)^2} \\ \frac{-1}{x_2+1} & 0 & \frac{1}{(x_2+1)^2} \\ 3 + \frac{x_1+1}{(x_2+1)^2} & \frac{1}{(x_2+1)^2} & \frac{-2(x_1+1)}{(x_2+1)^3} \end{bmatrix}$

and we have $D_1(x_1, x_2) = -\frac{1}{(x_2+1)^2} < 0$, $D_2(x_1, x_2) = -\frac{6}{(x_2+1)^3} < 0$, so that from Theorem 3.4.13, f is pseudoconvex.

3.28 (a).

3.29 (b).

3.30 The answer is positive.

3.31 $a \leq 0$, $b \leq 0$, $c \leq 0$, $ac - b^2 \leq 0$.

3.32 Apply Theorem 3.4.12 and its analogous for pseudoconcave functions.

3.33 Apply conditions (i) and (ii) given in Exercise 3.32.

3.34 If $f(x) \neq f(x_0)$, (3.28) implies (3.27). Assume that $f(x) = f(x_0)$ and that (3.27) is not verified. Then, there exists $t \in (0, 1)$ such that $y = x_0 + t(x - x_0) \in S$ with $f(y) > f(x) = f(x_0)$. The continuity of the function implies that the restriction of f on the line segment I of end points x_0, x attains its maximum value at an interior point $x^* \in riI$ so that we have $f(x^*) \geq f(y) > f(x_0)$. Since in the interval I^* of end points x^*, x, the function f assumes any intermediate value between $f(x^*)$ and $f(x)$, there exists a point $z \in riI^*$ such that $f(x) < f(z) < f(x^*)$. Consequently $f(x^*) > max\{f(z), f(x_0)\}$ and this contradicts (3.28) with $x = z$.

3.35 Consider the function $f(x) = \begin{cases} \frac{x}{1-x} & 0 \leq x < 1 \\ 0 & x = 1 \end{cases}$

The function f is not quasiconvex at $x_0 = 0$ according to (3.27), since $f(x_0) = f(1) = 0$ but $f(x) > 0$, $\forall x > 0$. On the other hand, for every x such that $f(x) \neq f(0)$, (3.28) is verified since f is increasing in $[0, 1)$.

Chapter 4

4.1 Set $V = \{(-c^T, Ax), \ x \in \Re^n\}$. System 1 is impossible if and only if $V \cap (int\Re_- \times \Re^m_-) = \emptyset$. Since $-e^1 \notin V \subset V^*$, Corollary 4.2.1 implies the existence of $\alpha > 0$, $\beta \in \Re^m_+$ such that $\alpha(-c^Tx) + \beta^T Ax = 0$, $\forall x \in \Re^n$ or, equivalently, $-\alpha c^T + \beta^T A = 0$. By setting $y = \frac{1}{\alpha}\beta$, the thesis is achieved.

4.2 System 1 is impossible if and only if

$$(W_1 \times W_2) \cap (int\Re^m_- \times \Re^s_-) = \emptyset \tag{9.1}$$

where $W_1 = \{Ax, \ x \in \Re^n\}$, $W_2 = \{Bx, \ x \in \Re^n\}$.
By setting $W = W_1 \times W_2$, from (9.1) we have $W \cap int\Re^{m+s}_- = \emptyset$ so that there exist $\alpha \in \Re^m_+$, $\beta \in \Re^s_+$, $(\alpha, \beta) \neq (0,0)$ such that $\alpha^T w_1 + \beta^T w_2 = 0$, $\forall w_1 \in W_1$, $\forall w_2 \in W_2$.
The existence of a separating hyperplane for which $\alpha \in \Re^m_+ \backslash \{0\}$ remains to be proven.
If $\alpha = 0$ for every separating hyperplane, then from (i) and (ii) of Theorem 4.2.1, there exists $\bar{w} \in W^* \cap (int\Re^m_- \times \Re^s_-)$ and this implies the existence of $w \in W \cap (int\Re^m_- \times \Re^s_-)$, and this contradicts the impossibility of system 1.
Assume now that system 2 has a solution. Then, system 1 is impossible, otherwise there exists $\bar{x} \in \Re^n$ such that $A\bar{x} < 0$, $B\bar{x} \leqq 0$ for which we have $\alpha^T A\bar{x} + \beta^T B\bar{x} < 0$.

4.3 System 1 is impossible if and only if

$$(W_1 \times W_2) \cap (int\Re^m_- \times \Re^s_- \backslash \{0\}) = \emptyset \tag{9.2}$$

where $W_1 = \{Ax, \ x \in \Re^n\}$, $W_2 = \{Bx, \ x \in \Re^n\}$.
By setting $W = W_1 \times W_2$, from (9.2) we have $W \cap int\Re^{m+s}_- = \emptyset$ so that there exist $\alpha \in \Re^m_+$, $\beta \in \Re^s_+$, $(\alpha, \beta) \neq (0,0)$ such that $\alpha^T w_1 + \beta^T w_2 = 0$, $\forall w_1 \in W_1$, $\forall w_2 \in W_2$.
It remains to be proven that if $\alpha = 0$ for every separating hyperplane, then necessarily we have $\beta > 0$.
Since $\alpha = 0$ implies $\beta \in \Re^s_- \backslash \{0\}$, from (i) and (ii) of Theorem 4.2.1, there exists $\bar{w} \in W^* \cap (int\Re^m_- \times \Re^s_- \backslash \{0\})$, and this implies the existence of $w \in W \cap (int\Re^m_- \times \Re^s_- \backslash \{0\})$, contradicting the impossibility of system 1.
The proof of the converse statement is similar to that given in Exercise 4.2.

4.4 The proof of each statement is given by contradicting (4.3), that is assuming the existence of a direction $d^* \in \Re^n$ such that $\nabla f(x_0)^T d^* < 0$, $\nabla g_i(x_0)^T d^* \leq 0$, $i \in I(x_0)$.
(i) Let $\hat{d} = d^* + \frac{1}{n}d$ where $d \in C^0$. We have $\nabla g_i(x_0)^T \hat{d} < 0$, $i \in I(x_0)$ and $\nabla f(x_0)^T \hat{d} < 0$ for n large enough, so that $\hat{d}$ is a feasible decreasing direction and this contradicts the optimality of x_0.
(ii) We have $d^* \in C = clF$ so that there exists a sequence of feasible directions $\{d_n\}$ converging to d^*. Consequently it is possible to choose a feasible

sequence of points $\{x_n\}$ converging to x_0 such that $\lim\limits_{n \to +\infty} \frac{x_n - x_0}{\|x_n - x_0\|} = \frac{d^*}{\|d^*\|}$.

Since $f(x_n) \geq f(x_0)$, we have $\lim\limits_{n \to +\infty} \frac{f(x_n) - f(x_0)}{\|x_n - x_0\|} = \frac{\nabla f(x_0)^T d^*}{\|d^*\|} \geq 0$ and this contradicts the assumption $\nabla f(x_0)^T d^* < 0$.

(iii) We have $d^* \in C = T$ so that $\exists \{x_n\} \subset S$, $\exists \{\alpha_n\} \subset \Re_+$, such that $x_n \to x_0$, $\alpha_n \to +\infty$, $\alpha_n(x_n - x_0) \to d^*$. Since $f(x_n) \geq f(x_0)$, we have $\nabla f(x_0)^T d^* \geq 0$ and this contradicts the assumption $\nabla f(x_0)^T d^* < 0$.

4.5 Assume, by contradiction, the existence of $\bar{x} \in S^*$ such that $f(\bar{x}) < f(x_0)$. The assumptions of generalized convexity imply $\nabla f(x_0)^T (\bar{x} - x_0) < 0$, $\nabla g_i(x_0)^T (\bar{x} - x_0) \leq 0$, $i = 1, ..., m$, and $\nabla h_j(x_0)^T (\bar{x} - x_0) = 0$, $j = 1, .., p$.

Consequently $(\nabla f(x_0) + \sum\limits_{i=1}^{m} \lambda_i \nabla g_i(x_0) + \sum\limits_{j=1}^{p} \mu_j \nabla h_j(x_0))^T (\bar{x} - x_0) < 0$ and this contradicts the Karush–Kuhn–Tucker conditions.

4.6 Let x_0 be a global maximum of S. From Theorem 1.2.8, x_0 can be expressed as a convex combination of finitely many extreme points $x^1, .., x^m$ of S. Applying the Jensen' Inequality (see (1.4)), we have $f(x_0) = f\left(\sum\limits_{i=1}^{m} \lambda_i x^i\right) \leq$

$\sum\limits_{i=1}^{n} \lambda_i f(x^i), \lambda_i \geq 0$, $i = 1, ..., n$, $\sum\limits_{i=1}^{m} \lambda_i = 1$. If $f(x^i) < f(x_0)$ for all i, then

$f(x_0) < \sum\limits_{i=1}^{n} \lambda_i f(x^i) = f(x_0)$ and this is absurd. It follows that $f(x^i) = f(x_0)$ for some i and the thesis is achieved.

4.7 Applying the Karush–Kuhn–Tucker conditions which are necessary and sufficient for optimality because of the pseudoconcavity of $U(x)$, we have $x_i(p, m) = \frac{1}{\lambda} \frac{\alpha_i}{p_i} U(x)$.

Taking into account that $\sum\limits_{i=1}^{n} p_i x_i = m = \frac{1}{\lambda} U(x) \sum\limits_{i=1}^{n} \alpha_i = \frac{1}{\lambda} U(x)$, the thesis is achieved.

(b) In this case, since $\sum\limits_{i=1}^{n} \alpha_i = 1$, we have $x_i(p, m) = \lambda u \frac{\alpha_i}{p_i}$, $U(x) = u =$

$\lambda u \prod\limits_{i=1}^{n} (\frac{\alpha_i}{p_i})^{\alpha_i}$, so that $\lambda = \prod\limits_{i=1}^{n} (\frac{\alpha_i}{p_i})^{-\alpha_i}$ and $e(p, u) = \sum\limits_{i=1}^{n} p_i x_i = u \prod\limits_{i=1}^{n} (\frac{\alpha_i}{p_i})^{-\alpha_i}$.

4.8 Since the objective function $p_0 y - p^T x$ is linear and F is concave, we can apply the Karush–Kuhn–Tucker conditions which become necessary and sufficient for optimality.

We have $x_i(p_0, p) = \frac{\alpha_i}{p_i} p_0 F(x)$ and $F(x) = y = \prod\limits_{i=1}^{n} x_i^{\alpha_i}$. This last equality implies (a) and by substitution, (b) follows.

4.9 Let x_1, x_2 be optimal solutions of problems $P(\alpha_1)$, $P(\alpha_2)$, respectively. We have $g(x_1) \leq \alpha_1$, $g(x_2) \leq \alpha_2$, so that the convexity of g implies $g(\lambda x_1 + (1 - \lambda)x_2) \leq \lambda g(x_1) + (1 - \lambda)g(x_2) \leq \lambda \alpha_1 + (1 - \lambda)\alpha_2$. By setting $\bar{\alpha} = \lambda \alpha_1 + (1 - \lambda)\alpha_2$, it follows that $\bar{x} = \lambda x_1 + (1 - \lambda)x_2$ is feasible for problem $P(\bar{\alpha})$ and consequently, $z(\bar{\alpha}) \leq f(\bar{x})$. From the convexity of f we have $z(\bar{\alpha}) \leq \lambda f(x_1) + (1 - \lambda)f(x_2) = \lambda \alpha_1 + (1 - \lambda)\alpha_2$.

4.10 Referring to Exercise 4.9, it is sufficient to note that $f(x) = c^T x$ and $g(x) = Ax - b - \theta u$ are convex functions.

Chapter 6

6.1 $Q(x) = \frac{1}{2}(x - s)^T Q(x - s) - \frac{1}{2}s^T Qs$. Since $\frac{1}{2}(x - s)^T Q(x - s) \leq 0$, we have $Q(x) \leq -\frac{1}{2}s^T Qs$.

6.2 See Corollary 4.6.2.

6.3 (a) $s = -Q^{-1}q = (1, 0)^T$, $s + T = \{x = (1, 0)^T + \alpha(1, 1)^T + \beta(-1, 2)^T$, $\alpha \geq 0, \beta \geq 0\}$.
(b) Note that $Q(x) = \frac{1}{2}(x - s)^T Q(x - s) - \frac{1}{2}s^T Qs$ with $\frac{1}{2}s^T Qs = 2$. Since $\frac{1}{2}(x - s)^T Q(x - s) \leq 0$, we have $Q(x) \leq -2$.
(c) The Hessian matrix of f is negative semidefinite on $s + T$.

6.4 $Q(x_1, x_2) = -2x_1^2 - 3x_1 x_2 - x_2^2$.

6.5 The function is quasiconvex on $s + T$ and on $s - T$ where $s = (-1, -4)^T$ and $T = \{(x_1, x_2) = \alpha(0, 1)^T + \beta(1, 2)^T, \alpha \geq 0, \beta \geq 0\}$; $Q(x_1, x_2)$ is pseudoconvex on $(s \pm T) \backslash \{s\}$.

6.6 $H = \{(x_1, x_2) \in \Re^2 : x_1 - 3x_2 + h_0 \geq 0\}$, $h_0 \leq -1$.

6.7 $Q(x_1, x_2) = -2(3x_1 + x_2)^2 + \beta(3x_1 + 2x_2)$, $\beta \leq -12$.

6.8 The answer is negative since every boundary point of H is necessarily a critical point of $Q(x)$.

6.9 (a) $Q(x_1, x_2) = -x_1 x_2 - x_2^2$; (b) $Q(x_1, x_2) = -x_1 x_2 + x_2^2$; (c) $Q(x_1, x_2) = x_1^2 - x_2^2$.

6.10 It is easy to verify that the maximal domains of quasiconvexity of $Q_0(x_1, x_2) = kx_1 x_2, k < 0$ are $\Re_+^2$ and $\Re_-^2$. Conversely, by setting $2Q_0(x_1, x_2) = \alpha x_1^2 + 2\beta x_1 x_2 + \gamma x_2^2$, we have, from Theorem 6.5.8, $\alpha \leq 0$, $\beta \leq 0, \gamma \leq 0$. If $\alpha < 0$, then $Q_0(1, 0) < 0$ so that, by continuity, $Q_0(x_1, x_2) < 0$ on a suitable neighbhourhood of $(1, 0)$ and this implies that $\Re_+^2 \neq T$. It follows that $\alpha = 0$ and, by similar arguments, that $\gamma = 0$.

6.11 It is easy to verify, by applying Theorem 6.5.7, that $Q_0(x) = \beta x_1 x_2 + q_1 x_1 + q_2 x_2$, $\beta < 0$ is merely pseudoconvex on $\Re_+^2$ for all $q = (q_1, q_2)^T \leq 0$. Conversely, note that Q is necessarily nonsingular, so that we must have

$q^T s = -q^T Q^{-1} q \geq 0$ for all $q \leq 0$ or, equivalently, (see also Theorem 6.5.8), $\gamma q_1^2 - 2\beta q_1 q_2 + \alpha q_2^2 \geq 0$ for all $q_1 \leq 0, q_2 \leq 0$. Since $\alpha \leq 0, \beta \leq 0, \gamma \leq 0$, necessarily we have $\alpha = \gamma = 0, \beta < 0$.

6.12 Apply Theorem 6.6.1 to the case $c = 0$.

6.13 Follow the same line of the proof given in Theorem 6.6.1.

Chapter 7

7.1 Case (ii) of Theorem 7.2.3 occurs; f is pseudoconvex if and only if $q_0 \geq -3$.

7.2 Case (iii) of Theorem 7.2.3 occurs; f is pseudoconvex if and only if $q_0 \geq 10$.

7.3 The proof is similar to the one given in Theorem 7.2.4.

7.4 Suppose, by contradiction, that $v_-(A) > 1$ and let v_1 and v_2 be two linearly independent eigenvectors associated with two negative eigenvalues of A. Let W be the linear subspace generated by v_1 and v_2. Let us note that $\dim(\ker A) \leq n - 2$, so that $\ker A \neq b^\perp$ and $W \cap b^\perp \neq \emptyset$. Let $v \in W \cap b^\perp$, $v \neq 0$. Since v is a linear combination of v_1 and v_2, we have $v^T A v < 0$. Consider the line $x = x_0 + tv$, $x_0 \in S$, $t \in \Re$ which is contained in S since $b^T x + b_0 = b^T x_0 + b_0 > 0$. It is easy to verify that the restriction $\varphi(t) = f(x_0 + tv)$ is of the kind $\varphi(t) = \alpha t^2 + \beta t + \gamma$ with $\alpha < 0$ and this contradicts the pseudoconvexity of f.

7.5 Consider $f(x, y) = \frac{x^2 + y^2 - 10}{(x+1)^2}$ and its restriction on the line $y = x + 3$, i.e., $\varphi(x) = \frac{2x^2 + 6x - 1}{(x+1)^2}$. We have $\varphi'(x) = \frac{-2x+8}{(x+1)^3}$, so that $x = 4$ is a strict local maximum point for $\varphi(x)$. Consequently, $\varphi(x)$ and in turns $f(x)$, is not pseudoconvex.

7.6 Let us note that cases (ii–iv) of Theorem 7.4.2 do not occur since the non-singularity of A implies the existence of α satisfying (i). In fact, we have $b - \frac{\|b\|^2}{b_0} A^{-1} a = \alpha A^{-1} b$, so that $\alpha = \frac{\|b\|^2 (b_0 - a^T A^{-1} b)}{b_0 b^T A^{-1} b}$. By substituting α in (i) of Theorem 7.4.2, together with $a^T b = \frac{\|b\|^2}{b_0} a^T A^{-1} a + \alpha b^T A^{-1} b$, the thesis is achieved.

7.7 Consider $f(x, y) = \frac{x^2 + y^2}{(x+y+1)^2}$ and its restriction on the line $x = 0$.

7.8 The non-singularity of A implies the validity of (i) of Theorem 7.4.2 with $\alpha = \frac{\|b\|^2 (b_0 - a^T A^{-1} b)}{b_0 b^T A^{-1} b}$. If $b^T A^{-1} b \neq 0$, (ii) follows (see also the proof given in Exercise 7.6). If $b^T A^{-1} b = 0$, from $b - \frac{\|b\|^2}{b_0} A^{-1} a = \alpha A^{-1} b$, we have $b^T A^{-1} a = b_0$. Substituting in (i) of Theorem 7.4.2 the thesis is achieved.

7.9 (i) and (ii) follow directly from Theorem 7.4.2 by setting $A = 0$. The Hessian matrix of f, given by $\nabla^2 f(x) = \frac{2k(b^T x + b_0) + 6(a_0 - k b_0)}{(b^T x + b_0)^4} b b^T$, is negative semidefinite in case (ii), while it is positive semidefinite in case (i) with $k > 0$. In case (i) with $k < 0$, the function f is not necessarily convex as can be verified by analyzing the function $f(x) = \frac{-x + 5}{(x+1)^2}$.

7.10 Case (i) of Theorem 7.4.2 occurs; f is pseudoconvex on S if and only if $a_0 \geq \frac{1}{2}$.

7.11 Case (ii) of Theorem 7.4.2 occurs; f is pseudoconvex if and only if $a_0 \geq 1$.

7.12 Case (iii) of Theorem 7.4.2 occurs; f is pseudoconvex if and only if $b_0 \in [-\frac{1}{2}, \frac{3}{2}]$, $b_0 \neq 0$.

7.13 Case (iv) of Theorem 7.4.2 occurs; $f(x)$ is pseudoconvex if and only if $a_2 \leq \frac{1}{2}$.

7.14 Firstly, consider the case where a and d are linearly independent. The function h is obtained from g of Theorem 7.2.4 by setting $b = d$, $b_0 = d_0$, $a_0 = 0$. Since $d = \delta_1 a + \delta_2 d$ if and only if $\delta_1 = 0$, $\delta_2 = 1$, condition (ii) of Theorem 7.2.4 never occurs. Furthermore, condition (i) of Theorem 7.2.4 is equivalent to having $d_0 = \gamma_1 + d_0$, $c_0 \geq (\gamma_1 + d_0)\gamma_2$, so that $\gamma_1 = 0$, $c_0 \geq \gamma_2 d_0$ and $c = \gamma_1 a + \gamma_2 d = \gamma_2 d$. Consequently, (ii) holds with $t = \gamma_2$.
Consider now the case where a and d are linearly dependent, i.e., $a = kd$. We have $h(x) = \frac{k(d^T x)^2 + (k d_0 d + c)^T x + c_0}{d^T x + d_0}$. If $k \geq 0$, the numerator is a convex function so that (i) holds (see also Theorem 3.2.10). If $k < 0$, by referring to Theorem 7.2.3, we have $Q = 2k d d^T$, $q = k d_0 d + c$, $q_0 = c_0$. Since $Q\bar{y} = d$ if and only if $d^T \bar{y} = \frac{1}{2k}$, case (ii) of Theorem 7.2.3 never occurs since $k < 0$. By referring to (iii) of Theorem 7.2.3, $Q\bar{x} = -q$ is equivalent to stating that $c = td$ with $t = -k(d_0 + 2 d^T \bar{x})$, while the condition $(d^T \bar{x} + d_0)^2 + 2n(\bar{x}) d^T \bar{y} \leq 0$ is equivalent to $c_0 \geq t d_0$.

7.15 (a) Case (iv) of Theorem 7.3.2 occurs;
$D = \{(x_1, x_2) \in \Re^2 : 3x_1 + 2x_2 + 6 > 0, \ x_1 + 4x_2 + 3 > 0\}$.
(b) Case (v) of Theorem 7.3.2 occurs;
$D_1 = \{(x_1, x_2) \in \Re^2 : 5x_1 - x_2 - 15 > 0, \ 2x_1 - x_2 + 8 > 2\}$,
$D_2 = \{(x_1, x_2) \in \Re^2 : 5x_1 - x_2 - 15 < 0, \ 0 < 2x_1 - x_2 + 8 < 2\}$.
(c) Case (iii) of Theorem 7.3.2 occurs;
$D_1 = \{(x_1, x_2) \in \Re^2 : 3x_1 + 5x_2 + 2 > 1\}$,
$D_2 = \{(x_1, x_2) \in \Re^2 : 0 < 3x_1 + 5x_2 + 2 < 1\}$.
(d) Case (i) of Theorem 7.3.2 occurs;
$D = \{(x_1, x_2) \in \Re^2 : -2x_1 + 6x_2 + 7 > 0\}$.

7.16 (a) When $\theta = 0$, case (i) of Theorem 7.3.2 occurs; $D = \{(x_1, x_2) \in \Re^2 : 2x_1 + 3x_2 + 5 > 0\}$.

When $\theta > 0$, case (iv) of Theorem 7.3.2 occurs; $D = \{(x_1, x_2) \in \Re^2 : 4x_1 + (6+\theta)x_2 - 8\theta + 10 > 0, \ 2x_1 + 3x_2 + 5 > 0\}$. When $\theta < 0$, case (v) of Theorem 7.3.2 occurs;
$D_1 = \{(x_1, x_2) \in \Re^2 : 4x_1 + (6+\theta)x_2 - 8\theta + 10 > 0, \ 2x_1 + 3x_2 + 5 + \frac{1}{\theta} > 0\}$,
$D_2 = \{(x_1, x_2) \in \Re^2 : 4x_1 + (6+\theta)x_2 - 8\theta + 10 < 0, \ 0 < 2x_1 + 3x_2 + 5 < -\frac{1}{\theta}\}$.
(b) When $c_0 = \frac{3}{2}$, f reduces to a linear function.
When $c_0 > \frac{3}{2}$, case (i) of Theorem 7.3.4 occurs;
$D = \{(x_1, x_2) \in \Re^2 : 2x_1 + 2x_2 + 3 > 0\}$.
When $c_0 < \frac{3}{2}$, case (ii) of Theorem 7.3.4 occurs;
$D_1 = \{(x_1, x_2) \in \Re^2 : 2x_1 + 2x_2 + 3 > \sqrt{\frac{3}{2} - c_0}\}$,
$D_2 = \{(x_1, x_2) \in \Re^2 : 0 < 2x_1 + 2x_2 + 3 < \sqrt{\frac{3}{2} - c_0}\}$.

7.17 (a) Case (iii) of Theorem 7.5.1 occurs, with $\beta = 1$, $\delta = -1$, $\lambda_1, \lambda_2 = 1$.
(b) Case (ii) of Theorem 7.5.1 occurs, with $\gamma = -1$.

7.18 f is pseudoconvex if and only if $m_0 \geq 4$. When $m_0 = 4$, case (i) of Theorem 7.5.1 occurs, with $\alpha = 1$. When $m_0 > 4$, case (iii) of Theorem 7.5.1 occurs, with $\beta = \frac{2}{m_0 - 4}$, $\delta = \frac{-m_0 + 2}{m_0 - 4}$, $\lambda_1 = 2, \lambda_2 = \frac{2}{m_0 - 4}$.

References

1. Abadie J. M., *On the Kuhn-Tucker theorem*, in Nonlinear Programming, Abadie J. (Ed.), North-Holland, Amsterdam, 21–36, 1967.
2. Aggarwal S. P., *Quadratic fractional functionals programming*, Cahiers du Centre d'Etude de Recherche Operationelle, 15, 157–165, 1973.
3. Aggarwal S. P., *Indefinite quadratic programming with a quadratic constraint*, Cahiers du Centre d'Etude de Recherche Operationelle, 15, 405–410, 1973.
4. Almogy Y., and Levin O., *Parametric analysis of a multi-stage stochastic problem*, Proceedings of the fifth IFORS Conference, 359–370, 1964.
5. Almogy Y., and Levin O., *A class of fractional programming problems*, Operations Research, 19, 57–67, 1971.
6. Arrow K. J., Hurwicz L., and Uzawa H., *Studies in Linear and Nonlinear Programming*, Stanford University Press, Stanford, 1958.
7. Arrow K. J. and Enthoven A. C., *Quasi-concave programming*, Econometrica, 29, 779–800, 1961.
8. Arrow K. J., Hurwicz L., and Uzawa H., *Constraint qualifications in maximization problems* , Naval Research Logistic Quarterly, 8, 175–181, 1961.
9. Avriel M., and Zang I., *Generalized convex functions with applications to nonlinear programming*, Chapter 2 in Mathematical Programs for Activity Analysis, Van Moeseke P. (Ed.), North-Holland Publishing Co., 1974.
10. Avriel M., *Non linear Programming - Analysis and Methods*, Prentice-Hall, Englewood Cliffs, N. J., 1976.
11. Avriel M., and Schaible S., *Second order characterizations of pseudoconvex functions*, Mathematical Programming, 14, 170–185, 1978.
12. Avriel M., Diewert W. E., Schaible S., and Ziemba W. T., *Introduction to concave and generalized concave functions*, in Generalized Concavity in Optimization and Economics, Schaible S., and Ziemba W. T. (Eds.), Academic Press, New York, 1981.
13. Avriel M., Diewert W. E., Schaible S. and Zang I., *Generalized Concavity*, Plenum Press, 1988.
14. Bajalinov E. B., *Linear-Fractional Programming: Theory, Methods, Applications, and Software*, Kluwer Academic Publishers, Boston MA, 2004.
15. Barros A. I., *Discrete and Fractional Programming Techniques for Location Models*, Kluwer Academic Publisher, 1998.

16. Bazaraa M. S., Goode J. J., and Shetty C. M., *Constraint qualifications revisited*, Management Sciences, 18, 567–573, 1972.
17. Bazaraa M. S. and Shetty C. M., *Foundations of Optimization*, Lectures Notes in Economics and Mathematical Systems, Vol. 122, Springer-Verlag, New York, 1976.
18. Bazaraa M. S., Sherali H. D., and Shetty C. M., *Nonlinear Programming*, John Wiley & Sons, Inc., 1993.
19. Bector C. R., and Singh C., *B-vex functions*, Journal of Optimization Theory and Applications, 71, 237–253, 1991.
20. Bector C. R., Suneja S. K., and Lalitha C. S., *Generalized B-vex functions and generalized B-vex programming*, Journal of Optimization Theory and Applications, 71, 561–576, 1993.
21. Ben-Israel A., and Mond B., *What is invexity?*, Journal of Australian Mathematical Society, Ser. B, 28, 1–19, 1986.
22. Benson H. P., *Generating Sum-of-Ratios Test Problems in Global Optimization*, Technical Note, Journal of Optimization Theory and Applications, 119, 615–621, 2003.
23. Benson H. P., *On the global optimization of sums of linear fractional functions over a convex set*, Journal of Optimization Theory and Applications, 121, 19–39, 2004.
24. Berge C., *Espace topologiques. Fonctions multivoques*, Dunod Paris, 1966.
25. Bitran G., and Novaes A. G., *Linear programming with a fractional objective function*, Operations Research, 21, 22–29, 1973.
26. Bitran G., *Experiments with linear fractional problems*, Naval Research Logistic Quarterly, 26, 689–693, 1979.
27. Brighi L., and John R., *Characterizations of pseudomonotone maps and economic equilibrium*, Journal of Statistics and Management Systems 5, 253–273, 2002.
28. Bykadorov, I. A., *On quasiconvexity in fractional programming*, in Generalized Convexity, Komlósi S., Rapcsak T., and Schaible S. (Eds.), Lecture Notes in Economics and Mathematical Systems, Vol. 405, Springer-Verlag, 281–293, 1994.
29. Cambini A., *Un algoritmo per il massimo del quoziente di due forme affini con vincoli lineari*, Technical Report A-42, Department of Mathematics, University of Pisa, 1977.
30. Cambini A., Martein L., and Pellegrini L., *A decomposition algorithm for a particular class of nonlinear programs*, First Meeting AFCET-SMF on Applied Mathematics, Ecole Polytechnique, Palaiseau, Tome II, 179–189, 1978.
31. Cambini A., *An algorithm for a special class of generalized convex programs*, in Generalized Concavity in Optimization and in Economics, Edited by Schaible and W. T. Ziemba, Academic Press, New York, 491–508, 1981.
32. Cambini A., Martein L., and Sodini C., *An algorithm for two particular nonlinear fractional programs*, Methods of Operations Research, 45, 61–70, 1983.
33. Cambini A., and Martein L., *A modified version of Martos's algorithm for the linear fractional problem*. Methods of Operations Research, 53, 33–44, 1986.
34. Cambini A., Martein L., and Schaible S., *On Maximizing a Sum of Ratios*, Journal of Information and Optimization Sciences, 10, pp. 65–79, 1989.

35. Cambini A.,, Martein L., *Linear fractional and bicriteria linear fractional programs*, in Generalized Convexity and Fractional Programming with Economic Applications, Cambini A., Castagnoli E., Martein L., Mazzoleni P., and Schaible S. (Eds.), Lecture Notes in Economics and Mathematical Systems, vol. 345, Springer-Verlag, Berlin Heidelberg, 155–166, 1990.

36. Cambini A., Castagnoli E., Martein L., Mazzoleni P., and Schaible S., (Eds.), *Generalized Convexity and Fractional Programming with Economic Applications*, in Lecture Notes in Economics and Mathematical Systems, vol. 345, Springer-Verlag, Berlin Heidelberg, 1990.

37. Cambini, A., and Martein, L., *Equivalence in linear fractional programming*, Optimization 23, 41–51, 1992.

38. Cambini A., Schaible S., and Sodini C., *Parametric linear fractional programming for an unbounded feasible region*, Journal of Global Optimization, 3, 157–169, 1991.

39. Cambini A., and Martein L., *Generalized concavity and optimality conditions in vector and scalar optimization*, in Generalized Convexity, Komlósi S., Rapcsák T., and Schaible S. (Eds.), Lectures Notes in Economics and Mathematical Systems, vol. 405, Springer-Verlag, Berlin, 337–357, 1994.

40. Cambini A, Marchi A., Martein L., and Schaible S. *An analysis of the Falk-Palocsay algorithm.* In "Scalar and vector optimization in economic and financial problems", Castagnoli E., and Giorgi G. (Eds.), Proceedings of the Workshop held in Milano, March 28, 1995. Also in Working Paper 96–05, The A. Gary Anderson Graduate School of Management, University of California, Riverside CA.

41. Cambini A., Martein L., and Stancu-Minasian I. M., *Some developments in bicriteria fractional programming*, in Operations Research and its Applications, Du D.-Z., Zhang X.-S., and Cheng K. (Eds.), Lecture Notes in Operations Research 2, Bejing World Publishing Corporation, 144–153, 1996.

42. Cambini A., Martein L., and Stancu-Minasian I. M., *A survey of bicriteria fractional problems*, Editura Academiei Romane, Mathematical Reports, 2, 127–162, 2000.

43. Cambini A., Carosi L., and Martein L., *On the Supremum in Fractional Programming*, in Generalized Convexity and Generalized Monotonicity, Hadjisavvas N., Martinez-Legaz J. E., Penot J.-P. (Eds.), Lecture Notes in Economics and Mathematical Systems, Vol. 502, 129–143, 2001.

44. Cambini A., Crouzeix J. P., and Martein L., *On the pseudoconvexity of a quadratic fractional function*, Optimization, 51, 677–687, 2002.

45. Cambini A., and Martein L., *Generalized convexity and optimality conditions in scalar and vector optimization*, in Rendiconti del Circolo Matematico di Palermo, Serie II, Suppl. 70, 135–153, 2002.

46. Cambini A., Dass B. K., and Martein L., (Eds.), *Generalized Convexity, Generalized Monotonicity, Optimality conditions and Duality in Scalar and Vector Optimization*, Taru Publications and Academic Forum, F-23 Model Town, Delhi, 2003.

47. Cambini A., and Martein L., *Pseudolinearity in scalar and vector optimization*, in Handbook of Generalized Convexity and Generalized Monotonicity, Hadjisavvas N, Komlósi S, and Schaible S. (Eds.), Nonconvex Optimization and its Applications, Vol. 76, Springer, 151–193, 2005.

48. Cambini A., Martein L., and Schaible S., *On the pseudoconvexity of the sum of two linear fractional functions*, in Generalized Convexity, Generalized Monotonicity and Applications, Eberhard A., Hadjisavvas N., and D. T. Luc. (Eds.), Series: Nonconvex Optimization and its Applications, Vol. 77, Springer Berlin Heidelberg New York, 161–172, 2005.

49. Cambini A., Martein L., and Schaible S., *Pseudoconvexity, pseudomonotonicity and the generalized Charnes-Cooper transformation*, Pacific Journal of Optimization, 1, 265–275, 2005.

50. Cambini A., and Martein L., *Pseudomonotonicity of a linear map on the interior of the positive orthant*, in Generalized Convexity and Related Topics, Konnov I. V., D. T. Luc, and Rubinov A. M. (Eds.), Series: Lectures Notes in Economics and Mathematical Systems, Vol. 583, Springer-Verlag, Berlin Heidelberg, 115–131, 2007.

51. Cambini R., *A class of nonlinear programs: theoretical and algorithmical results*, in Generalized Convexity, Komlósi S., Rapcsák T., and Schaible S. (Eds.), Lectures Notes in Economics and Mathematical Systems, Vol. 405, Springer-Verlag, Berlin, 294–310, 1994.

52. Cambini R., and Carosi L., *On generalized convexity of quadratic fractional functions*, in Rendiconti del Circolo Matematico di Palermo, Serie II, Suppl. 70, 155–176, 2002.

53. Cambini R., and Sodini C., *A finite algorithm for a particular D. C. quadratic programming problem*, in Annals of Operations Research, 117, 33–49, 2002.

54. Cambini R., and Sodini C., *A finite algorithm for a class of nonlinear multiplicative programs*, Journal of Global Optimization, 26, 279–296, 2003.

55. Cambini R., and Carosi L., *On generalized linearity of quadratic fractional functions*, Journal of Global Optimization, 30, 235–251, 2004.

56. Cambini R., and Sodini C., *A unifying approach to solve a class of parametrically-convexifiable problems*, in Generalized Convexity and Related Topics, Konnov I. V., D. T. Luc, and Rubinov A. M. (Eds.), Series: Lectures Notes in Economics and Mathematical Systems, Vol. 583, Springer-Verlag, Berlin Heidelberg, 149–166, 2007.

57. Cambini R., and Sodini C., *A sequential method for a class of box costrained quadratic programming problems*, Mathematical Methods of Operation Research, 67, 223–243, 2008.

58. Carosi L., and Martein L., *On the pseudoconvexity and pseudolinearity of some classes of fractional functions*, Optimization, 3, 385–398, 2007.

59. Carosi L., and Martein L., *Some classes of pseudoconvex fractional functions via the Charnes-Cooper transformation*, Lectures Notes in Economics and Mathematical Systems, Vol. 583, Springer, Berlin, 177–188, 2007.

60. Carosi L., and Martein L., *A sequential method for a class of pseudoconcave fractional problems*, Central European Journal of Operations Research, 16, 153–164, 2008.

61. Carter M., *Foundations of Mathematical Economics*, Cambridge (Mass.), MIT Press, 2001.

62. Chew K. L., and Choo E. U., *Pseudolinearity and efficiency*, Mathematical Programming, 28, 226–239, 1984.

63. Chabrillac Y., and Crouzeix J. P., *Definiteness and semi-definiteness of quadratic forms revised*, Linear Algebra and its Applications 63, 283–292, 1984.

64. Charnes A., and Cooper W. W., *Programming with linear fractional functionals*, Naval Research Logistic Quarterly, 9, 181–186, 1962.

65. Charnes A., and Cooper W. W., *An explicit general solution in linear fractional programming*, Naval Research Logistic Quarterly, 23, 161–167, 1976.

66. Charnes A., Cooper W. W., and Rhodes E., *Measuring the efficiency of decision making units*, European Journal of Operational Research 2, 429–444, 1978.

67. Charnes A., Cooper W. W., and Lewin A. Y., *Data Envelopment Analysis: Theory, Methodology and Applications*, Seiford L. M. (Ed.), Kluwer Academic Publishers, Boston MA, 1995.

68. Chen D. Z., Daescu O., Dai Y., Katoh N., Wu X., and Xu J., *Optimizing the sum of linear fractional functions and applications*, Proceedings of the eleventh annual ACM-SIAM Symposium on Discrete Algorithms, San Francisco, California, 707–716, 2000.

69. Colantoni C. S., Manes R. P., and Whinston A., *Programming, profit rates and pricing decisions*, Accounting Review, 467–481, 1969.

70. Cottle R. W., *Notes on a fundamental theorem in quadratic programming*, Siam Journal of Applied Mathematics, 12, 663–665, 1964.

71. Cottle R. W., *Nonlinear programs with positively bounded Jacobians*, Siam Journal of Applied Mathematics, 14, 147–158, 1966.

72. Cottle R. W., and Ferland J. A., *On pseudoconvex functions of nonnegative variables*, Mathematical Programming 1, 95–101, 1971.

73. Cottle R. W., and Ferland J. A., *Matrix-theoretic criteria for quasiconvexity and pseudoconvexity of quadratic functions*, Linear Algebra and its Applications, 5, 123–136, 1972.

74. Cottle R. W., *Complementarity and variational problems*, Istituto Nazionale di Alta Matematica, Vol. XIX, 177–208. 1976.

75. Craven B. D., *Duality for generalized convex fractional programs*, in Generalized Concavity in Optimization and Economics, Schaible S., and Ziemba W. T. (Eds.), Academic Press, 1981.

76. Craven B. D., *Invex functions and constrained local minima*, Bulletin of Australian Mathematical Society, 24, 357–366, 1981.

77. Craven B. D., and Glover B. M., *Invex functions and duality*, Journal of Australian Mathematical Society, Ser. A, 39, 1–20, 1985.

78. Craven B. D., *Fractional programming*, Sigma Series of Applied Mathematics 4, Heldermann Verlag, Berlin, 1988.

79. Crouzeix J.-P., *On second order conditions for quasiconvexity*, Mathematical Programming 18, 349–352, 1980.

80. Crouzeix J.-P., *Some differentiability properties of quasiconvex functions on $\mathfrak{R}^n$*, Optimization and Optimal Control, Auslender et al. (Eds), Springer-Verlag, Lecture Notes in Control and Information Sciences, Vol. 30, 9–20, 1981.

81. Crouzeix J.-P., *About differentiability properties of quasiconvex functions*, Journal of Optimization Theory and Applications, 36, 367–385, 1982.

82. Crouzeix J.-P., and Ferland J. A., *Criteria for quasiconvexity and pseudoconvexity: relationships and comparisons*, Mathematical Programming 23, 193–205, 1982.

83. Crouzeix J.-P., and Ferland J. A., *Criteria for differentiable generalized monotone maps*, Mathematical Programming, 75, 399–406, 1996.

84. Crouzeix J.-P., and Schaible S., *Generalized monotone affine maps*, SIAM Journal on Matrix Analysis and Applications, 17, 992–997, 1996.

85. Crouzeix J.-P., *Characterizations of generalized convexity and generalized monotonicity, a survey*, in Generalized Convexity, Generalized Monotonicity, Crouzeix J.-P., Martinez-Legaz J. E., and Volle M. (Eds.), Nonconvex Optimization and its Applications, Kluwer Academic Publishers, 237–256, 1998.

86. Crouzeix J.-P., Martinez-Legaz J. E., and Volle M., (Eds.), *Generalized Convexity, Generalized Monotonicity*, in Nonconvex Optimization and its Applications, Kluwer Academic Publishers, 1998.

87. Crouzeix J.-P., Marcotte P., and Zhu D. L., *Conditions ensuring the applicability of cutting-plane methods for solving variational inequalities*, Mathematical Programming Series A, 88, 521–539, 2000.

88. Crouzeix J.-P., Hassouni A., Lahlou A., and Schaible S., *Positive subdefinite matrices, generalized monotonicity and linear complementarity problems*, SIAM Journal on Matrix Analysis and Applications 22, 66–85, 2001.

89. Crouzeix J. P., *Criteria for generalized convexity and generalized monotonicity in the differentiable case*, in Handbook of Generalized Convexity and Generalized Monotonicity, Hadjisavvas N., Komlosi S., and Schaible S. (Eds.), Nonconvex Optimization and its Applications, Vol. 76, Springer, 89–119, 2005.

90. Dantzig G. B., and Cottle R. W. *Positive semi-definite programming*, in Nonlinear Programming, Abadie J. (Ed.), North-Holland, Amsterdam, 55–73, 1967.

91. Debreu G., *Definite and semi-definite quadratic form*, Econometrica 20, 285–300, 1952.

92. De Finetti B., *Sulle stratificazioni convesse*, Ann. Math. Pura Appl. 30,173–183, 1949.

93. Diewert W. E., *Generalized concavity and economics*, in Generalized Concavity in Optimization and Economics, Schaible S., and Ziemba W. T. (eds.), Academic Press, New York, 1981.

94. Diewert W. E., Avriel M., and Zang I., *Nine kinds of quasiconcavity and concavity*, Academic Press, 1993.

95. Drezner Z., Schaible S., and Simchi-Levi D., *Queuing-location problems on the plane*, Naval Research Logistic Quarterly, 37, 929–935, 1990.

96. Dür M., Horst R., and Thoai N. V., *Solving sum-of-ratios fractional programs using efficient points*, Optimization, 49, 447–466, 2001.

97. Eberhard A., Hadjisavvas N., and Dinh The Luc., (Eds.), *Generalized Convexity, Generalized Monotonicity and Applications*, Nonconvex Optimization and its Applications, Vol. 77, Springer, 2005.

98. Egudo R. R., and Mond B., *Duality with generalized convexity*, Journal of Austalian Mathematical Society, Ser. B, 28, 10–21, 1986.

99. Eicherberger J., *Game theory for Economists*, Journal of Economic Theory 25, 397–420, 1981.

100. Eichhorn W., *Generalized convexity in Economics: some Examples*, in Generalized Convexity and Fractional Programming with Economic Applications, Cambini A., Castagnoli E., Martein L., Mazzoleni P., and Schaible S. (Eds.), Lecture Notes in Economics and Mathematical Systems, vol. 345, Springer-Verlag, Berlin Heidelberg, 1990.

101. Ellero A., *The optimal level solutions method*, Journal of Information and Optimization Sciences, 17, 355–372, 1996.

102. Ellero A., and Moretti E., *A computational comparison between algorithms for linear fractional programming*, Journal of Information and Optimization Sciences, 13, 343–362, 1992.

103. Ellero A., and Moretti Tomasin E., *A parametric simplex-like algorithm for a fractional programming problem*, Rivista Matematica per le Scienze Economiche e Sociali, 2, 77–88, 1994.

104. Elkin R. M., *Convergence theorems for Gauss-Seidel and other minimization algorithms*, Ph. D. dissertation, University of Maryland, College Park, 1968.

105. Falk J. E., and Palocsay S. M. *Optimizing the sum of linear fractional functions*. Collection: Recent Advances in Global Optimization, Princeton University Press, 221–258, 1992.

106. Falk J. E., and Palocsay S. M. *Image space analysis of generalized fractional programs functions*, Journal of Global Optimization 4, 63–88, 1994.

107. Ferland J. A., *Quasi-convex and pseudo-convex functions on solid convex sets*, Department of Operations Research, Stanford University; Technical Report 71-4, 1971.

108. Ferland J. A., *Maximal domains of quasi-convexity and pseudo-convexity for quadratic functions*, Mathematical Programming 3, 178–192, 1972.

109. Ferland J. A., *Mathematical programming with quasiconvex objective functions*, Mathematical Programming 3, 296–301, 1972.

110. Ferland J. A., *Matrix-theoretic criteria for quasiconvexity of twice continuously differentiable functions*, Linear Algebra and its Applications, 38, 51–63, 1981.

111. Fenchel W., *Convex cones, sets and functions*, Mimeographed Lecture Notes, Princeton University, Princeton, New Jersey, 1951.

112. Finsler W., *Über das Vorkommen definiter Formen and Scharen quadratischer Formen*, Commentarii Math. Helvet., 9, 188–192, 1937.

113. Forgó F., *On the existence of Nash-equilibrium in n-person generalized concave games*, in Generalized Convexity, Komlósi S., Rapcsák T., and Schaible S (Eds.), Lecture Notes in Economics and Mathematical Systems, Vol. 405, Springer-Verlag, Berlin Heidelberg, 1994.

114. Frenk J. and Schaible S., *Fractional programming*, Handbook of Generalized Convexity and Generalized Monotonicity, Edited by (Hadjisavvas N., Komlósi S., and Schaible S., Nonconvex Optimization and its Applications, Vol. 76, Springer, 335–386, 2005.

115. Fudenberg D., and Tirole J., *Game Theory*, Massachussetts Institute of Technology, 1991.

116. Gerencsér L., *On a close relation between quasi-convex and convex functions and related investigations*, Mathematical Operationsforsh, Statistics 4, 201–211, 1973.

117. Giorgi G., *Inclusions among constraint qualifications in nonlinear programming*, Atti del Settimo Comvegno A.M.A.S.E.S., Itec Editrice Milano, 1983.

118. Giorgi G., and Molho E., *Generalized invexity: relationships with generalized convexity and application to the optimality and duality conditions*, in Generalized Concavity for Economic Applications, Mazzoleni P. (Ed.), Tecnoprint, Bologna, 53–70, 1992.

119. Giorgi G., and Guerraggio A., *Various types of nonsmooth invexity*, Journal of Information and Optimization Sciences, 17, 137–150, 1994.

120. Giorgi G., and Guerraggio A., *Constraints qualifications in the invex case*, Journal of Information and Optimization Sciences, 19, 373–384, 1998.

121. Giorgi G., and Guerraggio A., *The notion of invexity in vector optimization: smooth and nonsmooth case*, in Generalized convexity, Generalized monotonicity, Crouzeix J.-P., Martinez-Legaz J.-E., and Volle M. (Eds.), Nonconvex Optimization and its Applications, 27, Kluwer Academic Publishers, 389–405, 1998.

122. Giorgi G., and Thielfelder J., *Constrained quadratic forms and generalized convexity of C^2 − functions revisited*, in Generalized Convexity and Optimization for Economics and Financial Decisions, Giorgi G., and Rossi F. (Eds.), Pitagora Editrice Bologna, 1999.

123. Giorgi G., Jiménez B., and Novo V., *On constraint qualifications in directionally differentiable multiobjective optimization problems*, RAIRO Operations Research, 38, 255–274, 2004.

124. Ginsberg W., *Concavity and quasiconcavity in economics*, Journal of Economic Theory 6, 596–605, 1973.

125. Gotoh J., and Konno H., *Maximization of the ratio of two convex quadratic functions over a polytope*, Compute. Optim. Appl., 20, 43–60, 2001.

126. Gowda M. S., *Pseudomonotone and copositive star matrices*, Linear Algebra and Applications, 113, 107–118, 1989.

127. Gowda M. S., *Affine pseudomonotone mappings and the linear complementarity problems*, SIAM Journal of Matrix Analysis and Applications, 11, 373–380, 1990.

128. Gowda M. S., *On the transpose of a pseudomonotone matrix and the linear complementarity problem*, Linear Algebra and Applications, 140, 129–137, 1990.

129. Greenberg H. J., and Pierskalla W. P., *A review of quasi-convex functions*, Operations Research 19, 1553–1570, 1971.

130. Hadjisavvas N., and Schaible S., *On strong pseudomonotonicity and (semi)strict quasimonotonicity*, Journal of Optimization Theory and Applications, 79, 373–380, 1993.

131. Hadjisavvas N., Martinez-Legaz J. E., and Penot J.-P. (Eds.), *Generalized Convexity and Generalized Monotonicity*, Lecture Notes in Economics and Mathematical Systems, Vol. 502, Springer, Berlin Heidelberg, 2001.

132. Hadjisavvas N., Komlósi S., and Schaible S. (Eds.), *Handbook of Generalized Convexity and Generalized Monotonicity*, Nonconvex Optimization and its Applications, Vol. 76, Springer, 2005.

133. Hadjisavvas N., and Schaible S., *Generalized monotone maps*, in Handbook of Generalized Convexity and Generalized Monotonicity, Hadjisavvas N., Komlósi S., and Schaible S. (Eds.), Nonconvex Optimization and its Applications, Vol. 76, Springer, 389–420, 2005.

134. Hartman P., and Stampacchia G., *On some nonlinear elliptic differential functional equations*, Acta Math., 115, 153–188, 1966.

135. Hanson M. A., *On sufficiency of the Kuhn-Tucker conditions*, Journal of Mathematical Analysis and Applications, 80, 545–550, 1981.

136. Hanson M. A., and Mond B., *Further generalizations of convexity in mathematical programming*, Journal of Information and Optimization Sciences, 3, 25–32, 1982.

137. Hanson M. A., and Mond B., *Convex transformable programming problems and invexity*, Journal of Information and Optimization Sciences, 8, 201–207, 1987.

138. Hanson M. A., and Rueda N. G., *A sufficient condition for invexity*, Journal of Mathematical Analysis and Applications, 138, 193–198, 1989.

139. Hassouni A, *Sous-differentiels des fonctions quasiconvexes*, These de troisieme cicle, Université Paul Sabatier, Toulouse, 1983.

140. Hirche J., *On programming problem with a a linear-plus- linear fractional objective function*, Cahiers du Centre d'Etude de Recherche Operationelle, 26, 59–64, 1984.

141. Hirche J., *Optimizing of sums and products of linear fractional functions under linear constraints*, Fachbereich Mathematik und Informatik, Martin-Luther-Universitat Halle-Wittenberg, Halle, Germany, Report N. 3, 1995.

142. Ibaraki T., Ishii H., Iwase J., Hasegawa T., and Mine H., *Algorithms for fractional programming problems*, Journal of Operations Research, Society of Japan, 19, 2, 174–191, 1976.

143. Ibaraki T., *Solving mathematical programming problems with fractional objective functions*, in Generalized Concavity in Optimization and in Economics, Schaible S., and Ziemba W. T. (eds.), Academic Press, New York, 441–472, 1981.

144. Isbell J. R., and Marlow W. H., *Attrition games*, Naval Research Logistic Quarterly, 3, 71–93, 1956.

145. Jagannathan R., *On some properties of programming in parametric form pertaining to fractional programming*, Management Sciences, 12, 609–615, 1966.

146. Jeyakumar V., *Strong and weak invexity in mathematical programming*, Collection: Methods of Operations Research, 55, 109–125, 1980.

147. Jeyakumar V., and Yang X. Q., *On characterizing the solution sets of pseudolinear programs*, Journal of Optimization Theory and Applications, 87, 747–755, 1995.

148. John R., *Variational inequalities and pseudomonotone functions: some characterizations*, in Generalized Convexity, Generalized Monotonicity, Crouzeix J.-P., Martinez-Legaz J. E., and Volle M. (Eds.), Nonconvex Optimization and its Applications, Kluwer Academic Publishers, 291–301, 1998.

149. John R., *Quasimonotone individual demand*, Optimization, 47, 201–209, 2000.

150. John R., *A first order characterization of generalized monotonicity*, Mathematical Programming, 88, 147–155, 2000.

151. John R., *A note on Minty variational inequalities and generalized monotonicity*, in Generalized Convexity and Generalized Monotonicity, Hadjisavvas N., Martinez-Legaz J. E., and Penot J.-P. (Eds.), Lecture Notes in Economics and Mathematical Systems, Vol. 502, Springer, 240–246, 2001.

152. John R., *Uses of generalized convexity and generalized monotonicity in Economics*, in Handbook of Generalized Convexity and Generalized Monotonicity, Hadjisavvas N., Komlósi S., and Schaible S. (Eds.), Nonconvex Optimization and its Applications, Vol. 76, Springer, 619–666, 2005.

153. Joksch H. C., *Programming with fractional linear objective function*, Naval Research Logistic Quarterly, 197–204, 1964.

154. Karamardian S., *Duality in mathematical programming*, Journal of Mathematical Analysis and Applications, 20, 344–358, 1967.

155. Karamardian S., *Generalized complementarity problems*, Journal of Optimization Theory and Applications, 8, 161–168, 1971.

156. Karamardian S., *Complementarity problems over cones with monotone and pseudomonotone maps*, Journal of Optimization Theory and Applications, 18, 445–454, 1976.

157. Karamardian S., and Schaible S., *Seven kinds of monotone maps*, Journal of Optimization Theory and Applications, 66, 37–46, 1990.

158. Karamardian S., Schaible S., and Crouzeix J. P., *Characterizations of generalized monotone maps*, Journal of Optimization Theory and Applications, 76, 399–413, 1993.

159. Karush W., *Minima of functions of several variables with inequalities as side conditions*, M. S. dissertation, Department of Mathematics, University of Chicago, 1939.

160. Katzner D. W., *Static Demand Theory*, Macmillan, New York, 1970.

161. Kaul R. N., and Kaur S., *Generalizations of convex and related functions*, European Journal of Operational Research, 9, 369–377, 1982.

162. Kaul R. N., and Kaur S., *Optimality criteria in nonlinear programming involving nonconvex functions*, Journal of Mathematical Analysis and Applications, 105, 104–112, 1985.

163. Kaul R.N., Suneja S. K., and Lalitha C. S., *Generalized nonsmooth invexity*, Journal of Information and Optimization Sciences, 15, 1–17, 1994.

164. Khanh P. Q., *Invex-convexlike functions and duality*, Journal of Optimization Theory and Applications, 87, 141–165, 1995.

165. Khanh P. Q., *Sufficient optimality conditions and duality in vector optimization with invex-convexlike functions*, Journal of Optimization Theory and Applications, 87, 359–378, 1995.

166. Khurana A., and Arora S. R., *The sum of a linear and a linear fractional transportation problem with restricted and enhanced flow*, Journal of Interdisciplinary Mathematics, 9, 363–371, 2006.

167. Kim D. S., and Lee G. M., *Optimality conditions and duality theorems for multiobjective invex programs*, Journal of Information and Optimization Sciences, 12, 235–242, 1991.

168. Komlósi S., *Generalized convexity of quadratic functions*, Izv, Vyssh. Uchebn., Zaved Math, 9, 38–43, 1983.

169. Komlósi S., *Second-order characterizations of pseudoconvex and strictly pseudoconvex functions in terms of quasi-Hessians* , Contribution to the theory of optimization, Fórgo F. (Ed.), University of Budapest, 19–46, 1983.

170. Komlósi S., *First and second-order characterizationa of pseudolinear functions*, European Journal of Operations Research, 67, 278–286, 1993.

171. Komlósi S., *On pseudoconvex functions*, Acta Sciences Mathematics, 569–586, 1993.

172. Komlósi S., Rapcsák T., and Schaible S. (Eds.), *Generalized Convexity*, in Lecture Notes in Economics and Mathematical Systems, Vol. 405, Springer-Verlag, Berlin Heidelberg, 1994.

173. Konno H. and Kuno T., *Generalized linear multiplicative and fractional programming*, Annals of Operations Research, 25, 147–161, 1990.

174. Konno H., Yajima Y., and Matsui T., *Parametric simplex algorithms for solving a special class of nonconvex minimization problems*, Journal of Global Optimization, 1, 65–81, 1991.

175. Konno H., Kuno T., and Yajima Y., *Global minimization of a generalized convex multiplicative function*, Journal of Global Optimization, 4, 47–62. 1991.

176. Konno H., and Yajima Y., *Minimizing and maximizing the product of linear fractional functions*, Recent Advances in Global Optimization, Princeton University Press, Princeton, NJ, 259–273, 1992.

177. Konno H., and Watanabe H., *Bond portfolio optimization problems and their applications to index tracking*, Journal of the Operations Research Society of Japan, 39, 295–306, 1996.

178. Konno H., Thach P. T., and Tuy H., *Optimization on low rank nonconvex structures*, Kluwer Academic Publishers, Dodrecht, 1997.
179. Konno H., and Yamashita H., *Minimizing sums and products of linear fractional functions over a polytope*, Naval Research Logistic Quarterly, 46, 583–596, 1999.
180. Konno H., and Abe N., *Minimization of the sum of three linear fractional functions*, Journal of Global Optimization, 15, 419–432, 1999.
181. Konno H., and Fukaishi K., *A branch-and-bound algorithm for solving low-rank linear multiplicative and fractional programming problems*, Journal of Global Optimization, 18, 283–299, 2000.
182. Konno H., *Minimization of the sum of several linear fractional functions*, in Generalized Convexity and Generalized Monotonicity, Hadjisavvas N., Martinez-Legaz J. E., and Penot J.-P. (Eds.), Lectures Notes in Economics and Mathematical Systems, 502, Springer-Verlag, Berlin, 3–20, 2001.
183. Konnov I. V., *On quasimonotone variational inequalities*, Journal of Optimization Theory and Applications, Vol. 99, 165–181, 1998.
184. Konnov I. V., *Generalized monotone equilibrium problems and variational inequalities*, in Handbook of Generalized Convexity and Generalized Monotonicity, Hadjisavvas N., Komlósi S., and Schaible S. (Eds.), Nonconvex Optimization and its Applications, Vol. 76, Springer, 559–618, 2005.
185. Konnov I. V., Dinh The Luc, and Rubinov A. M., (Eds.), *Generalized Convexity and Related Topics*, Lecture Notes in Economics and Mathematical Systems, Vol. 583, Springer, 2007.
186. Kuhn H. W., and Tucker A. W., *Nonlinear Programming*, in Proceedings of the second Berkeley Symposium on Mathematical Statistics and Probability, Neimann J. (Ed.), University of California Press, Berkeley, California, 1951.
187. Kuno T., *A branch-and-bound algorithm for maximizing the sum of several linear ratios*, Journal of Global Optimization, 22, 155–174 2002.
188. Lee G. M., *Nonsmooth invexity in multiple programming*, Journal of Information and Optimization Sciences, 15, 127–136, 1994.
189. Li X. F., Dong J. L., and Liu Q. H., *Lipschitz B-vex functions and nonsmooth programming*, Journal of Optimization Theory and Applications, 93, 557–574, 1997.
190. Lo A., and Mackinlay C., *Maximizing predictability in stock and bond markets*, Microecon. Dyn., 1, 102–134, 1997.
191. Madden P., *Concavity and Optimization*, Basil Blackwell, 1986.
192. Mangasarian O. L., *Pseudoconvex-functions*, Journal of Siam Control Series A, 3, 281–290, 1965.
193. Mangasarian O. L., *Nonlinear programming*, McGraw-Hill, 1969.
194. Mangasarian O. L., *Convexity, pseudo-convexity and quasi-convexity of composite functions*, Cahiers du Centre d'Etude de Recherche Operationelle, 12, 114–122, 1970.
195. Majhii J., Janardan R., Schwerdt J., Smid M., and Gupta P., *Minimizing support structures and trapped area in two-dimensional layered manufacturing*, Technical Report TR-97-058, Department of Computer Science, University of Minnesota, 1997.
196. Majhii J., Janardan R., Smid M., and Gupta P., *On some geometric optimization problems in layered manufacturing*, in Proceedings of the fifth Workshop on Algorithms and Data Structures, Springer-Verlag, 136–149, 1997.

197. Marchi A., and Martein L., *Pseudomonotonicity of an affine map and the two-dimensional case*, to appear in Journal of Information and Optimization Sciences.

198. Martein L., and Pellegrini L., *Un algoritmo per la determinazione del massimo di una particolare funzione razionale fratta soggetta a vincoli lineari*, Technical Report A-45, Department of Mathematics, University of Pisa, 1977.

199. Martein L., and Pellegrini L., *Su una estensione di una particolare classe di problemi di programmazione frazionaria*, Technical Report A-47, Department of Mathematics, University of Pisa, 1977.

200. Martein L., and Pellegrini L., *Su una classe di problemi non lineari e non convessi*, Technical Report A-48, Department of Mathematics, University of Pisa, 1977.

201. Martein L., and Sodini C., *Un algoritmo per un problema di programmazione frazionaria non lineare e non convessa*, Technical Report A-93, Department of Mathematics, University of Pisa, 1982.

202. Martein L., *Maximum of the sum of a linear function and a linear fractional function*, Rivista Matematica per le Scienze Economiche e Sociali, 8, 13–20, 1985.

203. Martein L., and Schaible S., *On solving a linear program with one quadratic constraint*, Rivista Matematica per le Scienze Economiche e Sociali, 10, 75–90, 1987.

204. Martein L., *On the bicriteria maximization problem*, in Generalized Convexity and Fractional Programming with Economic Applications, Cambini A., Castagnoli E., Martein L., Mazzoleni P., and Schaible S. (Eds.), Lecture Notes in Economics and Mathematical Systems, Vol. 345, Springer-Verlag, Berlin Heidelberg, 77–84, 1990.

205. Martein L., and Bertolucci V., *A sequential method for a class of bicriteria problems*, Lectures Notes in Economics and Mathematical Systems, Vol. 583, Springer, Berlin, 347–358, 2007.

206. Martin D. H., *The essence of invexity*, Journal of Optimization Theory and Applications, 47, 65–76, 1985.

207. Martos B., *Hyperbolic programming*, Naval Research Logistic Quarterly, 11, 135–165, 1964.

208. Martos B., *The direct power of adjacent vertex programming methods*, Management Sciences, 12, 241–252, 1965.

209. Martos B., *Subdefinite matrices and quadratic forms*, Siam Journal of Applied Mathematics 17, 1215–1223, 1969.

210. Martos B., *Quadratic programming with a quasiconvex objective function*, Operations Research, 19, 82–97, 1971.

211. Martos B., *Nonlinear programming, Theory and methods*, Holland, Amsterdam, 1975.

212. Mas-Colell A., Whinston M. D., and Green J. R., *Microeconomic theory*, New York, Oxford, Oxford University Press, 1995.

213. Mititelu S., *Generalized invexities and global minimum properties*, Balkan Journal of Geometry and its Applications, 2, 61–72, 1997.

214. Mjelde K. M., *Allocation of resources according to a fractional objective*, European Journal of Operational Research 2, 116–124, 1978.

215. Mond B., and Weir T., *Generalized concavity and duality*, in Generalized Concavity in Optimization and Economics, Schaible S., and Ziemba W. (Eds.), Academic Press, 263–279, 1981.

216. Nikaido H., *Convex Structures and Economic Theory*, Academic Press, New York and London, 1972.
217. Ortega J. M., and Rheinboldt W. C., *Iterative solution of nonlinear equations in several variables*, Academic Press, New York, 1970.
218. Osuna-Gomez R., Rufian-Lizana A., and Ruiz-Canales P., *Invex functions and generalized convexity in multiobjective programming*, Journal of Optimization Theory and Applications, 98, 651–661, 1998.
219. Otani K., *A characterization of quasiconcave functions*, Journal of Economic Theory, 31, 194–196, 1983.
220. Ottaviani M., and Pacelli G., *Fractional programming and characterization of some vertices of the feasible region*, Journal of Optimization Theory and Applications 79, 333–344, 1993.
221. Pini R., *Invexity and generalized convexity*, Optimization, 22, 513–525, 1991.
222. Pini R., and Schaible S., *Invariance properties of generalized monotonicity*, Optimization 28, 211–222, 1994.
223. Ponstein J., *Seven kinds of convexity*, Siam Revue, 9, 115–119, 1967.
224. Preda V., *On duality with generalized convexity*, Bull. U.M.I., 291–305, 1991.
225. Preda V., Stancu-Minasian I. M., and Batatorescu A., *Optimality and duality in nonlinear programming involving semilocally preinvex and related functions*, Journal of Information and Optimization Sciences, 17, 585–596, 1996.
226. Preda V., and Stancu-Minasian I. M., *Duality in multiple objective programming involving semilocally preinvex and related functions*, Glasnik Matematicki, 32, 153–165, 1997.
227. Radzik T., *Newtons' method for fractional combinatorial optimization*, in Proceedings of 33rd Annual IEEE Symposium on Foundations of Computer Science, 659–669, 1992.
228. Rao M. R., *Cluster analysis and mathematical programming*, Journal of the American Statistical Association, 66, 622–626, 1971.
229. Rapcsák T., *On pseudolinear functions*, European Journal of Operational Research, 50, 353–360, 1991.
230. Reiland T. W., *Generalized invexity for nonsmooth vector-valued mappings*, Numerical Functional Analysis and Optimization, 10, 1191–1202, 1989.
231. Reiland T. W., *Nonsmooth invexity*, Bulletin of Australian Mathematical Society, 42, 437–446, 1990.
232. Ritter K., *A parametric method for solving certain nonconcave maximization problems*, Journal of Computational Systems Sciences 1, 11–51, 1967.
233. Roberts A. W., and Varberg D. E., *Convex Functions*, Academic Press, New York, 1973.
234. Rockafellar R. T., *Convex Analysis*, Princeton University Press, 1970.
235. Rueda N. G., and Hanson M. A., *Optimalty criteria in mathematical programming involving generalized invexity*, Journal of Mathematical Analisys and Applications, 130, 375–385, 1988.
236. Schaible S., *Beiträge zur quasikonvexen programming*, Doctoral Dissertation, Köln, Germany, 1971.
237. Schaible S., *Quasiconcave, strictly quasiconcave and pseudoconcave functions*, Methods of Operations Research, 17, 308–316, 1973.
238. Schaible S., *Quasiconcavity and pseudoconcavity of cubic functions*, Mathematical Programming, 5, 243–247, 1973.
239. Schaible S., *Second order characterizations of pseudoconvex quadratic functions*, Journal of Optimization Theory and Applications 21, 15–26, 1977.

240. Schaible S., *Minimization of ratios*, Journal of Optimization Theory and Applications, 19, 347–352, 1976.
241. Schaible S., *A note on the sum of a linear and linear-fractional function*, Naval Research Logistics Quarterly, 24, 691–693, 1977.
242. Schaible S., and Cottle R. W., *On pseudoconvex quadratic forms*, in General Inequalities II, Beckenbach E. F. (Ed.), Birkäuser-Verlag, Basel, 1980.
243. Schaible S., *Generalized convexity of quadratic functions*, in Generalized Concavity in Optimization and Economics, Schaible S., and Ziemba W. T. (Eds.), Academic Press, New York, 183–197, 1981.
244. Schaible S., *A survey of fractional programming*, in Generalized Concavity in Optimization and Economics, Schaible S., and Ziemba W. T. (Eds.), Academic Press, New York, 417–440, 1981.
245. Schaible S., *Fractional programming - State of the art*, Operational Research '81, North-Holland, Amsterdam, 479–493, 1981.
246. Schaible S., *Fractional programming: applications and algorithms*, European Journal of Operational Research 7, 111–120, 1981.
247. Schaible S., *Quasiconvex, pseudoconvex and strictly pseudoconvex quadratic functions*, Journal of Optimization Theory and Applications 35, 303–338, 1981.
248. Schaible S., and Ziemba W. T. (Eds.), *Generalized Concavity in Optimization and Economics*, Academic Press, New York, 1981.
249. Schaible S., *Bibliography in fractional programming*, Zeitschrift für Operations research. Ser. A, 26, 211–241, 1982.
250. Schaible S., *Fractional programming with several ratios*, Methods of Operations Research 49, 77–83, 1985.
251. Schaible S., *Special Functional Forms II: Quadratic Functions*, in *Generalized Concavity*, Avriel M., Diewert W. E., Schaible S. and Zang I. (Eds.), Plenum Press, 1988.
252. Schaible S., *Fractional programming-Some recent developments*, Journal of Information and Optimization Sciences, 10, 1–14, 1989.
253. Schaible S., *Generalized Monotonicity*, Technical Report N. 61, Department of Statistics and Applied Mathematics, 1992.
254. Schaible S., *Generalized Monotone Maps*, in Nonsmooth Optimization, Giannessi F. (Ed.), Gordon and Breach Science Publishers, Amsterdam, 392–408, 1992.
255. Schaible S., *Fractional programming with sums of ratios*, University of Pisa, Department of Statistics and Applied mathematics, Report N. 13, 1994.
256. Schaible S., *Fractional programming*, Handbook of Global Optimization, Horst R., and Pardalos P. M. (Eds.), Kluwer Academic Publishers, Dordrecht, 495–608, 1995.
257. Schaible S., and Sodini C., *Finite algorithm for generalized linear multiplicative programming*, Journal of Global Optimization, 87, 441–455, 1995.
258. Schaible S., and Shi J., *Fractional programming: the sum-of-ratios case*, Optimization Methods and Software, Vol. 18, 219–229, 2003.
259. Schechter M., *An extension of the Charnes-Cooper method in linear fractional programming*, in Continuous-time, Fractional and Multiobjective Programming, Singh C., and Dass B. K. (Eds.), Analytic Publishing Company, Delhi, 97–104, 1989.
260. Singh C., and Dass B. K. (Eds.), *Continuous-time, Fractional and Multiobjective Programming*, Analytic Publishing Company, Delhi, 1989.

261. Sodini C., *Minimizing the sum of a linear function and the square root of a convex quadratic form*, Methods of Operations Research, Vol. 53, 171–182, 1986.

262. Sodini C., *Equivalence and parametric analysis in linear fractional programming*, in Generalized Convexity and Fractional Programming with Economic Applications, Cambini A., Castagnoli E., Martein L., Mazzoleni P., and Schaible S. (Eds.), Lecture Notes in Economics and Mathematical Systems, Vol. 345, Springer-Verlag, Berlin Heidelberg, 143–154, 1990.

263. Stancu-Minasian I. M., *Applications of the fractional programming*, Economic Compututation and Economic Cybernetics Studies and Research, 14, 69–86, 1980.

264. Stancu-Minasian I. M., *Bibliography of fractional programming: 1960–1976*, Pure Appl. Math. Sci. 13, 1–2, 35–69, 1981.

265. Stancu-Minasian I. M., *A second bibliography of fractional programming: 1977–1981*, Pure Appl. Math. Sci. 17, 1–2, 87–102, 1983.

266. Stancu-Minasian I. M., *A third bibliography of fractional programming*, Pure Appl. Math. Sci. 22, 1–2, 109–122, 1985.

267. Stancu-Minasian I. M., and Tigan S., *On some fractional models occurring in minimum-risk problems*, in Generalized Convexity and Fractional Programming with Economic Applications, Cambini A., Castagnoli E., Martein L., Mazzoleni P., and Schaible S. (Eds.), Lecture Notes in Economics and Mathematical Systems, Vol. 345, Springer-Verlag, Berlin Heidelberg, 295–324, 1990.

268. Stancu-Minasian I. M., *A fourth bibliography of fractional programming*, Optimization, 23, 53–71, 1992.

269. Stancu-Minasian I. M., *Fractional Programming - Theory, Methods and Applications*, Kluwer Academic Publishers, Vol. 409, 1997.

270. Stoer J., and Witzgall C., *Convexity and Optimization in finite dimension I*, Springer-Verlag, Berlin-Hidelberg-New York, 1970.

271. Suneja S. K., Singh C., and Bector C. R., *Generalization of preinvex and B-vex functions*, Journal of Optimization Theory and Applications, 76, 577–587, 1993.

272. Swarup K., *Linear fractional functionals programming*, Operations Research, 1029–1036, 1965.

273. Szilágyi P., *A class of differentiable generalized convex functions*, in Generalized Convexity, Komlósi S., Rapcsák T., and Schaible S. (Eds.), Lecture Notes in Economics and Mathematical Systems, vol. 405, Springer-Verlag, Berlin Heidelberg, 104–115, 1994.

274. Takayama A., *Analytical Methods in Economics*, Harvester Wheatsheaf, 1994.

275. Thompson W. A., and Parke D. W., *Some properties of generalized concave functions*, Operations Research 21, 305–313, 1973.

276. Ujvári M., *Simplex-type algorithm for optimizing a pseudolinear quadratic fractional function over a polytope*, Report 2006-01, Department of Operations Research, Budapest, 2006.

277. Wagner H. M., and Yuan J. S. C., *Algorithmic equivalence in linear fractional programming*, Management Sciences, 14, 301–306, 1968.

278. Weber R. J., *Attainable Sets of Markets: An Overview*, in Generalized Concavity in Optimization and Economics, Schaible S., and Ziemba W. T. (Eds.), Academic Press, New York, 1981.

279. Weir T., and Mond B., *Pre-invex functions in multiple objective optimization*, Journal of Mathematical Analysis and Applications, 136, 29–38, 1988.

280. Wolf H., *A parametric method for solving the linear fractional programming problem*, Operations Research, 33, 835–841, 1985.

281. Wolf H., *Solving special nonlinear fractional programming problems via parametric linear programming*, European Journal of Operational Research, 23, 396–400, 1986.

282. Xu B., and Zhu D. L., *New classes of generalized invex monotonicity*, Journal of Inequalities and Applications, 1–19, 2006.

283. Yamamoto R., and Konno H., *An efficient algorithm for solving convex-convex quadratic fractional programs*, Journal of Optimization Theory and Applications, 133, 241–255, 2007.

284. Yao J. C., and Chadli O., *Pseudomonotone complementarity problems and variational inequalities*, in Handbook of Generalized Convexity and Generalized Monotonicity, Hadjisavvas N., Komlósi S., and Schaible S. (Eds.), Nonconvex Optimization and its Applications, Vol.76, Springer, 501–558, 2005.

285. Zangwill W. I., *Nonlinear Programming: A Unified Approach*, Prentice Hall, Enhlewoods Cliffs, 1969.

286. Zhu D. L., and Marcotte P., *New classes of generalized monotonicity*, Journal of Optimization Theory and Applications, 87, 457–471, 1995.

287. Zhu D. L., *The demand functions that satisfy the weak axiom of revealed preference and generalized monotonicity*, Economic Letters, 3, 369–374, 2001.

288. Zionts S., *Programming with linear fractional functionals*, Naval Research Logistics Quarterly, 449–452, 1968.

A

Mathematical Review

In this Appendix, we shall collect definitions, notation, and some basic results from Linear Algebra and Analysis that are used frequently in the book. More details can be found in the standard text-books on these topics.

A.1 Sets

If S is a set and x is an element of S, we write $x \in S$. We write $x \notin S$ if x is not an element of S.

The *union* of two sets S and T, denoted by $S \cup T$, is the set consisting of the elements which belong to either S or T.

The *intersection* of two sets S and T, denoted by $S \cap T$, is the set consisting of the elements which belong to both S and T.

If every element of a set S belongs to a set X, we say that S is a subset of X and we write $S \subset X$ if the inclusion is proper (that is there exists at least one element x of X such that $x \notin S$), otherwise we write $S \subseteq X$.

The *complement* of a set $S \subset X$, denoted by S^c, is the set $\{x \in X : x \notin S\}$.

The *Cartesian product* of the sets S, T, denoted by $S \times T$, is defined as $S \times T = \{(s, t) : s \in S, \ t \in T\}$.

Maps

Let S, T be sets. A map F from S to T is an association with to every element of S associates an element of T; if $x \in S$, then we denote by $F(x)$ the element of T associated with x by F. We call $F(x)$ the *value* of F at x or also the *image* of x under F. The set of all elements $F(x)$, for all $x \in S$, denoted by $F(S)$, or ImS, is called the image of F.

Sets of Real Numbers

The set of real numbers (also referred to as scalars) is denoted by $\Re$.

If a and b are real numbers, $[a, b]$ denotes the set of real numbers x satisfying $a \le x \le b$. A rounded, instead of a square, bracket denotes strict inequality in the definition.

For any real number x, the *absolute value* of x, denoted by $| \, x \, |$, is defined as follows:

$$| \, x \, | = \begin{cases} x & x > 0 \\ -x & x < 0 \\ 0 & x = 0 \end{cases}$$

If S is a subset of $\Re$ bounded below, that is there exists $m \in \Re$ such that $x \geq m$, $\forall x \in S$, then there is a greatest real number ℓ such that $x \geq \ell$ for all $x \in S$. The number ℓ is called the *infimum* of S and is denoted by $\inf S$. We write $\ell = -\infty$, if S is not bounded below.

Similarly, the least upper bound or *supremum* L of a set S is denoted by $\sup S$ and we write $L = +\infty$, if S is not bounded above.

A.2 The Euclidean Space $\Re^n$

An n-vector is an ordered n-tuple of real numbers and it will be viewed as column vector, that is $x = \begin{pmatrix} x_1 \\ x_2 \\ \vdots \\ x_n \end{pmatrix}$.

The set of all n-vectors is denoted by $\Re^n$. For any $x \in \Re^n$, x^T denotes the transpose of x, which is an n-dimensional row vector. For typographical reasons, we shall write $x = (x_1, ..., x_n)^T$, or $x^T = (x_1, ..., x_n)$.

In $\Re^n$ vector addition and scalar multiplication are defined by the formulas $x + y = (x_1 + y_1, ..., x_n + y_n)^T$, $\alpha x = (\alpha x_1, ..., \alpha x_n)^T$.

The *zero* vector, denoted by 0, is a vector consisting entirely of zeros.

The i-th *unit vector*, denoted by e^i, consists of zeros except for a 1 at the i-th position.

One of the important features of $\Re^n$ is that it is an n-dimensional *linear space* (also referred as a vector space) which is a set of elements $x, y, z, ...$ called vectors, for which the operations of addition of vectors and multiplications of vectors by scalars $\alpha, \beta, \gamma, ..$ are defined and are subject to the following basic rules from which all others can be derived:

• $x + y = y + x$, $x + (y + z) = (x + y) + z$;
• there exists a vector 0 such that $x + 0 = x$ for all x. For each x there is an element $(-x)$ such that $x + (-x) = 0$;
• $\alpha x = x \alpha$, $\alpha(\beta x) = (\alpha\beta)x$, $(\alpha + \beta)x = \alpha x + \beta x$, $1 \cdot x = x$, $0x = 0$, where 0 in the left member denotes the scalar zero.

A *subspace* V of $\Re^n$ is a subset of $\Re^n$ with itself constitutes a linear space. V is a subspace of $\Re^n$ if and only if it is closed with respect to the addition and with respect to the multiplication by a scalar, that is $\alpha x + \beta y \in V$ for every $x, y \in V$ and for every $\alpha, \beta \in \Re$.

A *linear manifold* W of $\Re^n$ is a translated subspace, that is a set of the form $W = w + V = \{w + x, \ x \in V\}$.

A set of m vectors $x^1, ..., x^m$ is said to be a set of *linearly independent* vectors if a relation of the form $\sum_{i=1}^{m} \alpha_i x^i = 0$ holds if and only if $\alpha_1 = ... = \alpha_m = 0$. Vectors are *linearly dependent* if they are not linearly independent.

If S is a nonempty subset of $\Re^n$, the set of all linear combinations $x = \sum_{i=1}^{s} \alpha_i x^i$ of vectors $x^1, ..., x^s$ in S is a subspace called the subspace *generated or spanned* by S.

Given a subspace V of $\Re^n$, a *basis* for V is a collection of vectors of V that are linearly independent and that spanned V. Every basis of a given subspace has the same number of elements. This number is said to be the *dimension* of V, denoted $dimV$, and it is equal to the maximum number of linearly independent vectors in V. The unit vectors $e^1, .., e^n$ constitutes a basis of $\Re^n$. A subspace of $\Re^n$ is proper if and only if its dimension is less than n. The dimension of a linear manifold $W = w + V$ is the dimension of V and we write $dimW = dimV$.

Linear manifolds of dimension 0, 1 and 2 are called *points*, *lines* and *planes*, respectively. An $n - 1$-dimensional linear manifold is called a *hyperplane*. Let U, V be subspaces of $\Re^n$. Then, $U \cap V$ and $U + V = \{u + v, \ u \in U, \ v \in V\}$ are subspaces verifying the relation $dim(U+V) = dimU + dimV - dim(U \cap V)$.

The Scalar Product

The *scalar or inner product* $x^T y$ of two vectors $x, y \in \Re^n$ is defined as

$$x^T y = \sum_{i=1}^{n} x_i y_i$$

The basic properties of the scalar product are as follows:
- $x^T x \geq 0$ for every x; $x^T x = 0$ if and only if $x = 0$;
- $x^T y = y^T x$;
- $(\alpha x + \beta y)^T z = \alpha x^T z + \beta y^T z$.

The vectors x, y are said to be *orthogonal* if $x^T y = 0$.

If V is a subspace of $\Re^n$, the *orthogonal complement* of V, denoted by $V^\perp$, consists of all vectors that are orthogonal to every vector on V; $V^\perp$ is a subspace and together V and $V^\perp$ span $\Re^n$ in the sense that every $x \in \Re^n$ can be written uniquely in the form $x = v + w$ with $v \in V$, $w \in V^\perp$. We have $dimV + dimV^\perp = n$.

A set of vectors $x^1, ..., x^s$ is said to be *orthogonal* if $(x^i)^T x^j = 0, i, j = 1, ...s, i \neq j$; if in addition $(x^i)^T x^i = 1, i = 1, ...s$, the set of vectors is called *orthonormal*. An important fact is that every subspace $\Re^n$ has an orthogonal basis which reduces to an orthonormal basis substituting x^i with $\dfrac{x^i}{(x^i)^T x_i}$.

Several times, in the book the vectors $x^1, ..., x^s$ are denoted by $x_1, ..., x_s$; however, this variation in notation will be clear in the context.

By means of the scalar product, a hyperplanes may be characterized as follows:

$H \subset \Re^n$ *is a hyperplane if and only if there exist* $\alpha \in \Re^n \setminus \{0\}, \beta \in \Re$, *such that* $H = \{x \in \Re^n : \alpha^T x = \beta\}$, *with* α, β *unique up to a common non-zero multiple.*

Corresponding to a hyperplane H there are the *positive and negative closed half-spaces*

$$H^+ = \{x \in \Re^n : \alpha^T x \geq \beta\}, \ H^- = \{x \in \Re^n : \alpha^T x \leq \beta\}$$

and the *positive and negative open half-spaces*

$$int H^+ = \{x \in \Re^n : \alpha^T x > \beta\}, \ int H^- = \{x \in \Re^n : \alpha^T x < \beta\}$$

Moreover, *every linear manifold of* $\Re^n$ *is an intersection of a finite number of hyperplanes.*

In particular any line may be expressed as an intersection of $n-1$ hyperplanes. A line may be also usefully expressed in parametric form.

The line through $x_1 \in \Re^n$ and $x_2 \in \Re^n$, with $x_1 \neq x_2$, is

$$\{x \in \Re^n : x = x_1 + \lambda(x_2 - x_1), \ \lambda \in \Re\}$$

The line through x_0 and direction u with $x_0, u \in \Re$, is

$$\{x \in \Re^n : x = x_0 + tu, \ t \in \Re\}$$

The set $r = \{x \in \Re^n : x = x_0 + tu, \ t \geq 0\}$ is the *half-line* through x_0 and direction u, while the set

$$[x_1, x_2] = \{x \in \Re^n : x = x_1 + t(x_2 - x_1), \ t \in [0,1]\}$$

is the *line segment* of end points x_1, x_2.

Matrices

A *matrix* is a rectangular array of numbers, called *elements*. We write

$$A = \begin{bmatrix} a_{11} & a_{12} & \cdots & a_{1n} \\ \vdots & \vdots & \ddots & \vdots \\ a_{m1} & a_{m2} & \cdots & a_{mn} \end{bmatrix}$$

for a matrix A having m rows and n columns. Such a matrix is referred as an $m \times n$ matrix. The notation $A = [a_{ij}]$ is used.

The *transpose* of A, denoted by A^T, is the $n \times m$ matrix A^T whose ij element is a_{ji}.

An $m \times n$ matrix whose elements are all zero is called a *zero matrix* and denoted by O.

The *sum* of two $m \times n$ matrices $A = [a_{ij}]$ and $B = [b_{ij}]$ is the matrix

$A + B = [a_{ij} + b_{ij}]$. The *product* λA of a matrix A and a scalar λ, is obtained by multiplying each element of A by λ. The product AB of an $m \times n$ matrix A and an $n \times p$ matrix B is the $m \times p$ matrix C with elements $c_{ij} = \sum_{k=1}^{n} a_{ik} b_{kj}$.

Let A be an $m \times n$ matrix. The *range space* or *image* of A, denoted by ImA, is the set of all vectors $z \in \Re^m$ such that $z = Ax$ for some $x \in \Re^n$, i.e. $ImA = \{z \in \Re^m : z = Ax, \; x \in \Re^n\}$. ImA is the subspace of $\Re^m$ spanned by the columns of A and its dimension is the maximum numbers of the columns of A which are linearly independent. The *kernel* of A, denoted by $kerA$, is the set of all vectors $x \in \Re^n$ such that $Ax = 0$, i.e., $kerA = \{x \in \Re^n : Ax = 0\}$. $kerA$ is a subspace of $\Re^n$ and we have $dim(kerA) + dim(ImA) = n$.

The *rank* of an $m \times n$ matrix A, denoted by $rankA$, is equal to the maximum number of linearly independent columns in A. This number is also equal to the maximum number of linearly independent rows in A. A is said to be of *full rank* if the rank of A is equal to the minimum of m and n.

Square Matrices, Eigenvalues, and Quadratic Forms

Let A be a *square* matrix (a matrix with $m = n$). We use I to denote the *identity matrix* that is the matrix $[a_{ij}]$ with $a_{ij} = 0$ for $i \neq j$, and $a_{ii} = 1$ for $i = 1, 2, ..., n$. A is said to be *symmetric* if $A^T = A$, i.e., $a_{ij} = a_{ji}$ for all i, j. A is *diagonal* if $a_{ij} = 0$ whenever $i \neq j$. A is *nonsingular* or *invertible* if there is a matrix A^{-1}, called the inverse of A, such that $A^{-1}A = I = AA^{-1}$. A is nonsingular if and only if its *determinant*, denoted by $detA$, is non-zero. If A and B are square matrices we have:
$(AB)^T = B^T A^T$, $(A^{-1})^T = (A^T)^{-1}$, $(AB)^{-1} = B^{-1}A^{-1}$.
Corresponding to an $n \times n$ square matrix A, a scalar λ and a non-zero vector x satisfying the equation $Ax = \lambda x$ are said to be, respectively, an *eigenvalue* and *eigenvector* of A; λ is an eigenvalue of A if and only if $A - \lambda I$ is singular or, equivalently, if and only if $det(A - \lambda I) = 0$. This last result, when expanded, yields an n-th order polynomial equation which can be solved for n complex roots λ which are the eigenvalues of A.

When A is symmetric we have the following special properties:
(a) the eigenvalues of A are real numbers;
(b) eigenvectors associated with distinct eigenvalues are orthogonal;
(c) there is an orthogonal basis for $\Re^n$, each element of which is an eigenvector.

A symmetric $n \times n$ matrix A is said to be *positive definite* if $x^T A x > 0$ for all $x \in \Re^n$, $x \neq 0$. It is called *positive semidefinite* if $x^T A x \geq 0$ for all $x \in \Re^n$. We have the following properties:
• For any $m \times n$ matrix A, the matrix $A^T A$ is symmetric and positive semidefinite. $A^T A$ is positive definite if and only if $rankA = n$;
• A square symmetric matrix is positive semidefinite (positive definite) if and only if all of its eigenvalues are non-negative (positive);
• The inverse of a positive definite matrix is positive definite.

A.3 Topological Concepts

The Euclidean *norm* of a vector x is defined by

$$\| x \| = \sqrt{x^T x}$$

From the properties of the scalar product there follows the following basic properties:

1. $\| x \| \geq 0$ for all $x \in \Re^n$; $\| x \| = 0$ if and only if $x = 0$;
2. $\| \alpha x \| = \alpha \| x \|$ for every $\alpha \in \Re$ and every $x \in \Re^n$;
3. $\| x + y \| \leq \| x \| + \| y \|$ for all $x, y \in \Re^n$ (triangle inequality);

Another important inequality, called the *Cauchy-Schwartz inequality*, states that for any two vectors x, y, we have

4. $| x^T y | \leq \| x \| \| y \|$ with equality holding if and only if $y = \alpha x$ for some scalar α.

The Euclidean norm allow us to define some topological concepts.

A ball around x_0 and radius ϵ is the set $B(x_0, \epsilon) = \{ x \in \Re^n : \| x - x_0 \| < \epsilon \}$. Such a ball is also referred to as the *neighborhood* of x_0 of radius ϵ.

Let S be a set of $\Re^n$. A point x_0 is an *interior point* of S if there exists $\epsilon > 0$ such that $B(x_0, \epsilon) \subset S$. A point x_0 is an *accumulation point* or a *limit point* of S if every ball around x_0 contains a point $x \neq x_0$ belonging to S. A point x_0 is an *isolated point* of S if $x_0 \in S$ but it is not a limit point of S. A point x_0 is a *boundary point* of S if every ball around x_0 contains points in S and points not in S. Finally a point x_0 is an *exterior point* of S if it is interior to the *complement* of S.

A set S is:
- *open* if all of its points are interior points;
- *closed* if it contains all of its limit points;
- *bounded* if it is contained in a ball;
- *compact* if it is closed and bounded.

The complement of an open set is closed, and the complement of a closed set is open.

The *interior* of S, denoted by $int S$, is the set of its interior points. The interior of a set is always open. The *boundary* of S, denoted by ∂S, is the set of its boundary points.

A set S is closed if and only if $\partial S \subseteq S$. The *closure* of S, denoted by $cl S$, is the smallest closed set containing S.

The union of open sets is open. The intersection of finitely many open sets is open.

The union of finitely many closed sets is closed. The intersection of closed sets is closed.

Every linear manifold in $\Re^n$ is closed.

A sequence of vectors $x_1, ..., x_k, ...$, denoted $\{x_k\}$, is said to *converge* to the limit x if $\| x_k - x \| \to 0$ as $k \to +\infty$, that is if for any given $\epsilon > 0$, there is a N such that $k \geq N$ implies $\| x_k - x \| < \epsilon$. If $\{x_k\}$ converges to x, we write $x_k \to x$ or $\lim_{k \to +\infty} x_k = x$.

A set S is closed if and only if the limits of convergent sequences in S are in S, that is $\{x_k\} \subset S$, $x_k \to x$ *implies* $x \in S$.

A point x is a limit point of the sequence $\{x_k\}$ if there is a subsequence of $\{x_k\}$ convergent to x.

An important result, known as the Bolzano-Weierstrass Theorem, states that *a set S is compact if and only if every sequence of elements of S has a subsequence that converges to an element of S.*

A.4 Functions

A map f from $S \subset \Re^n$ to $\Re$ is referred as a *real-valued function* defined on S.
A map F from $S \subset \Re^n$ to $\Re^m$ is referred as a *vector-valued function* defined on S. F can be equivalently expressed by $F(x) = (f_1(x), ..., f_m(x))$, where $f_i(x), i = 1, ..., m$, are real-valued functions defined on S.
In this book, a real-valued function is referred as a function. This variation will be clear in the context.

Continuity
Let S be a nonempty set of $\Re^n$ and let f be a function defined on S.
• f is said to be *lower semicontinuous* at $x_0 \in S$, if for each $\epsilon > 0$ there exists $\delta > 0$ such that $x \in S$ and $\| x - x_0 \| < \delta$ implies $f(x) > f(x_0) - \epsilon$.
The lower semicontinuity of f at $x_0 \in S$ is equivalent to the following statement: f is lower semicontinuous at $x_0 \in S$, if for any sequence $\{x_n\} \subset S$ such that $x_n \to x_0$ and $f(x_n) \to z$, we have $z \geq f(x_0)$.
• f is said to be *upper semicontinuous* at $x_0 \in S$, if for each $\epsilon > 0$ there exists $\delta > 0$ such that $x \in S$ and $\| x - x_0 \| < \delta$ implies $f(x) < f(x_0) + \epsilon$.
The upper semicontinuity of f at $x_0 \in S$ is equivalent to the following statement: f is upper semicontinuous at $x_0 \in S$, if for any sequence $\{x_n\} \subset S$ such that $x_n \to x_0$ and $f(x_n) \to z$, we have $z \leq f(x_0)$.
• f is said to be *continuous* at $x_0 \in S$, if it is both lower semicontinuous and upper semicontinuous at x_0 or, equivalently, if for each $\epsilon > 0$ there exists $\delta > 0$ such that $x \in S$ and $\| x - x_0 \| < \delta$ implies $\| f(x) - f(x_0) \| < \epsilon$.
The continuity at $x_0 \in S$ is equivalent to the following statement: f is continuous at $x_0 \in S$, if for any sequence $\{x_n\} \subset S$ such that $x_n \to x_0$ we have $f(x_n) \to f(x_0)$.
• f is said to be continuous (lower semicontinuous, upper semicontinuous) on S if f is continuous (lower semicontinuous, upper semicontinuous) at every element of S.
• The Euclidean norm and the scalar product are continuous functions.
Sums and products of continuous functions are continuous functions; a composite of continuous functions is a continuous function.
• Let S be a nonempty set of $\Re^n$ and let $F(x) = (f_1(x), ..., f_m(x))$ be a vector-valued function on S. F is said to be continuous (respectively, lower

semicontinuous, upper semicontinuous) on S if each function f_i, $i = 1, ..., m$, is continuous (respectively, lower semicontinuous, upper semicontinuous) on S.

Infimum of a Function

Let f be a real-valued function on $S \subseteq \Re^n$. The infimum of the image $f(S)$, i.e., the largest number ℓ (with $\ell = -\infty$ admitted) such that $f(x) \geq \ell$ for all $x \in S$ is called the *infimum* of f on S. It is denoted by $\inf_{x \in S} f(x)$. In general, the value $\inf_{x \in S} f(x)$ may or may not be attained. If there is a point $x_0 \in S$ such that $f(x_0) = \inf_{x \in S} f(x)$, this is expressed by writing $f(x_0) = \min_{x \in S} f(x)$.

Such a point x_0 is called a *global minimum* for f on S and $f(x_0)$ is the minimum value of f on S.

Similarly, the supremum of the image $f(S)$, i.e., the smallest number L (with $L = +\infty$ admitted) such that $f(x) \leq L$ for all $x \in S$ is called the *supremum* of f on S. It is denoted by $\sup_{x \in S} f(x)$. If there is a point $x_0 \in S$ such that $f(x_0) = \sup_{x \in S} f(x)$, this is expressed by writing $f(x_0) = \max_{x \in S} f(x)$.

Such a point x_0 is called a *global maximum* for f on S and $f(x_0)$ is the maximum value of f on S.

It is clear that on S

$$\inf_{x \in S} f(x) = -\sup_{x \in S}(-f(x)); \quad \sup_{x \in S} f(x) = -\inf_{x \in S}(-f(x))$$

One of the basic problems in Optimization is to determine conditions which imply the existence of the minimum and/or the maximum value of a function. With this regards, the following theorem holds.

The Generalized Weierstrass Theorem

Let S be a nonempty compact set of $\Re^n$. Then:

(i) A lower semicontinuous function f on S attains its minimum value on S, i.e., there exists $x_0 \in S$ such that $f(x_0) = \min_{x \in S} f(x)$.

(ii) An upper semicontinuous function f on S attains its maximum value on S, i.e., there exists $x_0 \in S$ such that $f(x_0) = \max_{x \in S} f(x)$.

(iii) A continuous function on S attains its minimum and maximum values on S, i.e., there exists $x_1 \in S$, $x_2 \in S$ such that $f(x_1) = \min_{x \in S} f(x)$ and $f(x_2) = \max_{x \in S} f(x)$.

Derivatives and Differentiability

Let S be an open set of $\Re^n$, and let f be a function defined on S. We define its *partial derivative* at a point $x \in S$ by

$$\frac{\partial f(x)}{\partial x_i} = \lim_{t \to 0} \frac{f(x + te^i) - f(x)}{t}$$

if the limit exists. Assuming all of these partial derivatives exist, the *gradient* of f at x_0 is defined as the column vector

$$\nabla f(x) = \begin{pmatrix} \frac{\partial f(x)}{\partial x_1} \\ \frac{\partial f(x)}{\partial x_2} \\ \vdots \\ \frac{\partial f(x)}{\partial x_n} \end{pmatrix}.$$

We shall say that f is *differentiable* at $x_0 \in S$ if the gradient of f at x_0 exists and

$$f(x) = f(x_0) + \nabla f(x_0)^T (x - x_0) + o(x - x_0)$$

where $o(x - x_0)$ is such that $\lim_{x \to x_0} o(x - x_0) = 0$.

The function f is said to be differentiable on an open set S if it is differentiable at every point of S.

If f is differentiable on an open set S it is obviously differentiable on an arbitrary subset of S (not necessarily open). For such a reason, we shall say that f *is differentiable on a subset* $X \subset \Re^n$ *if it is differentiable on an open set containing* X.

The gradient of f thus takes the place of a derivative. The rules for derivative hold as usual: if f, g are differentiable on S, and $\alpha \in \Re$, then

$$\nabla \left(f(x) + g(x) \right) = \nabla f(x) + \nabla g(x); \quad \nabla \alpha f(x) = \alpha \nabla f(x)$$

$$\nabla \left(f(x) g(x) \right) = g(x) \nabla f(x) + f(x) \nabla g(x)$$

$$\nabla \left(\frac{f(x)}{g(x)} \right) = \frac{g(x) \nabla f(x) - f(x) \nabla g(x)}{(g(x))^2}$$

If f is differentiable on an open set S and the gradient $\nabla f(x)$ is a continuous function of x, f is said to be *continuously differentiable on* S.

If in addition, each partial derivative is a continuously differentiable function, we say that f is *twice continuously differentiable* on S. We use the notation $\frac{\partial^2 f(x)}{\partial x_i \partial x_j}$ to indicate the i-th partial derivative of $\frac{\partial f(x)}{\partial x_j}$ at a point $x \in S$.

The matrix whose ij-th entry is $\frac{\partial^2 f(x)}{\partial x_i \partial x_j}$, is called the *Hessian* matrix, or Hessian, of f and it is denoted by $\nabla^2 f(x)$. If f is twice continuously differentiable on S, the Hessian of f is symmetric.

If $F(x) = (f_1(x), ... f_m(x))$ is a vector-valued function defined on an open set $S \subseteq \Re^n$, it is called differentiable (respectively, continuously differentiable) if each component f_i of F is differentiable (respectively, continuously differentiable). The *Jacobian matrix* of F, denoted by $J(F(x))$, is the $m \times n$ matrix whose i-th row is the transpose of the gradient $\nabla f_i(x)$, i.e.,

$$J(F(x)) = \begin{bmatrix} \dfrac{\partial f_1(x)}{\partial x_1} & \cdots & \dfrac{\partial f_1(x)}{\partial x_n} \\ \dfrac{\partial f_2(x)}{\partial x_1} & \cdots & \dfrac{\partial f_2(x)}{\partial x_n} \\ \vdots & \ddots & \vdots \\ \dfrac{\partial f_m(x)}{\partial x_1} & \cdots & \dfrac{\partial f_m(x)}{\partial x_n} \end{bmatrix}$$

The Directional Derivative

Let S be an open set of $\Re^n$, and let f be a function defined on S. For any $x_0 \in S$ and for any $u \in \Re^n$, setting $\varphi(t) = f(x_0 + tu)$, we define the one-sided *directional derivative* of f at x_0 in the direction u, to be

$$\varphi'(0) = \lim_{t \to 0^+} \frac{\varphi(t) - \varphi(0)}{t} = \lim_{t \to 0^+} \frac{f(x_0 + tu) - f(x_0)}{t}$$

provided that the limit exists.

If f is differentiable on S, we have

$$\varphi'(t) = \nabla(f(x_0 + tu))^T u; \quad \varphi'(0) = \nabla f(x_0)^T u.$$

The Second Directional Derivative

If f is a twice continuously differentiable function on S, we have

$$\varphi''(t) = u^T \nabla^2 f(x_0 + tu)u; \quad \varphi''(0) = u^T \nabla^2 f(x_0)u.$$

The Mean Value Theorem

Let S be an open set of $\Re^n$, and let f be a continuously differentiable function on S. If $x_1, x_2 \in S$, then there exists $\theta \in [0,1]$ such that

$$f(x_2) = f(x_1) + \nabla f(x_1 + \theta(x_2 - x_1))^T (x_2 - x_1)$$

Furthermore, if f is a twice continuously differentiable function on S, then then there exists $\theta \in [0,1]$ such that

$$f(x_2) = f(x_1) + \nabla f(x_1)^T (x_2 - x_1) + \frac{1}{2}(x_2 - x_1)^T \nabla^2 f(x_1 + \theta(x_2 - x_1))(x_2 - x_1)$$

Implicit Function Theorem

Consider a set of m equations in n variables

$$h_i(x) = 0, \quad i = 1, 2, ..., m.$$

The implicit function theorem addresses the question as to whether if $n - m$ of the variables are fixed, the equations can be solved for the remaining m variables. Thus, selecting m variables, say $x_1, ..., x_m$, we wish to determine if these may be expressed in terms of the remaining variables in the form

$$x_i = \phi_i(x_{m+1}, ..., x_n), \quad i = 1, ..., m.$$

Theorem Let $x^0 = (x_1^0, ..., x_n^0) \in \Re^n$ be such that:
(i) $h_i(x) = 0$, $i = 1, ..., m$;
(ii) The functions h_i, $i = 1, ..., m$, have continuous partial derivatives of order p in some neighborhood of x^0 for some $p \geq 1$;
(iii) The $m \times m$ Jacobian matrix

$$J = \begin{bmatrix} \frac{\partial h_1(x^0)}{\partial x_1} & \cdots & \frac{\partial h_1(x^0)}{\partial x_m} \\ \frac{\partial h_2(x^0)}{\partial x_1} & \cdots & \frac{\partial h_2(x^0)}{\partial x_m} \\ \vdots & \ddots & \vdots \\ \frac{\partial h_m(x^0)}{\partial x_1} & \cdots & \frac{\partial h_m(x^0)}{\partial x_m} \end{bmatrix}$$

is nonsingular.
Then, there exists a neighborhood of $\hat{x}^0 = (x_{m+1}^0, ..., x_n^0) \in \Re^{n-m}$ such that for $\hat{x} = (x_{m+1}, ..., x_n)$ in this neighborhood there are functions $\phi_i(\hat{x})$, $i = 1, 2, ..., m$, such that
(i) ϕ_i has continuous partial derivatives of order p, $i = 1, 2, .., m$;
(ii) $x_i^0 = \phi_i(\hat{x}^0)$, $i = 1, ..., m$;
(iii) $h_i(\phi_i(\hat{x}), \phi_2(\hat{x}), ..., \phi_m(\hat{x}), \hat{x}) = 0$, $i = 1, ..., m$.

Local Minima and Maxima
Let S be a subset of $\Re^n$, and let f be a function defined on S.
A point $x_0 \in S$ is said to be a *local minimum* for f on S if there exists a ball $B(x_0, \epsilon)$ around x_0 and radius ϵ such that

$$f(x) \geq f(x_0), \ \forall x \in B(x_0, \epsilon) \cap S$$

If $B(x_0, \epsilon)$ can be chosen so that the strict inequality $f(x) > f(x_0)$ holds for all $x \in B(x_0, \epsilon) \cap S$, $x \neq x_0$, the point x_0 is said to be a *strict local minimum*. It is clear that a global minimum is a local minimum. The converse, however, is not true.
The corresponding definitions for a *local maximum* and for a *strict local maximum* are obtained from the definitions just given by reversing the inequalities and replacing "minimum" by "maximum".

Necessary Optimality Conditions
Let x_0 be an interior point of S. If x_0 *is a local minumum or a local maximum and f is differentiable at x_0, then $\nabla f(x_0) = 0$.*
The above condition is called a *first-order optimality condition* since it uses the first partial derivatives of f.
A point x_0 such that $\nabla f(x_0) = 0$ is referred as a *critical point*.
Necessary optimality conditions stated in terms of the Hessian of f are called *second-order optimality conditions*. One such condition is given below.
Let x_0 be an interior point of S. If x_0 *is a local minumum (respectively, local maximum) and f is twice continuously differentiable on S, then $\nabla f(x_0) = 0$ and $\nabla^2 f(x_0)$ is positive semidefinite (respectively, negative semidefinite).*

Sufficient Optimality Conditions

Let f be a twice continuously differentiable function on S. If $\nabla f(x_0) = 0$ and $\nabla^2 f(x_0)$ is positive definite (respectively, negative semidefinite), then x_0 is a strict local minimum point (respectively, strict local maximum).

B

Concave and Generalized Concave Functions

In this Appendix, for the sake of completeness and taking also into account that concave and quasiconcave functions are more frequently encountered in Economics, we shall present the definitions and the main properties of these functions.

Definitions
Let $S \subseteq \Re^n$ be a convex set. A function f defined on S is said to be:
(a) *concave* on S if if for all $x_1, x_2 \in S$,

$$f(\lambda x_1 + (1 - \lambda)x_2) \geq \lambda f(x_1) + (1 - \lambda)f(x_2), \quad \forall \lambda \in [0, 1].$$

(b) *strictly concave* on S if for all $x_1, x_2 \in S$, with $x_1 \neq x_2$,

$$f(\lambda x_1 + (1 - \lambda)x_2) > \lambda f(x_1) + (1 - \lambda)f(x_2), \quad \forall \lambda \in (0, 1).$$

(c) *quasiconcave* on S if for all $x_1, x_2 \in S$,

$$f(\lambda x_1 + (1 - \lambda)x_2) \geq \min\{f(x_1), f(x_2)\}, \quad \forall \lambda \in [0, 1].$$

(d) *strictly quasiconcave* on S if for all $x_1, x_2 \in S$, $x_1 \neq x_2$,

$$f(\lambda x_1 + (1 - \lambda)x_2) > \min\{f(x_1), f(x_2)\}, \quad \forall \lambda \in (0, 1).$$

(e) *semistrictly quasiconcave* on S if for every $x_1, x_2 \in S$, with $f(x_1) \neq f(x_2)$,

$$f(\lambda x_1 + (1 - \lambda)x_2) > \min\{f(x_1), f(x_2)\}, \quad \forall \lambda \in (0, 1)$$

(f) *pseudoconcave* on S (under differentiability assumption) if for all x_1, $x_2 \in S$,
$$f(x_1) < f(x_2) \Rightarrow \nabla f(x_1)^T (x_2 - x_1) > 0.$$

(g) *strictly pseudoconcave* on S (under differentiability assumption) if for all $x_1, x_2 \in S$, $x_1 \neq x_2$,

$$f(x_1) \leq f(x_2) \Rightarrow \nabla f(x_1)^T (x_2 - x_1) > 0.$$

A semistrictly quasiconcave function on the convex set S is not necessarily a quasiconcave funtion. This is true by assuming the upper semicontinuity of f on S.

An important case where quasiconcavity reduces to concavity is related to homogeneity:

if f is a homogeneous positive function of degree α with $0 < \alpha \leq 1$ on the convex set $S \subseteq \Re^n$, then f is quasiconcave if and only if it is concave.

In the differentiable case, quasiconcavity reduces to concavity when the function has not critical points:

if f is a differentiable function on the open convex set $S \subseteq \Re^n$ such that $\nabla f(x) \neq 0$, $\forall x \in S$, then f is pseudoconcave on S if and only it is quasiconcave on S.

The inclusion relationships among the various classes of concave and generalized concave functions are illustrated in Figure B.1 assuming differentiability. All inclusions are proper.

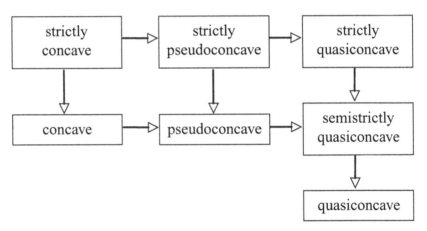

Fig. B.1. Relationships between various types of concavity

Properties

Let $S \subseteq \Re^n$ be a convex set.

• If f is a concave function on S, then the upper-level set $S_{\geq \alpha} = \{x \in S : f(x) \geq \alpha\}$ is convex for every $\alpha \in \Re$.

• A non-negative linear combination of concave functions on S is a concave function on S.

• A non-negative linear combination of strictly concave functions on S is a strictly concave function on S.

• If $f_i, i = 1, ..., m$ are concave functions on S, then $z(x) = \min_{i \in \{1,..,m\}} \{f_i(x)\}$ is concave on S.

- If $g : \Re \to \Re$ is a non-decreasing concave function and $h : S \to \Re$ is a concave function on S, then the composite function $f(x) = g(h(x))$ is concave on S.
- If $g : \Re \to \Re$ is an increasing concave function and $h : S \to \Re$ is a strictly concave function on S, then the composite function $f(x) = g(h(x))$ is strictly concave on S.
- If f is a quasiconcave (respectively, strictly quasiconcave, semistrictly quasiconcave) function on S and $g : \Re \to \Re$ be an increasing function, then the composite function $g \circ f$ is quasiconcave (respectively, strictly quasiconcave, semistrictly quasiconcave) on S. This result still holds for a quasiconcave function even if g is a non-decreasing function.
- If f is pseudoconcave on S and $\phi : \Re \to \Re$ is a differentiable function such that $\phi'(z) > 0, \forall z \in \Re$, then the composite function $\phi \circ f$ is pseudoconcave.
- Let $g(x) = Ax + b$ where A is an $m \times n$ matrix, $b \in \Re^m$ and let f be a quasiconcave (pseudoconcave) function on S. Then, $z(x) = f(Ax + b)$ is quasiconcave (pseudoconcave) on S.

Characterizations

Let $S \subseteq \Re^n$ be a convex set and let f be a function defined on S.
- f is concave on S if and only if its hypograph $hypof = \{(x, z) : x \in S,\ z \leq f(x)\}$ is a convex set.
- f is strictly concave on S if and only if its hypograph is a convex set and does not contain any line segments.
- f is quasiconcave on S if and only if all of its upper level sets $S_{\geq \alpha} = \{x \in S : f(x) \geq \alpha\}$ are convex.
- f is quasiconcave (respectively, strictly quasiconcave, semistrictly quasiconcave) on S if and only if the restriction of f on each line segment contained in S is a quasiconcave (respectively, strictly quasiconcave, semistrictly quasiconcave) function.
- f is strictly quasiconcave on S if and only f is quasiconcave and every restriction on a line segment is not constant.
- f is semistrictly quasiconcave on S if and only f is quasiconcave and every local maximum is also global for every restriction on a line segment.

Under differentiabilty assumptions:
- f is concave if and only if

$$f(y) \leq f(x) + \nabla f(x)^T (y - x), \quad \forall x, y \in S.$$

and it is strictly concave if and only the inequality is strict.
- f is concave if and only if its Hessian matrix is negative semidefinite at each point of S. If the Hessian matrix of f is negative definite at each point of S, then f is strictly concave.
- f is quasiconcave on S if and only if the following implication holds:

$$x_1,\ x_2 \in S,\ f(x_1) \leq f(x_2) \Rightarrow \nabla f(x_1)^T (x_2 - x_1) \geq 0.$$

Note that both for a strictly and a semistrictly quasiconcave function there is not a first-order characterization.

- f is quasiconcave on S if and only
(i) $x_0 \in S$, $u \in \Re^n$, $u^T \nabla f(x_0) = 0$ imply $u^T \nabla^2 f(x_0) u \leq 0$;
(ii) $x_0 \in S$, $x_1 \in S$, $f(x_1) > f(x_0)$, $\nabla f(x_0) = 0$, $u^T \nabla^2 f(x_0) u = 0$ with $u = x_0 - x_1$, imply that for every $\epsilon > 0$ there exists $k \in (0, \epsilon)$ such that $x_0 + ku \in S$ and $f(x_0) \geq f(x_0 + ku)$.

- f is (strictly) pseudoconcave on S if and only if f is quasiconcave on S and every critical point is a (strict) local maximum for f on S.
- f is (strictly) pseudoconcave if and only if for every $x_0 \in S$ and $u \in \Re^n$ such that $u^T \nabla f(x_0) = 0$, the function $\varphi(t) = f(x_0 + tu)$ attains a (strict) local maximum at $t = 0$.
- f is (strictly) pseudoconcave on S if and only if for every $x_0 \in S$ and $u \in \Re^n$ such that $u^T \nabla f(x_0) = 0$, either $u^T \nabla^2 f(x_0) u < 0$ or $u^T \nabla^2 f(x_0) u = 0$ and the function $\varphi(t) = f(x_0 + tu)$ attains a (strict) local maximum at $t = 0$.
- f is (strictly) pseudoconcave if and only if
(i) $x \in S$, $u \in \Re^n$, $u^T \nabla f(x_0) = 0 \Rightarrow u^T \nabla^2 f(x_0) u \leq 0$;
(ii) $x \in S$, $\nabla f(x) = 0 \Rightarrow f$ has a (strict) local maximum at x.

- f is pseudoconcave if and only if
(i) $(-1)^{|R|} D_R(x) \geq 0$, $\forall x \in S$, $\forall R \subseteq \{1, 2, .., n\}$, $R \neq \emptyset$;
(ii) $x \in S$, $\nabla f(x) = 0 \Rightarrow f$ has a local maximum at x
where $D_R(x)$, $R \subseteq \{1, 2, .., n\}$ are the bordered principal minors of the bordered Hessian $D(x) = \begin{bmatrix} 0 & \nabla^T f(x) \\ \nabla f(x) & \nabla^2 f(x) \end{bmatrix}$.
- A sufficient condition for f to be pseudoconcave is

$$(-1)^k D_k(x) > 0, \quad \forall x \in S, \quad \forall k = 1, 2, .., n$$

where $D_k(x)$, $k = 1, .., n$ are the bordered leading principal minors of the bordered Hessian.

Maxima and Minima
The set of all maximizers is convex for all classes of generalized concave functions and the Karush-Kuhn-Tucker conditions become sufficient for global optimality under suitable assumptions of generalized concavity.

Sufficiency of the Karush–Kuhn–Tuckers Conditions
Consider the problem

$$P: \quad \max f(x), \quad x \in S = \{x \in X : g_i(x) \geq 0, \ i = 1, 2, ..., m\}$$

where $f, g_i, i = 1, ..., m$ are functions defined on the open set $X \subseteq \Re^n$, and let x_0 be a feasible point. Suppose that f is pseudoconcave at $x_0 \in S$, and the g_i, $i = 1, .., m$, are differentiable and quasiconcave at x_0. If there exist $\lambda_i \in \Re$, $i = 1, .., m$, such that

$$\begin{cases} \nabla f(x_0) + \sum_{i=1}^{m} \lambda_i \nabla g_i(x_0) = 0 \\ \lambda_i \geq 0, \quad i = 1, ..., m \\ \lambda_i g_i(x_0) = 0, \quad i = 1, ..., m \end{cases}$$

then x_0 is a global maximum point.

• Let f be a continuous and semistrictly quasiconcave function on the convex and closed set S. If f attains its minimum value on S, then it is reached at some boundary point.

• Let f be a continuous and semistrictly quasiconcave function on the convex and closed set S containing no lines. If f attains its minimum value on S, then it is reached at an extreme point.

Some Classes of Generalized Concave Functions

Let f and g be functions defined on a convex set $S \subseteq \Re^n$, and let

$$z(x) = \frac{f(x)}{g(x)}$$

Then, the following properties hold.

(i) If f is non-positive and concave on S, g is positive and concave on S, then z is semistrictly quasiconcave on S; z is pseudoconcave if f and g are differentiable on S;

(ii) If f is non-negative and concave on S, g is positive and convex on S, then z is semistrictly quasiconcave on S; z is pseudoconcave on S if f and g are differentiable on S;

(iii) If f concave on S and g is positive and affine on S, then z is semistrictly quasiconcave on S; z is pseudoconcave on S if f is differentiable on S.

• The product

$$z(x) = f(x) \cdot g(x).$$

is quasiconcave if f is non-negative and concave, and g is positive and concave.

• The function $z(x) = \prod_{i=1}^{k} (f_i(x))^{\alpha_i}$, $\alpha_i > 0$ where $f_i(x)$, $i = 1, ..., k$, are positive-valued concave functions on the convex set $S \subseteq \Re^n$ is quasiconcave. In particular, the Cobb-Douglas function:

$$f(x) = A x_1^{\alpha_1} x_2^{\alpha_2} .... x_n^{\alpha_n}, \quad A > 0, \ x_i > 0, \ \alpha_i > 0, \ i = 1, ..., n$$

is pseudoconcave.

• The constant elasticity of substitution

$$f(x) = (a_1 x_1^\alpha + a_2 x_2^\alpha + ... + a_n x_n^\alpha)^{\frac{1}{\alpha}}, \ a_i > 0, \ x_i > 0, \ i = 1, ..., n, \ \alpha \neq 0$$

is quasiconcave if and only if $\alpha \leq 1$.

Index

Printed in the United States
By Bookmasters